Lecture Notes in Control and Information Sciences

Volume 467

About this Series

This series aims to report new developments in the fields of control and information sciences—quickly, informally and at a high level. The type of material considered for publication includes:

1. Preliminary drafts of monographs and advanced textbooks
2. Lectures on a new field, or presenting a new angle on a classical field
3. Research reports
4. Reports of meetings, provided they are

    (a) of exceptional interest and

    (b) devoted to a specific topic. The timeliness of subject material is very important.

More information about this series at http://www.springer.com/series/642

Selim S. Hacısalihzade

Control Engineering
and Finance

 Springer

Selim S. Hacısalihzade
Department of Electrical and Electronics
    Engineering
Boğaziçi University
Bebek, Istanbul
Turkey

ISSN 0170-8643   ISSN 1610-7411   (electronic)
Lecture Notes in Control and Information Sciences
ISBN 978-3-319-87805-8   ISBN 978-3-319-64492-9   (eBook)
https://doi.org/10.1007/978-3-319-64492-9

Printed on acid-free paper

This Springer imprint is published by Springer Nature
The registered company is Springer International Publishing AG
The registered company address is: Gewerbestrasse 11, 6330 Cham, Switzerland

O D E

Hark!

   All ye, listen to this ode

     Never heard before abroad.

      Dear me, this sounds flawed,

       Elementary, someone called.

        You said it smacks of fraud

         Eer, well, it *is* a bit odd...

Telling as this short story might be,

   Unsurprisingly it all began by the sea.

     Twain had met just for a cup of tea.

      Keen and eager was our wannabe,

       Unimpeachable, the other, in actuality.

        Yet, it had to be, it was nature's decree:

         Lovely and smiling ever so sweetly,

          Ablaze, she became, forever, the addressee!

Acknowledgements

There is a long list of people to acknowledge and thank for their support in preparing this book. I want to begin by thanking Jürg Tödtli who supported my interest in the field of quantitative finance—even though I did not know the term at that time—while we were at the Institute of Automatic Control at the ETH Zurich many decades ago. This interest was triggered and then re-triggered in countless discussions over the years with my uncle Ergün Yüksel. Special thanks are certainly due to Manfred Morari, the former head of the Institute of Automatic Control at the ETH Zurich, who encouraged me to write this book and to publish it in the Lecture Notes in Control and Information Science series of Springer Verlag and who also offered me infrastructure and library access during the preparation of the manuscript.

Very special thanks go to Florian Herzog of Swissquant who supported me while I was writing this book by offering the use of his lecture notes of Stochastic Control, a graduate course he held at the ETH Zurich. I did so with gratitude in Chapters 5 and 9. The data for the empirical study reported in Chapter 8 were graciously supplied by Özgür Tanrverdi of Access Turkey Opportunities Fund for several years, to whom I am indebted. I also want to thank Jens Galschiøt, the famous Danish sculptor for allowing me to use a photograph of his impressive and inspiring sculpture "Survival of the Fattest" to illustrate the inequality in global wealth distribution.

Parts of this book evolved from a graduate class I gave at Boğaziçi University in Istanbul during the last years and from project work by many students there, notably Efe Doğan Yılmaz, Ufuk Uyan, Ceren Sevinç, Mehmet Hilmi Elihoş, Yusuf Koçyiğit and Yasin Çotur. I am most grateful to Yasin, a Ph.D. candidate at Imperial College in London now, who helped with calculations and with valuable feedback on earlier versions of the manuscript.

I am indebted to my former student Yaşar Baytın, my old friend Sedat Ölçer, Head of Computer Science and Engineering at Bilgi University in Istanbul, Bülent Sankur, Professor Emeritus of Electrical and Electronics Engineering at Boğaziçi

University, and especially my dear wife Hande Hacısalihzade for proofreading parts of the manuscript and their most valuable suggestions.

I am grateful to Petra Jantzen and Shahid S. Mohammed at Springer for their assistance with the printing of this volume.

Hande, of course, also deserves special thanks for inspiring me (and certainly not only for writing limericks!), hours of lively discussions, her encouragement, and her endless support during the preparation of this book.

Selim S. Hacısalihzade
Istanbul 2017

Contents

Chapter 1
Introduction

What we need is some financial engineers.

— Henry Ford

Diversifying sufficiently among uncorrelated risks can reduce portfolio risk toward zero. But financial engineers should know that's not true of a portfolio of correlated risks.

— Harry Markowitz

1.1 Control Engineering and Finance

At first glance, the disciplines of finance and control engineering may look as unrelated as any two disciplines could be. However, this is true only to the uninitiated observer. For trained control engineers, the similarities of the underlying problems are striking. One, if not the main, intent of control engineering is to control a process in such a way that it behaves in the desired manner in spite of unforeseen disturbances acting upon it. Finance, on the other hand, is the study of the management of funds with the objective of increasing them, in spite of unexpected economical and political events. Once formulated this way, the similarity of control engineering to finance becomes obvious.

Perhaps the most powerful tool for reducing the risk of investments is diversification. If one can identify the risks specific to a country, a currency, an instrument class and an individual instrument, the risk conscious investor—as they should all be—ought to distribute her wealth among several countries, several currencies and among different instrument classes like, for instance, real estate, bonds of different issuers, equity of several companies and precious metals like gold.

© Springer International Publishing AG 2018
S. S. Hacısalihzade, *Control Engineering and Finance*, Lecture Notes in Control and Information Sciences 467, https://doi.org/10.1007/978-3-319-64492-9_1

Clearly, within equity investments, it makes sense to invest in several companies, ideally in different countries. In this context, as discussed in detail in Chapter 8, Modern Portfolio Theory attempts to maximize the expected return of a portfolio for a given amount of risk, or equivalently minimize the amount of risk for a given level of expected return by optimizing the proportions of different assets in the portfolio.

Chapters 4, 5, and 8 show that Optimal Stochastic Control constitutes an excellent tool for constructing optimal portfolios. The use of financial models with control engineering methods has become more widespread with the aim of getting better and more accurate solutions. Since optimal control theory is able to deal with deterministic and stochastic models, finance problems can often be seen as a mixture of the two worlds.

A generic feedback control system is shown in Figure 1.1. The system is composed of two blocks, where P denotes the process to be controlled and C the controller. In this representation r stands for the reference, e the error, u the control input, d the disturbance, and y the output.

The structure in Figure 1.2 can be used as a theoretical control model for dynamic portfolio management which suggests three stages: the *Estimator* estimates the return and the risk of the current portfolio and its constituents, the *Decider* determines the timing of the portfolio re-balancing by considering relevant criteria and the *Actuator* changes the current portfolio to achieve a more desirable portfolio by solving an optimization problem involving the model of the portfolio. In many cases, also due to regulatory constraints, the Actuator involves a human being but it can compute and execute buy/sell transactions without human involvement as well.

Looking at these two figures, one can observe that they have a similar structure in the sense of a feedback loop. Therefore, finance or investing, and more specifically the problem of portfolio management can be regarded as a control problem where r is the expected return of the portfolio, e is the difference between the expected and

Fig. 1.1 A generic feedback control system

Fig. 1.2 A theoretical control model for dynamic portfolio management

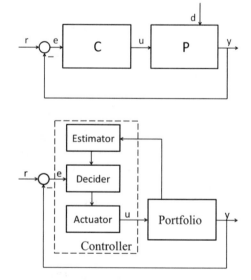

the actual portfolio returns, P is the portfolio, C is the algorithm which maximizes the portfolio return under certain constraints and u are the buy/sell instructions to re-balance the portfolio.

1.2 Outline

Numerous books have been published about the solution of various financial problems using control engineering techniques as researchers conversant in both fields became aware of their similarities. Many such specific examples can be found in the literature, especially in the areas of optimal control and more recently in stochastic control. Therefore, even a superficial review of this particular interdisciplinary field covering all subjects would have to weigh several volumes. Hence, any textbook in this field has either to take one specific application of control theory in finance and explore it in depth or be an eclectic collection of several problems. This Book chooses to take the latter path also because several excellent textbooks of the former type are already available.

This Volume presents a number of different control engineering applications in finance. It is intended for senior undergraduate or graduate students in electrical engineering, mechanical engineering, control engineering, industrial engineering and financial engineering programs. For electrical/mechanical/control/industrial engineering students, it shows the application of various techniques they have already learned in theoretical lectures in the financial arena. For financial engineering students, it shows solutions to various problems in their field using methods commonly used by control engineers. This Book should also appeal to students and practitioners of finance who want to enhance their quantitative understanding of the subject.

There are no *sine qua non* prerequisites for reading, enjoying and learning from this textbook other than basic engineering mathematics and a basic understanding of control engineering concepts. Nevertheless, the first half of the Book can be seen as a refresher of or an introduction to several tools like mathematical modeling of dynamic systems, analysis of stochastic processes, calculus of variations and stochastic calculus which are then applied to the solution of some financial problems in the second part of the Book. It is important to remember that this is by no means a mathematics book even though it makes use of some advanced mathematical concepts. Any definitions of these concepts or any derivations are neither rigorous nor claim to be complete. Therefore, where appropriate, the reader looking for mathematical precision is referred to standard works of mathematics.

After this introductory Chapter, the Book begins by discussing a very important topic which is often neglected in most engineering curricula, namely, mathematical modeling of physical systems and processes. Chapter 2 discusses what a model is and gives various classification methods for different types of models. The actual modeling process is illustrated using the popular inverted pendulum and the stock prices. The generic modeling process is explained, highlighting parameter identification using various numerical optimization algorithms, experiment design, and model validation. A small philosophical excursion in this Chapter is intended to alert the reader to the differences between reality and models.

Chapter 3, Probability and Stochastic Processes, begins by introducing Kolmogorov's axioms and reviewing some basic definitions and concepts of random variables such as moments, probability distributions, multidimensional distribution functions and statistical independence. Binomial and normal distributions are discussed and related to each other through the central limit theorem. Mathematical description and analysis of stochastic processes with emphasis on several relevant special classes like stationary and ergodic processes follows. Some special processes such as the normal process, the Wiener process, the Markov process and white noise are discussed. Students not comfortable with the concepts in this Chapter should study them very carefully, since they constitute an important mathematical foundation for the rest of the Book.

Chapter 4, Optimal Control, is thought as an introduction to this vast field. It begins with calculus of variations and goes through the fixed and variable endpoint problems as well as the variation problem with constraints. Application of these techniques to dynamic systems leads to the solution of the optimal control problem using the Hamilton-Jacobi method for closed-loop systems where all the states have to be fed back to assure optimality independent of initial conditions. Pontryagin's minimum principle is discussed in connection with optimal control in the case of the control variables being limited (as they always are in practice). Special emphasis is given to optimal control of linear systems with a quadratic performance index leading to the Riccati equation. Dynamic programming is briefly presented. The Final Section of this Chapter gives an introduction to Differential Games, thus establishing a firm link between Optimal Control and Finance.

Chapter 5, Stochastic Analysis, constitutes, perhaps, the heart of the Book. It begins with a rigorous analysis of white noise and introduces stochastic differential equations (SDE's). Stochastic integration and Itô integrals are introduced and their properties are scrutinized. Stochastic differentials are discussed and the Itô lemma is derived. Methods for solving SDE's using different techniques including Itô calculus and numerical techniques are shown for scalar and vector valued SDE's both in the linear and the non-linear cases. Several stochastic models used in financial applications are illustrated. The connection between deterministic partial differential equations and SDE's is pointed out.

Chapter 6, Financial Markets and Instruments is written in an informal and colloquial style, because it might be the first instance where an engineering student encounters the capital markets and various financial instruments. The Chapter begins with the concept of time value of money and introduces the main classes of financial instruments. It covers the fixed income instruments like savings accounts, certificates of deposit, and introduces Bonds. The Chapter then moves on to talk about a number of other financial instruments, including common stocks, various types of funds and derivative instruments. Fundamental concepts relating to risk, return and utility are discussed in this Chapter. Finally, the role and importance of the banks for the proper functioning of the economy are presented.

Special emphasis is given to Bonds in Chapter 7. Bond returns and valuations are derived heuristically based on the return concept defined in the previous Chapter. Fundamental determinants of interest rates and bond yields, together with the Macaulay duration are discussed with examples. The yield curve is presented together with its implications for the economy.

Chapter 8, Portfolio Management, is dedicated to the problem of maximizing the return of a portfolio of stocks over time while minimizing its risk. The fundamental concept of the Efficient Frontier is introduced within the context of this quintessential control engineering problem. The problem of choosing the sampling frequency is formulated as the question of choosing the frequency of re-balancing a stock portfolio by selling and buying individual stocks. Empirical studies from the Swiss, Turkish and American stock markets are presented to address this question. Additional algorithms for managing stock portfolios based on these empirical studies are proposed. Management of bond portfolios using the Vašíček model and static bond portfolio management methods finish off this Chapter.

Chapter 9, Derivative Financial Instruments, begins by reviewing forward contracts, futures and margin accounts. Options are discussed in detail in this Chapter. The Black-Scholes options pricing model is presented and the pricing equation is derived. Selected special solutions of this celebrated equation for the pricing of European and American options are shown. Finally, the use of options in constructing structured products to enhance returns and cap risks in investing is demonstrated.

There are three Appendices which contain various mathematical descriptions of dynamic systems including the concept of state space; matrix algebra and matrix calculus together with some commonly used formulas; and the requisite standardized normal distribution tables.

Every Chapter after this one includes three types of exercises at its conclusion: Type A exercises are mostly verbal and review the main points of the Chapter; they aim to help the reader gauge her[1] comprehension of the topic. Type B exercises require some calculation and they are intended to deepen the understanding of the methods discussed in the Chapter. Type C exercises involve open ended questions and their contemplation typically requires significant time, effort and creativity. These questions can qualify as term or graduation projects, and may indicate directions for thesis work.

Five books are enthusiastically recommend, especially to those readers who have an appetite for a less technical take on the workings of the financial markets. This short reading list should accompany the reader during his perusal of this Book and the odds are that he will return to these classics for many years to come.

- "The Physics of Wall Street: A Brief History of Predicting the Unpredictable" by James Owen Weatherall, 2014,
- "The Drunkard's Walk: How Randomness Rules Our Lives" by Leonard Mlodinow, 2008,

---

[1]To avoid clumsy constructs like "his/her", where appropriate, both male and female personal pronouns are used throughout the Book alternately and they are interchangeable with no preference for or a prejudice against either gender.

- "The Intelligent Investor" by Benjamin Graham,[2] 2006, [47]
- "The Black Swan: The Impact of the Highly Improbable" by Nassim Nicholas Taleb, 2007, and
- "Derivatives: The Tools that Changed Finance" by the father and son Phelim and Feidhlim Boyle, 2001.

It is humbling to see how inherently challenging subjects like randomness, investing, and financial derivatives are so masterfully explained in these books for laypersons in excellent prose.

---

[2]Benjamin Graham, British-American economist and investor (1894– 1976).

Chapter 2
Modeling and Identification

"That's another thing we've learned from your Nation," said
Mein Herr, "map-making. But we've carried it much further
than you. What do you consider the largest map that would
be really useful?"
"About six inches to the mile."
"Only six inches!" exclaimed Mein Herr. "We very soon got to six
yards to the mile. Then we tried a hundred yards to the mile.
And then came the grandest idea of all! We actually made a
map of the country, on the scale of a mile to the mile!"
"Have you used it much?" I enquired.
"It has never been spread out, yet," said Mein Herr: "the farmers
objected: they said it would cover the whole country, and shut
out the sunlight! So we now use the country itself, as its own
map, and I assure you it does nearly as well."

— Lewis Carroll, Sylvie and Bruno Concluded

2.1 Introduction

Most control engineering exercises begin with the phrase "Given is the system with the transfer function ...". Alas, in engineering practice, a mathematical description of the plant which is to be controlled is seldom available. The plant first needs to be modeled mathematically. In spite of this fact, control engineering courses generally do not spend much time on modeling. Perhaps this is because modeling can be said to be more of an art than a science. This Chapter begins by defining what is meant by the words model and modeling. It then illustrates various types of models, studies the process of modeling and concludes with the problem of parameter identification and related optimization techniques.

© Springer International Publishing AG 2018
S. S. Hacısalihzade, *Control Engineering and Finance*, Lecture Notes in Control and Information Sciences 467, https://doi.org/10.1007/978-3-319-64492-9_2

2.2   What Is a Model?

Indeed, what is a model? Probably one gets as many different responses as the number of persons one poses this question.[1] Therefore, it is not surprising that Merriam-Webster Dictionary offers 23 different definitions under that one entry. The fourth definition reads "a usually miniature representation of something" as in a "model helicopter". The ninth definition reads, rather unglamorously, "one who is employed to display clothes or other merchandise" as in Heidi Klum, Naomi Campbell or Adriana Lima depending on which decade you came off age. Well, if your interest in models is limited to these definitions you can close this Book right now.

Definitions 11 and 12 read "a description or analogy used to help visualize something (as an atom) that cannot be directly observed" and "a system of postulates, data and inferences presented as a mathematical description of an entity or state of affairs; also: a computer simulation based on such a system <climate models>". These definitions are closer to the sense of the word model that is used throughout this Book.

These definitions might be considered too general to be of any practical use. Let us therefore think of a model as an *approximate representation of reality*. One obvious fact, forgotten surprisingly often, is that a model is an abstraction and that any model is, by necessity, an approximation of reality. Consequently, there is no one "true model", rather there are models which are better than others. But what does "better" mean? This clearly depends on the context and the problem at hand.

Example: An illustrative example which can be found in many high school physics books is dropping an object from the edge of a table and calculating the time it will take for the object to hit the floor. Assuming no air friction, here one can write the well-known Newtonian[2] equation of motion which states that the object will move with a constant acceleration caused by the weight of the object, which again is caused by the Earth's gravitational attraction:

$$m\ddot{x}(t) = mg \,. \tag{2.1}$$

$\ddot{x}(t)$ denotes the second derivative of the distance x (acceleration) of the object from the edge of the table as a function of time after the drop; g is the Earth's acceleration constant. m is the mass of the object but it is not relevant, because it can be canceled away. Solving the differential Equation (2.1) with the initial conditions $x(0) = 0$ and $\dot{x}(0) = 0$ results in

$$x(t) = \frac{1}{2}gt^2 \,. \tag{2.2}$$

---

[1]This Chapter is an extended version of Chapter 3 in [51].

[2]Isaac Newton, English astronomer, physicist, mathematician (1642–1726); widely recognized as one of the most influential scientists of all time and a key figure in the scientific revolution; famous for developing infinitesimal calculus, classical mechanics, a theory of gravitation and a theory of color.

Solving (2.2) for the impact time t_i with the height of the table denoted as h results in the well-known formula

$$t_i = \sqrt{\frac{2h}{g}} .$$ (2.3)

Let us now consider the problem of dropping an object attached to a parachute from an aircraft. Here, one can no longer assume there is no air friction. The very reason for deploying a parachute is to make use of the air friction to slow down the impact velocity of the drop. Hence, to calculate the impact time with better accuracy, (2.1) needs to be modified to account for the air friction, a force which increases with the square of velocity:

$$\ddot{x}(t) = g - kS\dot{x}^2(t) .$$ (2.4)

$\dot{x}(t)$ denotes the first derivative of the distance x (velocity) of the object from the drop height, k is the viscous friction coefficient and S is the effective surface area of the parachute. It is no longer possible to have a simple analytical solution of this non-linear differential equation and one has to revert to numerical means to solve it. When the terminal velocity is reached (*i.e.*, $kS\dot{x}^2 = g \Rightarrow \ddot{x} = 0$) the object stops accelerating and keeps falling with a constant velocity. □

This example shows that the same physical phenomenon can be modeled in different ways with varying degrees of complexity. It is not always possible to know as in this case which model will give better results a priori. Therefore, one often speaks of the "art" of modeling. An engineer's approach to modeling might be to begin with certain restrictive assumptions leading to a simple model. At further steps, these assumptions can be relaxed to modify the initial simple model and to account for further complications until a model is attained which is appropriate for the intended purpose.

There are many different types of models employed in science and philosophy. The reader might have heard of mental models, conceptual models, epistemological models, statistical models, scientific models, economic models or business models just to name a few. This Book limits itself to scientific and mathematical models. Scientific modeling is the process of generating abstract, conceptual, graphical or mathematical models using a number of methods, techniques and theories. The general purpose of a scientific model is to represent empirical phenomena in a logical and objective way. The process of modeling is presented in the next section.

A mathematical models helps to describe a system using mathematical tools. The use of mathematical models is certainly not limited to engineering or natural sciences applications. Mathematical models are increasingly being used in social sciences like economics, finance, psychology and sociology[3,4].

---

[3]Isaac Asimov, American writer (1920–1992).

[4]Which science fiction enthusiast is unaffected by the Asimovian character Hari Seldon's "psychohistory", which combines history, sociology and statistics to make general predictions about the

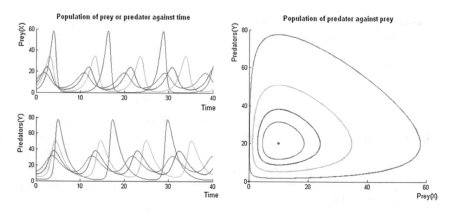

Fig. 2.1 Prey and predator populations as a numerical solution of the set of equations (2.5). Different shades of gray indicate different initial conditions [59]

Mathematical models can be classified under the following dichotomic headings: *Inductive* versus *deductive* models: A deductive model is based on a physical theory. In the example above, Newton's laws of motion were used to model the movement of a falling object. An inductive model, on the other hand, is based on empirical findings and their generalizations without forming a generally applicable law of nature. A well-known example is the set of Lotka-Volterra equations used in modeling the dynamics of biological systems in which two species interact as predator and prey [78]. Here the rate of change of the number of preys, x (say, rabbits) is proportional to the number of preys who can find ample food and who can breed, well, like rabbits. This exponential growth is corrected by prey-predator encounters (say, with foxes). Dually, the number of predators y decreases in proportion to the number of predators either because they starve off or emigrate. This exponential decay is corrected by the encounters of predators with preys. This can be modeled mathematically with the following differential equation system with $\alpha, \beta, \gamma, \delta > 0$ (Figure 2.1).

$$\frac{dx}{dt} = \alpha x - \beta xy \,,$$
$$\frac{dy}{dt} = -\gamma y + \delta xy \,. \tag{2.5}$$

Deterministic versus *stochastic* models: In a deterministic model no randomness is involved in the development of future states of the modeled system. In other words, a deterministic model will always produce the same output starting with a given set of initial conditions as in the example above. A stochastic model, on the other hand, includes randomness. The states of the modeled system do not have unique values.

future behavior of large groups of populations—like the Galactic Empire with a quintillion citizens [7]?

They must rather be described by probability distribution functions. A good example in physics is the movement of small particles in a liquid—Brownian motion—resulting from their bombardment by a vast number of fast moving molecules [36]. Finance uses mostly stochastic models for predictions. These are discussed in detail in Chapter 5.

Random or Chaotic?

Webster Dictionary defines "random" as "a haphazard course; without definite aim, direction, rule, or method". The same source defines "chaos" as "complete confusion and disorder; a state in which behavior and events are not controlled by anything". Those descriptions sound quite similar. However, in the mathematical or engineering context the word "chaos" has a very specific meaning. Chaos theory is a field in mathematics which studies the behavior of dynamic systems that are extremely sensitive to initial conditions. Tiny differences in initial conditions result in completely different outcomes for such dynamic systems. As the frequently told anecdote goes, Edward Norton Lorenz, a meteorologist with a strong mathematical background in non-linear systems, was making weather predictions using a simplified mathematical model for fluid convection back in 1961. The outcomes of the simulations were wildly different depending on whether he used three digits or six digits to enter the initial conditions of the simulations. Although it was known for a long time that non-linear systems had erratic or unpredictable behavior, this experience of Lorenz became a monumental reminder that even though these systems are deterministic, meaning that their future behavior is fully determined by their initial conditions, with absolutely no random elements involved, they cannot be used to predict the future for any meaningful purpose. Lorenz is reputed to have quipped "Chaos is when the present determines the future, but the approximate present does not approximately determine the future".

A double pendulum is made of two rods attached to each other with a joint. One of the pendulums is again attached with a joint to a base such that it can revolve freely around that base. Anyone who watches this contraption swing for some time cannot help being mesmerized by its unpredictable slowing downs and speeding ups. The double pendulum can be modeled very accurately by a non-linear ordinary differential equation (see, for instance, [135] for a complete derivation). However, when one tries to simulate (or solve numerically) this equation one will observe the sensitive dependency of its behavior on the initial conditions one chooses.

There is even a measure for chaos. The Lyapunov exponent of a dynamic system is a quantity that characterizes the rate of separation of infinitesimally close trajectories. Two trajectories in state space with initial separation δx_0 diverge at a rate given by $|\delta \mathbf{x}(t)| \approx e^{\lambda t}|\delta \mathbf{x}_0|$ where λ denotes the Lyapunov exponent.

Static versus *dynamic* models: In static models, as the name suggests, the effect of time is not considered. Dynamic models specifically account for time. Such models make use of difference or differential equations with time as a free variable. Looking at the input-output relationship of an amplifier far from saturation, a static model will simply consist of the amplification factor. A dynamic model, on the other hand, will use a time function to describe how the output changes dynamically, including its transient behavior, as a consequence of changes in the input [26].

Discrete versus *continuous* models: Some recent models of quantum gravity [122] notwithstanding, in the current *Weltbild* time flows smoothly and continuously. Models building on that—often tacit—assumption make use of differential equations, solutions of which are time continuous functions. However, the advent of the digital computer which works with a clock and spews out results at discrete time points made it necessary to work with discrete time models which are best described using difference equations. Whereas the classical speed control by means of a fly ball governor is based upon a continuous time model of a steam engine [130], modern robot movement controllers employing digital processors use discrete time models [106].

Lumped parameter models versus *distributed parameter* models: Distributed parameter or distributed element models assume that the attributes of the modeled system are distributed continuously throughout the system. This is in contrast to lumped parameter or lumped element models, which assume that these values are lumped into discrete elements. One example of a distributed parameter model is the transmission line model which begins by looking at the electrical properties of an infinitesimal length of a transmission line and results in the telegrapher's equations. These are partial differential equations involving partial derivatives with respect to both space and time variables [92]. Equations describing the elementary segment of a lossy transmission line developed by Heaviside[5] in 1880 as shown in Figure 2.2 are

$$\frac{\partial}{\partial x} V(x, t) = -L \frac{\partial}{\partial t} I(x, t) - R I(x, t), \qquad (2.6)$$

$$\frac{\partial}{\partial x} I(x, t) = -C \frac{\partial}{\partial t} V(x, t) - G V(x, t). \qquad (2.7)$$

$V(x, t)$ and $I(x, t)$ denote the position and time dependent voltage and current respectively. The parameters R, G, L, C are the distributed resistance, conductance, inductance and capacitance per unit length. A lumped parameter model simplifies the description of the behavior of spatially distributed physical systems into a topology consisting of discrete entities that approximate the behavior of the distributed system. The mathematical tools required to analyze such models are ordinary differential equations involving derivatives with respect to the time variable alone. A simple example for a lumped parameter models is an electrical circuit consisting of a resistor and an inductor in series. Such a circuit can adequately be described by the ordinary differential equation

Fig. 2.2 Schematic representation of the elementary components of a transmission line with the infinitesimal length dx

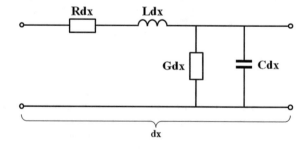

---

[5]Oliver Heaviside, English mathematician and engineer (1850–1925).

$$V(t) = L\frac{dI(t)}{dt} + RI(t). \tag{2.8}$$

Linear models versus *non-linear* models: It can be said that a model is linear if the net response caused by two stimuli $u_1(t)$, $u_2(t)$ is the sum of the responses $f(u_1(t))$, $f(u_2(t))$ which would have been caused by each stimulus individually [64]. Formally,

$$f(\alpha_1 u_1(t) + \alpha_2 u_2(t)) = \alpha_1 f(u_1(t)) + \alpha_2 f(u_2(t)). \tag{2.9}$$

Most physical phenomena are inherently non-linear. However, they may legitimately be modeled linearly either in limited operational ranges (*e.g.,* if an amplifier, as in Figure 2.3a, working far from saturation amplifies 1 to 5, it amplifies 2 to 10) or linearized around an operating point (*e.g.,* a diode, as in Figure 2.3b). While writing this Chapter, it was suggested by a student that dividing all models in linear and non-linear models is like dividing the world in bananas and non-bananas. Such a remark is clearly facetious, because there is a closed theory for linear systems and many tools and methods exist for dealing with them. On the other hand, non-linear systems are analyzed mostly with approximative numerical methods specific to the problem at hand [136].

It can be said that nature is inherently dynamic, stochastic, continuous, and non-linear with distributed parameters, hence difficult to model accurately. However, depending on the application area, many simplifying assumptions can be made to make the model more amenable to analytical techniques as indicated above.

To sum it up, a good model should provide some insight which goes beyond what is already known from direct investigation of the phenomenon being studied. The more observations it can explain and the more accurate predictions it can make, the better is the model. Combine that with the quote "Everything should be made as simple as possible, but not simpler" attributed to Einstein,[6] one has all the elements of a good model. A corollary to this statement is that the level of detail incorporated in the model is determined by the intended purpose of the model itself.

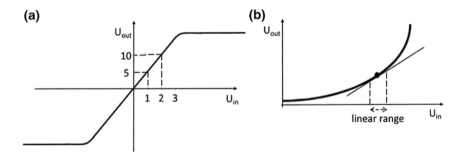

Fig. 2.3 **a** Linear amplifier with saturation. **b** Diode linearized around an operating point

---

[6] Albert Einstein, German-American physicist (1879–1955); famous for his explanation of the photoelectric effect and the relativity theory which radically changed the understanding of the universe.

The four distinctly different types of models depending on their purposes can be categorized as descriptive, interpretive, predictive or explanatory models. *Descriptive* models represent relationships between various variables using a concise mathematical language. Hooke's law stated as

$$F = -kx \, , \tag{2.10}$$

is a well-known example here, with F representing the force which lengthens a spring by the distance x. One could describe this law in words or using diagrams, but the expression in (2.10) is a pithy description.

Suppose one wants to model the purchasing power of money in an inflationary economy. One might plot the number of loaves of bread one can buy with a constant salary every month and see that this number drops with time and that the rate of the drop is constant as in Figure 2.4. One can use this exponential decay model to *interpret* the decay rate as a measure of the inflation.

A *predictive* model is required if the model will be used to see how a system will react to a certain stimulus. A typical example is a linear electrical circuit with a resistor, an inductor and a capacitor. The input-output transfer function of such a circuit can be described using the Laplace transform, for example, as

$$G(s) = \frac{2}{s^2 + 3s + 8} \, . \tag{2.11}$$

The response of the system to a unit step at its input can be simulated using this model as shown in Figure 2.5.

An *explanatory* model is a useful description and explanation why and how something works. It has no claim on accuracy or completeness. An example from finance is how the exchange rate between US Dollars and Turkish Lira affects the inflation in Turkey due to the changing price of imported goods.

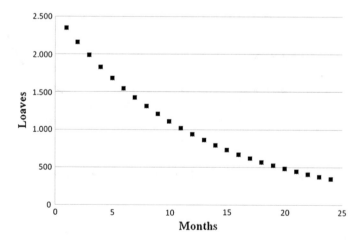

Fig. 2.4 Number of loaves of bread one can buy with a fixed salary in an inflationary economy

A general mathematical model can be represented with a block diagram[7] as shown in Figure 2.6.

> **A Philosophical Excursion**
>
> Let us consider the Newtonian model of gravity. Newton posits an invisible force able to act over distances which attracts bodies. This force is proportional to the masses of the bodies and inversely proportional to the square of the distance between the bodies. Note that Newton made no attempt at explaining the cause of that attraction. This model is immensely successful. One can even argue that, together with his three laws of motion,

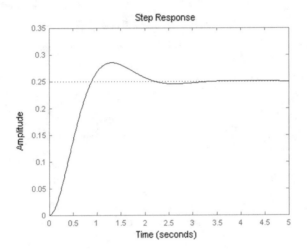

Fig. 2.5 Response of the system in (2.11) to a unit step at its input

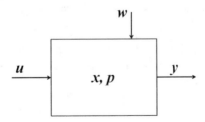

Fig. 2.6 Block diagram representation of a mathematical model has different types of variables: input variables which can be controlled **u**, input variables which cannot be controlled (often random and called disturbances) **w**, output variables **y**, internal state variables **x**, and model parameters **p**. Since there can be several of each type of variables, those are aggregated in vectors. The general state space representation of such a system is given by the equations $\dot{\mathbf{x}}(t) = \mathbf{f}(\mathbf{x}, \mathbf{u}, \mathbf{w}, \mathbf{p}, t)$ and $\mathbf{y}(t) = \mathbf{g}(\mathbf{x}, \mathbf{u}, \mathbf{w}, \mathbf{p}, t))$

---

[7]Lower case boldface characters are used for vectors and upper case boldface characters are used for matrices throughout the Book unless otherwise specified.

Newtonian mechanics constitutes one of the pillars on which the industrial revolution and the ensuing increase in the welfare of the human population during the last three centuries builds. Construction machines, the houses these machines build, the cars that travel among these houses, the aircraft that move people from one continent to another, the spacecraft that brought the Apollo astronauts to the moon, just to give a couple of examples, are all built using Newtonian mechanics. So, does Newtonian mechanics constitute a correct model of reality?

On the other hand, if one wants to use the satellite based global positioning system (GPS) to determine a position on Earth accurately, Newtonian mechanics will fail. Einstein's general relativity theory does not talk about invisible forces of attraction between bodies but posits the very fabric of space being curved due to the presence of matter. A consequence of this theory is that "time flows faster" in the vicinity of matter. Concretely, this means that the clocks on the surface of the earth and on a satellite in earth orbit do not tick to the same rhythm. Since GPS makes use of accurate time intervals between signals received and sent by satellites, not taking this relativistic effect into account would reduce its accuracy. So, does Einstein's mechanics constitute a correct model of reality?

In case of building an aircraft, one might argue that Newtonian mechanics is demonstrably correct since aircraft fly all the time. In case of building an accurate GPS, one might argue that general relativity is demonstrably correct since, using GPS, hikers or soldiers in unfamiliar terrain know exactly where they are. It follows that the question about which model is better or correct depends on the problem at hand. Clearly, one does not need to deal with the complexity of Einstein's field equations when a Newtonian model suffices to go to the moon.

The pesky question arises inevitably: yes, but how is it in reality? Do objects attract each other or is the space curved due to the presence of matter or is it another yet to be developed theory? The answer is probably none of the above. Depending on the problem at hand it will be more appropriate to use one model of reality or other. The point to remember is that a model of reality is just that: a model. Like all models, that model also has its limitations. If one wants to have a model which incorporates *all* aspects of reality that model will have to be the reality itself as Mein Herr points out in Lewis Carroll's Sylvie and Bruno Concluded: "We actually made a map of the country, on the scale of a mile to the mile!"

Then pops up the next question: which model is closer to reality? Well, even if there is an absolute reality out there, it is highly questionable that science will ever know how it is in reality. As observations and intellectual capability improve, scientists will most likely be able to develop models which will explain more phenomena, more of the observations and make more precise predictions. However, they will probably never know in the epistemological sense whether they are any closer to the reality. Besides, and here comes a rather provocative statement, science is not interested in finding the reality, it is content to find progressively better models of it; search for reality is outside the scope of science!

2.3 Modeling Process

Maybe it will relax the reader to hear that there are no exact rules in mathematical modeling and there are no "correct answers" as should be apparent by now. But how does one get a mathematical model of a process? Let us begin answering this question with a couple of examples.

Example: Consider the inverted pendulum problem: there is a pole mounted with a ball bearing fulcrum on a cart as shown in Figure 2.7. The pole's vertical equilibrium point is unstable and it will fall down due to the smallest of disturbances. Therefore, the task is to move the cart left and right in such a way as to keep the pole stabilized at its vertical position, much like a jongleur at a country fair who keeps a sword balanced at the tip of his finger by moving his hand back and forth. How could one achieve this target?

The engineers among the readers will immediately recognize this as a typical control problem and will want to apply various techniques they have learned at control engineering classes. They will also remember that a mathematical model of the "plant" they should control is required. Therefore, they will begin to solve the problem by building a mathematical model of the system composed of the cart and the pole. To build the model, it makes sense firstly to analyze the forces involved. A diagram as in Figure 2.8 is very helpful here. Summing the forces in the diagram of the cart in the horizontal direction gives

$$M\ddot{x} + b\dot{x} + N = F, \qquad (2.12)$$

with M as the mass of the cart and b as the friction coefficient. Summing the forces in the diagram of the pole in the horizontal direction gives

$$N = m\ddot{x} + ml\ddot{\theta}\cos\theta - ml\dot{\theta}^2\sin\theta, \qquad (2.13)$$

with m as the mass of the pole, l as the length of the pole and θ as the angle of the pole from the vertical. Substituting (2.13) in (2.12) yields the first equation of motion for the system:

$$F = (M + m)\ddot{x} + b\dot{x} + ml\ddot{\theta}\cos\theta - ml\dot{\theta}^2\sin\theta. \qquad (2.14)$$

Fig. 2.7 How to move the cart left and right such that the pole on the cart stays vertical?

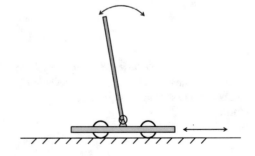

Fig. 2.8 Analyzing the
forces in the cart-pole system

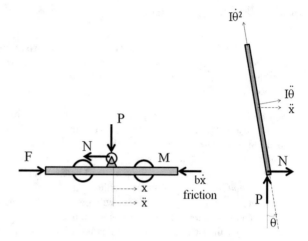

Summing up the forces acting on the longitudinal axis of the pole:

$$P sin\theta + N cos\theta - mg sin\theta = ml\ddot{\theta} + m\ddot{x}cos\theta . \tag{2.15}$$

Then, summing the moments around the center of gravity of the pole to get rid of the terms N and P:

$$- Pl sin\theta - Nl cos\theta = I\ddot{\theta} , \tag{2.16}$$

with I as the moment of inertia of the pole and combining (2.16) with (2.15) yields the second equation of motion as

$$(I + ml^2)\ddot{\theta} + mgl sin\theta = -ml\ddot{x}cos\theta . \tag{2.17}$$

At this stage it is assumed that the pole does not deviate too much from the vertical and the equations are linearized about $\theta = \pi$. With $\theta = \pi + \phi$ and ϕ small means $cos\theta \approx -1$, $sin\theta \approx -\phi$ and further $(\frac{d\theta}{dt})^2 = (\frac{d\phi}{dt})^2 \approx 0$ results in the linearized equations of motion

$$(I + ml^2)\ddot{\phi} - mgl\phi = ml\ddot{x} , \tag{2.18}$$
$$F = (M + m)\ddot{x} + b\dot{x} - ml\ddot{\phi} . \tag{2.19}$$

Defining the state vector as $\mathbf{x} = [x_1 \ x_2 \ x_3 \ x_4]^T = [x \ \dot{x} \ \phi \ \dot{\phi}]^T$ and assuming that the position of the cart (x) as well as the deviation from the equilibrium point of the pole (ϕ) are measured, the state space representation (see Appendix A) of the model can be written in the form

$$\dot{\mathbf{x}} = \mathbf{A}\mathbf{x} + \mathbf{b}u \qquad (2.20)$$
$$\mathbf{y} = \mathbf{C}\mathbf{x} + \mathbf{d}u$$

where

$$
\mathbf{A} = \begin{bmatrix} 0 & 1 & 0 & 0 \\ 0 & \dfrac{-(I+ml^2)b}{I(M+m)+Mml^2} & \dfrac{m^2gl^2}{I(M+m)+Mml^2} & 0 \\ 0 & 0 & 0 & 1 \\ 0 & \dfrac{-mlb}{I(M+m)+Mml^2} & \dfrac{mgl(M+m)}{I(M+m)+Mml^2} & 0 \end{bmatrix}, \quad
\mathbf{b} = \begin{bmatrix} 0 \\ \dfrac{(I+ml^2)}{I(M+m)+Mml^2} \\ 0 \\ \dfrac{ml}{I(M+m)+Mml^2} \end{bmatrix}, \quad (2.21)
$$

$$
\mathbf{C} = \begin{bmatrix} 1 & 0 & 0 & 0 \\ 0 & 0 & 1 & 0 \end{bmatrix}, \quad \mathbf{d} = \begin{bmatrix} 0 \\ 0 \end{bmatrix} \qquad (2.22)
$$

Comparing these equations with Figure 2.6 one can see that F, the force moving the cart left and right, is the control vector $\mathbf{u}$ (a scalar in this case), $\mathbf{f}$ and $\mathbf{g}$ are linear functions given by equation (2.20), the parameter vector $\mathbf{p}$ comprises the physical parameters M, m, l, I, g and b, and the disturbance $\mathbf{w}$ is ignored.

These equations represent a natural end of the modeling exercise. The system is modeled with a linear 4th order ordinary differential equation. The rest is a straight forward exercise in control engineering. Possible ways of calculating F are discussed in Chapter 4 and many impressive demonstrations can be found on the YouTube. $\square$

2.3.1 Stock Prices

An interesting problem that arises in financial engineering is how to determine the relative weights of stocks in a diversified portfolio [83] (Chapter 8 addresses this problem in detail). This requires estimates of future stock prices. Most of the time, these estimates are based on past prices of the stocks.[8]

Example: Imagine that the closing prices of a stock during the last 24 months are as shown in Table 2.1.

In order to be able to estimate the closing prices of this stock for the next few months one can naively model the price as a linear function of time and find the equation of the line that best fits the data of the past 24 months. A simple linear regression analysis based on Gaussian[9] least squares method results in

---

[8] Louis Jean-Baptiste Alphonse Bachelier, French mathematician (1870–1946) is credited with being the first person who tried modeling stock prices mathematically [134].

[9] Carl Friedrich Gauss, German mathematician (1777–1855); contributed significantly to many fields in mathematics and physics.

Fig. 2.9 Closing prices of a
stock during 24 months and
their regression line given by
equation (2.23)

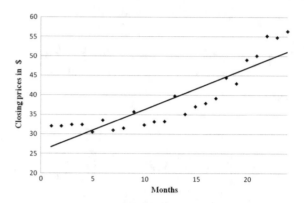

$$p(t) = 1.0596\,t + 25.635\,, \tag{2.23}$$

with p as the stock price and t as the time in months.

Plotting the equation (2.23) together with the past data is shown in Figure 2.9.

Equation (2.23) is a very rough model and the closeness of its fit to available data measured by Pearson product-moment correlation coefficient is $r^2 = 0.7935$ (the closer r^2 to 1, the better the fit [111]). When (2.23) is used to estimate the closing prices of this stock for months 25–28 (in other words, extrapolation) it results in the values in Table 2.2.

Now, observing the general tendency of the prices in Figure 2.9 to increase faster with time, one can hope to get a better fit of the data by using an exponential rather than a linear model. The regression analysis results in

$$p(t) = 27.593\,e^{0.0258t}\,. \tag{2.24}$$

Table 2.1 Closing prices of a stock during 24 months

Month	Closing price	Month	Closing price	Month	Closing price	Month	Closing price
1	32.12	7	31.03	13	39.85	19	43.01
2	32.12	8	31.56	14	35.21	20	49.05
3	32.45	9	35.75	15	37.11	21	50.07
4	32.48	10	32.39	16	38.00	22	55.22
5	30.37	11	33.19	17	39.25	23	54.82
6	33.61	12	33.29	18	44.52	24	56.46

Table 2.2 Closing prices
estimated using (2.23)

Month	Closing price
25	52.13
26	53.18
27	54.24
28	55.30

Fig. 2.10 Closing prices of a stock during 24 months and their regression line given by equation (2.24)

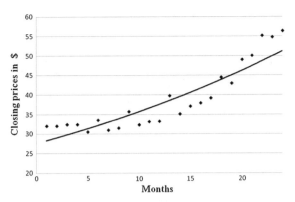

Table 2.3 Closing prices estimated using equation (2.24) are very similar to the result of the previous estimation

Month	Closing price
25	52.59
26	53.97
27	55.38
28	56.82

Table 2.4 Closing prices estimated using equation (2.25) look very different from the previous estimations

Month	Closing price
25	53.20
26	45.36
27	30.33
28	5.21
29	−33.41

Plotting the equation (2.24) together with the past data is shown in Figure 2.10.

Just by looking at their respective graphs, equation (2.24) seems to be a better model than equation (2.23). Also, here $r^2 = 0.8226$. When equation (2.24) is now used to estimate the closing prices of this stock for months 25–28, the values in Table 2.3 are obtained.

Can these results be improved? One can try to model the stock prices with a polynomial to get a better fit. Indeed, for a sixth order polynomial the best curve fit results in the equation

$$p(t) = -0.000008\,t^6 + 0.0006\,t^5 - 0.0149\,t^4 + 0.1829\,t^3 - 1.0506\,t^2 + 2.489\,t + 30.408\,. \tag{2.25}$$

The fit is impressive with $r^2 = 0.9666$. Since the fit is better, one might expect to get better estimates for the future values of the stocks. Therefore, (2.25) is now used to estimate the closing prices of this stock for months 25–28 which results in the values in Table 2.4.

Plotting the equation (2.25) together with the past data is shown in Figure 2.11.

Fig. 2.11 The model given
by equation (2.25) fits
available data very well.
However, this model results
in forecasts with negative
values for the stock and is
therefore, completely useless
for making predictions

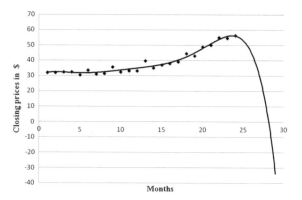

Clearly, the model in equation (2.25) is not an appropriate model for forecasting future stock prices. Actually, one can see that there is always at least one $n - 1$ or higher degree polynomial that goes through n points. This means that one can always find a polynomial model that fits the given data points perfectly ($r^2 = 1$ or zero residuals). However, such models are seldom useful for interpolating and hardly ever useful for extrapolating. Actually, the modeling of stock prices has been of great interest to scientists and researchers from fields as diverse as actuary and physics [134]. Stock prices are modeled using stochastic models based on Brownian motion with impressive success as shown in Chapters 5 and 9 [1]. □

2.3.2 Lessons Learned

Two modeling problems were presented and they were attacked with two very different approaches. In the first example, Newton's laws of motion were applied to the problem, deriving the differential equations governing the physical system in a fairly straight forward manner. This approach can be generalized to problems concerning the modeling of physical systems. One can analyze the given system using known physical laws like continuity or conservation of energy and obtain equations pertaining to the behavior of the system. Whenever possible, this is the preferred approach to modeling, because it also gives insight into the workings of the system. This way, what is actually going on can be "understood" and the model can be used to make fairly reliable predictions. The parameters have physical meanings like the mass of a body or the diameter of a pipe which can either be measured or estimated.

In the second example, the relationship between the input of the system (time) and its output (closing price of the stock) was considered as a black-box without any knowledge of its inner workings. This approach does not give any insights and is really an act of desperation. The modeling problem is quickly reduced to a curve fitting problem which can be solved mechanically. For instance, empirical evidence gained over centuries in the stock markets indicates the use of an exponential model [2]. Neither in this case nor in the case of polynomial models, the parameters of the model have any physical meaning. Also, as should be apparent by now, a good curve fit does not necessarily imply a good model which can be used to make predictions.

While modeling, either way, one always has to keep in mind the underlying assumptions. In the first example the equations of motion were linearized close to the vertical position of the pole. That means the model describes the behavior of the system for small deviations of the pole from the vertical in a fairly accurate way. On the other hand, as the pole moves away from its vertical position, the predictive power of the model deteriorates rapidly. Also, a controller based on that linearized model fails miserably. A common error in modeling and while using models is to forget the underlying assumptions and not be aware of the limitations or the applicability range of the model and try to use it to make predictions.

2.4   Parameter Identification

When physical laws are used to model a system, the parameters of the model (*e.g.,* the length of the pole) are often readily measurable. If not directly measurable, they can often be calculated with sufficient precision (*e.g.,* the mass of the pole can be calculated by first measuring its dimensions, using these dimensions to calculate its volume and multiplying the volume with the density of the material used to produce the pole). However, one might still end up with parameters that are neither directly measurable nor calculable (*e.g.,* the friction coefficient). Such parameters must be determined experimentally. The process of determining those model parameters is called *parameter identification.* Actually, while using black-box models, identification is often the only way to determine the model parameters, because the parameters have no physical meaning.

In order to be able to perform a reasonable parameter identification, a sufficiently large number of observations are necessary. Let us consider again the second example above. Imagine that the data available is limited to the first five data points. The result of a linear regression analysis on the available data is shown in Figure 2.12. A comparison with Figure 2.9 shows a radically different course of the regression line. Whereas the regression based on five data points predicts falling stock prices, a similar regression analysis using 24 data points predicts a rising course of the stock prices. This example demonstrates the necessity of using a sufficiently large number of data points for being able to perform parameter identification reliably. In most practical cases the consequence of using a small number of data points is not as dramatic as in this example. Rather, assuming that the model structure is correct, increasing the number of data points will increase the accuracy of the estimated parameters. Most widely used identification methods make use of linear or non-linear least squares methods. For instance [32] gives a detailed analysis of this problem.

How can one identify the parameters of a model when there are no or insufficient data for identification? The solution lies in conducting experiments to obtain sufficient data. Those experiments generally consist of exciting the system with a sufficiently rich set of inputs and observing its output. For that strategy to be successful, the model must be identifiable. A model is said to be identifiable if it is theoretically possible to infer the true value of the model's parameters after obtaining an infinite number of observations from it. In other words there must be a unique solution to the problem of finding the mapping how the input is mapped to the output of the model (see, for instance, [74] for a formal definition of identifiability).

Fig. 2.12 Regression line based on the first five data points predicts falling stock prices

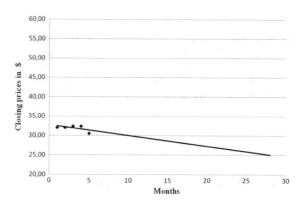

If a system is modeled as a linear system, in an experiment to identify the parameters of the model, in theory, it is sufficient to excite the system with an impulse or a step function and to observe its output at different times.[10] Since measurements cost time and money, one cannot make an infinite number of measurements. Therefore, one has to agree on a number of measurements. The next question to be resolved is when to conduct those measurements. Clearly, it does not help to make many measurements once the transient behavior is over. Last two points make sense intuitively. The general problem of experiment design is more complicated and treated in detail in *e.g.,* [9, 43, 108].

2.5   Mathematics of Parameter Identification

Figure 2.13 illustrates how to identify the parameters of a process. The procedure begins by applying a test input u to the system and measure its output y. The same signal is used to simulate the output of the model, $\hat{y}$. The difference between the measured and simulated outputs is built as $e = y - \hat{y}$ and squared to get $J(\mathbf{p})$ which has to be minimized with respect to the model parameters $\mathbf{p}$. This process can be repeated until the model output approximates the measured output closely enough.

How is $J(\mathbf{p})$ minimized? For that purpose, it is convenient to look at sampled values of the signals such that

$$e_k = y(t_k) - \hat{y}(t_k) \,, \tag{2.26}$$

$$J(\mathbf{p}) = \sum_{k=1}^{n} e_k^2 \,, \tag{2.27}$$

where t_k are the sampling time points and n is the number of samples. Thus, the problem becomes finding the optimal values of m parameters $p_1, p_2, \ldots, p_m$ which are aggregated in the parameter vector $\mathbf{p}$ as

---

[10]A unit impulse function contains all frequencies, because its Laplace transform is simply 1. Therefore, theoretically, it is a sufficiently rich input for identification purposes.

$$\mathbf{p}^* = \arg \min_{\mathbf{p}} J(\mathbf{p}).$$

(2.28)

But how does one numerically find the optimal values of the parameters which minimizes a function? There are scores of different ways to solve that problem. Let us first review some basic concepts before discussing several different algorithms.

2.5.1 Basics of Extremes

Given a continuous function with a single variable, $f(x)$ where x is unbounded and the derivatives $f'(x), f''(x), \ldots$ exist and are continuous, the Taylor series expansion can be given as

$$f(x + \Delta x) = f(x) + f'(x)\Delta x + \frac{f''(x)}{2!}(\Delta x)^2 + \ldots$$

$$\Rightarrow \Delta f = f(x + \Delta x) - f(x) = f'(x)\Delta x + \frac{f''(x)}{2!}(\Delta x)^2 + \ldots.$$

Looking at Figure 2.14 one can see that the points A and B result when $f'(x) = 0$ is solved for x. However, at A $f''(x) < 0$, therefore A is a maximum point and at B $f''(x) > 0$, therefore B is a minimum point.

If, on the other hand, $f(x)$ is monotonous and x is bounded as in Figure 2.15, second derivatives become irrelevant. The extremes are at the boundaries of the definition domain of the function.

When functions with two variables, $f(x, y)$, are now considered again with unbounded x, y, the necessary conditions for extreme points become

$$f_x := \frac{\partial f}{\partial x} = 0, \quad f_y := \frac{\partial f}{\partial y} = 0.$$

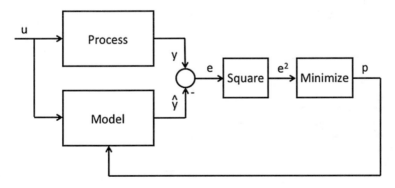

Fig. 2.13 Schematic representation of parameter identification

Furthermore, for a maximum point the conditions

$$f_{xx} < 0, \qquad f_{xx} f_{yy} - f_{xy}^2 > 0,$$

and for a minimum point the conditions

$$f_{xx} > 0, \qquad f_{xx} f_{yy} - f_{xy}^2 > 0,$$

must be satisfied [19].

Extending these conditions for functions with n variables $x_1, x_2, \ldots, x_n$, the necessary conditions for extrema become

$$f_{x_i} = 0, \qquad \forall i = 1, 2, \ldots, n.$$

Fig. 2.14 Continuous function with an unbounded variable

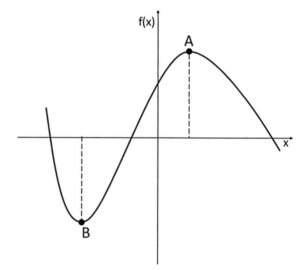

Fig. 2.15 The function $f(x)$ is defined for $a \leq x \leq b$. Absolute minimum is at $x = a$ and absolute maximum is at $x = b$. If the definition domain were open, i.e., $a < x < b$, then the function would not have a minimum or a maximum in the domain. In that case one speaks of an infimum and a supremum

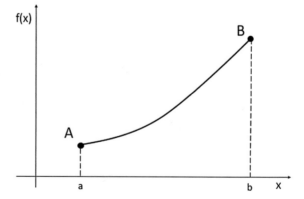

Fig. 2.16 Where to look for extreme values of a function

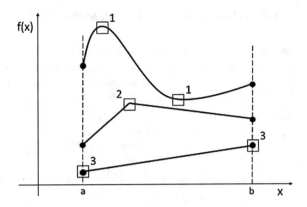

Furthermore, one has to build the Hessian matrix $\mathbf{H}$ composed of the second-order partial derivatives as

$$\mathbf{H} = \begin{vmatrix} f_{x_1x_1} & f_{x_1x_2} & \cdots & f_{x_1x_n} \\ f_{x_2x_1} & f_{x_2x_2} & \cdots & f_{x_2x_n} \\ \cdots & \cdots & \cdots & \cdots \\ f_{x_nx_1} & f_{x_nx_2} & \cdots & f_{x_nx_n} \end{vmatrix},$$

calculate the leading first minors[11] of $\mathbf{H}$ as D_i and check their signs.[12] For minima

$$D_i > 0 \qquad \forall i = 1, 2, \ldots, n,$$

and for maxima

$$(-1)^i D_i > 0 \qquad \forall i = 1, 2, \ldots, n.$$

It is important to keep the Weierstrass extreme value theorem in mind whilst seeking the extreme points of a function. This theorem loosely says that if a function f is continuous in the closed and bounded interval $[a, b]$, then f must have at least a maximum and at least a minimum. Those extreme points can be found either *1*) at the stationary points ($f_{x_i} = 0$), or *2*) at points where one or more partial derivatives become discontinuous, or *3*) at the boundaries of the interval. Figure 2.16 illustrates this theorem for functions with a single variable.

---

[11] A minor of a matrix $\mathbf{A}$ is the determinant of some smaller square matrix, cut down from $\mathbf{A}$ by removing one or more of its rows or columns. Minors obtained by removing just one row and one column from square matrices are called first minors. First minors that are obtained by successively removing the last row and the last column of a matrix are called leading first minors.

[12] A similar test checks the positive or negative definiteness of the Hessian matrix via its eigenvalues to determine whether a critical point ($f_{x_i} = 0$) is a minimum or a maximum [90].

2.5.2 *Optimization with Constraints*

Engineers very often encounter the problem of minimizing a function while at the same time a functional equality constraint must be fulfilled. Formally, for a function with two variables x and y, one is seeking the extreme points of $z = f(x, y)$ with the constraint $g(x, y) = 0$. Using the following total differentials for extrema

$$df = \frac{\partial f}{\partial x}dx + \frac{\partial f}{\partial y}dy = 0 \quad \Rightarrow \quad \frac{dy}{dx} = -\frac{f_x}{f_y},$$

$$dg = \frac{\partial g}{\partial x}dx + \frac{\partial g}{\partial y}dy = 0 \quad \Rightarrow \quad \frac{dy}{dx} = -\frac{g_x}{g_y}.$$

Therefore,

$$\frac{dy}{dx} = -\frac{f_x}{f_y} = -\frac{g_x}{g_y}.$$

Introducing the so-called Lagrange[13] multiplier

$$\lambda = -\frac{f_x}{g_x} = -\frac{f_y}{g_y},$$

or

$$f_x + \lambda g_x = 0, \qquad f_y + \lambda g_y = 0.$$

This means

$$\frac{\partial}{\partial x}(f + \lambda g) = 0, \qquad \frac{\partial}{\partial y}(f + \lambda g) = 0.$$

With the new principal function, called the Lagrangian, $F = f + \lambda g$ and F being stationary for optimality, the optimization with constraints problem is now reduced to a simple algebraic problem with three equations

$$F_x = 0,$$
$$F_y = 0,$$
$$g(x, y) = 0,$$

with the three unknowns x, y, λ. This result can readily be extended to optimizing a function $z = f(x_1, x_2, \ldots, x_n)$ with n variables subject to r constraints

---

[13] Joseph-Louis Lagrange, French mathematician (1736–1813).

$$g_1(x_1, x_2, \ldots, x_n) = 0,$$
$$g_2(x_1, x_2, \ldots, x_n) = 0,$$
$$\vdots$$
$$g_r(x_1, x_2, \ldots, x_n) = 0,$$

by constructing the Lagrangian $F = f + \lambda_1 g_1 + \lambda_2 g_2 + \ldots + \lambda_r g_r$ and the condition for optimality (all derivatives of F must vanish) results in n equations $F_{x_1} = F_{x_2} = \ldots = F_{x_n} = 0$. $(n + r)$ algebraic equations with $(n + r)$ unknowns need to be solved now.

Example: Where is the minimum of $f(\mathbf{x}) = 2x_1^2 + x_2^2$ subject to the condition $g(\mathbf{x}) = 1 - x_1 - x_2$?

Firstly, building the Lagrangian as

$$L(x_1, x_2, \lambda) = 2x_1^2 + x_2^2 + \lambda(1 - x_1 - x_2),$$

and taking the derivatives of the Lagrangian and equating them to zero as

$$\frac{\partial L}{\partial x_1}(x_1^*, x_2^*, \lambda^*) = 4x_1^* - \lambda^* = 0,$$

$$\frac{\partial L}{\partial x_2}(x_1^*, x_2^*, \lambda^*) = 2x_2^* - \lambda^* = 0,$$

$$\frac{\partial L}{\partial \lambda}(x_1^*, x_2^*, \lambda^*) = 1 - x_1^* - x_2^* = 0,$$

and solving these equations results in $x_1^* = \frac{1}{3}$, $x_2^* = \frac{2}{3}$ and $\lambda^* = \frac{4}{3}$. Therefore, the optimal value of the objective function is $\frac{2}{3}$. Since the objective function is convex and the constraint is linear, this optimum solution minimizes the objective function subject to the constraint.

The geometric interpretation of this problem becomes apparent when the objective function and the constraint are plotted as shown in Figure 2.17. The gradients of f and g at the optimum point must point in the same direction but they may have different lengths. In other words, $\nabla f(\mathbf{x}^*) = \lambda \nabla g(\mathbf{x}^*)$. This is exactly the condition for optimality $\nabla L(\mathbf{x}^*, \lambda^*) = 0$.

It is not always easy to find interpretations for Lagrange multipliers in constrained optimization problems. However, in economic problems of the type "Maximize $p(x)$ subject to $g(x) = b$" where $p(x)$ is a profit to maximize and b is a limited amount of resource, then the optimal Lagrange multiplier λ^* is the marginal value of the resource. In other words, if b were increased by Δ, the profit would increase by $\lambda^* \Delta$. Similarly, in problems of the type "Minimize $c(x)$ subject to $d(x) = b$" where $c(x)$ is the cost to minimize and b is the demand to be met, then the optimal Lagrange multiplier λ^* is the marginal cost of meeting the demand.

Coming back to the example, the minimum value of $f(\mathbf{x}) = 2x_1^2 + x_2^2$ subject to the condition $x_1 + x_2 = 1$ was found to be $\frac{2}{3}$ at $x_1^* = \frac{2}{3}$, $x_2^* = \frac{2}{3}$. If the right hand side is now changed from 1 to 1.05 (*i.e.*, $\Delta = 0.05$), then the optimum objective function value goes from $\frac{2}{3}$ to roughly $\frac{2}{3} + \frac{4}{3} \cdot 0.05 = \frac{2.2}{3}$. $\qquad\qquad$ □

Example: A common problem in finance is the allocation of funds among different investment instruments. How should an investor divide her savings among three mutual funds with expected returns 10%, 10% and 15% respectively, to minimize her risk while achieving an expected return of 12%? The risk is defined as the variance of the return of an investment (this is discussed in detail in Chapter 6). With the fraction x of the total invested amount being invested in Fund 1, y in Fund 2 and z in Fund 3, where necessarily $x + y + z = 1$, the variance of the return in this example has been calculated to be

$$r(x, y, z) = 400x^2 + 800y^2 + 200xy + 1600z^2 + 400yz.$$

So, the problem is to minimize $r(x, y, z)$ subject to

$$x + y + 1.5z = 1.2,$$
$$x + y + z = 1.$$

Using the Lagrangian method above, the optimal solution results in

$$x^* = 0.5, \quad y^* = 0.1, \quad z^* = 0.4, \quad \lambda_1^* = 1800, \quad \lambda_2^* = -1380,$$

where λ_1 is the Lagrange multiplier associated with the first constraint and λ_2 with the second constraint. The optimal value of the objective function, in other words the minimum variance of the return, is 390.

It is now possible to answer the following question approximately: If an expected return of 12.5% was desired instead of 12%, what would the corresponding variance of the return be? Δ is 0.05 and the variance would increase by $\Delta\lambda_1 = 0.05 \cdot 1800 = 90$. Therefore, the answer is $390 + 90 = 480$. Intuitively, it is not surprising that a higher risk has to be accepted in order to achieve a higher return. This thinking is the basis of Modern Portfolio Theory which is treated in Chapter 8.   □

Fig. 2.17 Geometric interpretation of minimizing $f(x_1, x_2) = 2x_1^2 + x_2^2$ subject to the constraint $g(x_1, x_2) = 1 - x_1 - x_2$

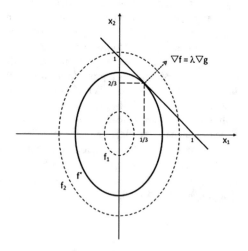

Sometimes the constraints which need to be considered are inequalities. The procedure for finding the constrained optimum in this case becomes a bit more tedious and is conducted by checking so-called Karush-Kuhn-Tucker conditions, often abbreviated as KKT [73].

2.6 Numerical Methods for Parameter Identification

It became clear so far, that parameter identification ultimately boils down to an optimization problem with constraints. The objective function to minimize is $Z = Z(p_1, p_2, \ldots, p_m)$ where $p_1, p_2, \ldots, p_m$ are the m parameters to be identified and Ω denotes the permissible region in the parameter space. As Figure 2.18 shows for a case with two parameters, the lowest point of Z in Ω is sought. In most cases, an analytical solution of the problem is not possible and solving it numerically is the only option. There are a myriad of different algorithms in the literature. Some of the best books in this area are [14, 16, 21, 100]. Only a couple of representative examples of numerical optimization algorithms are described in this section.

There are two important caveats here: (*a*) no single numerical method is applicable for all problems, and (*b*) where the objective function is not convex (multi-modal), meaning that it has several local minima, one can never be sure that the solution the numerical search algorithm delivers is the absolute minimum. Therefore, for most of the numerical optimization algorithms to be successful, several assumptions about the objective function are necessary:

- Z is convex (unimodal) in Ω,
- Z is "sufficiently smooth" in Ω, meaning that the parameter landscape has no precipitous ravines,
- Z is differentiable with respect to all parameters p_i in Ω,
- The region Ω is simply connected, meaning that it has no "holes".

The numerical search algorithms must necessarily have termination criteria. These can be (a) the maximum number of steps exceeded, (b) change in the value of Z from one step to the next is below a threshold, (c) change in parameter values from one step to the next is below a threshold.

Fig. 2.18 The curves denote the counter lines of the objective function Z (points where Z is constant). Ω is the permissible region for the parameters p_1, p_2. A is a relative minimum in Ω. B is the absolute minimum in Ω

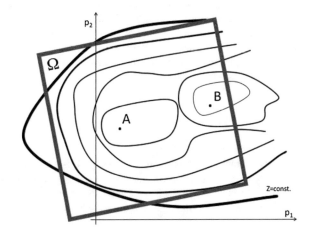

2.6.1 Golden Section

A popular method for numerically minimizing a unimodal function with a single variable derives from the Fibonacci[14] search method. Assuming that the interval (b^1, u^1) entails the minimum, the algorithm can be defined as follows:

1. Find the points $\lambda_1^1 = (1 - \beta)(u^1 - b^1) + b^1$ and $\lambda_2^1 = \beta(u^1 - b^1) + b^1, k = 1$; go to (3)
2. Find the points $\lambda_1^k = (1 - \beta)(u^k - b^k) + b^k$ and $\lambda_2^k = \beta(u^k - b^k) + b^k$; calculate $Z(\lambda_1^k)$ and $Z(\lambda_2^k)$
3. If $Z(\lambda_1^k) > Z(\lambda_2^k) b^{k+1} = \lambda_1^k, \lambda_1^{k+1} = \lambda_2^k, u^{k+1} = u^k, k = k + 1$, go to (2) and calculate λ_2^k.
   Else $u^{k+1} = \lambda_2^k, \lambda_2^{k+1} = \lambda_1^k, b^{k+1} = b^k, k = k + 1$, go to (2) and calculate λ_1^k
4. Repeat above steps until $\lambda_2^k - \lambda_1^k < \epsilon$.

This method is called golden section,[15] because

$$\frac{b^1 \lambda_2^1}{b^1 u^1} = \frac{b^1 \lambda_1^1}{b^1 \lambda_2^1} = \beta, \qquad \frac{b_1 \lambda_1^1}{b^1 u^1} = 1 - \beta$$

$$\Rightarrow \beta^2 = 1 - \beta \qquad \Rightarrow \beta = \frac{-1 + \sqrt{5}}{2} \approx 0.618,$$

which is known as the golden ratio.

2.6.2 Successive Parameter Optimization

So, what does one do when one has several parameters which must be identified? One possibility is to keep all the parameters constant except for one and optimize it, possibly using thegolden section method or something similar. Then go to the next parameter, keep all the rest of the parameters constant and optimize that one. One keeps doing this successively for each parameter.[16] The optimum combination of parameters which minimizes the objective function will result pretty quickly, provided that the objective function is unimodal. However, most of the time, one has no way of knowing whether the objective function is unimodal. Therefore, it is best to assume that it is not and repeat the optimization process, starting with a new set of initial values of parameters until the results are satisfactory. Figure 2.19 illustrates successive parameter optimization.

---

[14]Leonardo Fibonacci, Italian mathematician (c.1175–c.1250).

[15]Already the Ancient Greek mathematicians, most notably Euclid, were fascinated by the so-called golden ratio which is recurrent in the nature. This ratio is given as $\frac{a+b}{a} = \frac{a}{b}$.

[16]In Germanic literature this method is known as "*achsenparallele Suche*" or "search parallel to the axes".

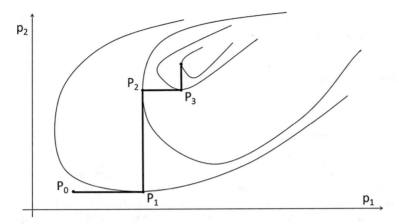

Fig. 2.19 Optimizing one parameter after the other. Note that the contour lines are not known

Let us call the starting set of parameters P_0 and the step size Δ.

$$
\begin{aligned}
P_0: \quad & Z_0 && = Z(p_{10}, p_{20}, \ldots, p_{n0}); \\
P_1: \quad & Z_{01} && = Z(p_{10} + N_1\Delta, p_{20}, \ldots, p_{n0}); \quad p_{11} = p_{10} + N_1\Delta \\
P_2: \quad & Z_{02} && = Z(p_{11}, p_{20} + N_2\Delta, \ldots, p_{n0}); \\
& && = Z(p_{11}, p_{21}, p_{30}, \ldots, p_{n0}); \\
& && \quad\vdots \\
P_n: \quad & Z_{0n} && = Z(p_{11}, p_{21}, p_{31}, \ldots, p_{n0} + N_n\Delta); \\
& && = Z(p_{11}, p_{21}, p_{31}, \ldots, p_n); \\
& && = Z_1 .
\end{aligned}
$$

P_n is then the starting point of the second search phase. It makes sense to reduce the step size Δ as one comes closer to the minimum.

Another way for optimizing an objective function is to start somewhere, look for the steepest gradient and move a step in that direction. At this new point again look for the steepest gradient and move a step in that direction and so on. This method is known as the *steepest descent*. Several variations of the steepest descent method are explained in detail, for instance, in [123]. It is also possible to proceed without computing any gradients. The so-called *direct search* method looks around the immediate vicinity of the starting point to make a picture of the topography and makes a long step in a descent direction. The procedure is then repeated [71].

The use of so-called *linear programming* is indicated if the objective function and the constraints are all linear [121]. This method of optimization was developed in the middle of the previous century to solve logistics problems with the then nascent digital computers.

A more modern approach is the so-called *particle swarm optimization* which finds an optimum by iteratively improving the result. It begins with a swarm of particles as candidate results and moves these around the parameter space based on their positions and velocities. Each particle moves towards a better solution based not only on their own positions and velocities but also on those of other particles [68].

2.7   Model Validation

Model validation is the process of checking whether a model is appropriate for the intended purpose. Remembering that a model is, by its very nature, but an approximation of reality, it becomes clear that it cannot replicate all the features of the actual system. Therefore, to validate a model means to verify that it can reproduce the features that are essential for the purpose for which the model was developed. One has to keep in mind that model validation is an integral part of the modeling process. In other words, one has to verify the model's appropriateness both during the modeling process and at the end of it. This is done primarily by simulating the system behavior by solving the model equations with the identified parameters. In most cases, the model consists of a set of differential equations which must be integrated numerically. Many techniques and algorithms and even dedicated simulation languages are available [26, 27].

Another important requirement of a model is its parsimony. It is quite possible to develop models that describe the same phenomena appropriately but with varying degrees of complexity. Among such models, the one with the lowest order or smallest number of parameters, the most parsimonious one, is to be preferred. This principle is a direct application of Occam's razor.[17]

A common problem in connection with modeling and particularly model validation is model reduction. In many applications, the available model of a system, although appropriate for most purposes of modeling, might be too complex for controller design purposes. It then makes sense to reduce the order of the model. There are several ways of achieving model reduction. Some are obvious, straight forward and simple. Others are more sophisticated and can be rather tedious to perform (*e.g.,* balanced model reduction [50]).

2.8   Summary

This Chapter began by discussing what is meant by a model. A model is always an approximation of reality. Therefore, there is no one "true model". Depending on the context and purpose, there are models which are better than others. Using the example of a falling object it was seen that depending on the context, it might make sense to neglect certain aspects of reality in the modeling process.

---

[17]Occam's razor is the law of parsimony, economy or succinctness. It is a principle urging one to select from among competing hypotheses the one which makes the fewest assumptions. Although the principle was known earlier, Occam's razor is attributed to the 14th-century English Franciscan friar William of Ockham.

Fig. 2.20 Complete
modeling process as adapted
from [30]. Darker shades
indicate more advanced
phases of modeling

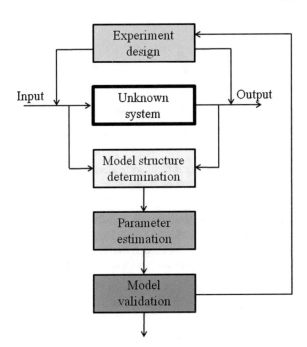

Although there are many different types of models, the coming Chapters focus on mathematical models. Mathematical models can be classified under several dichotomic headings: inductive versus deductive models, deterministic versus stochastic models, static versus dynamic models, discrete versuscontinuous models, lumped parameter models versus distributed parameter models and linear models versus non-linear models. Nature is inherently dynamic, stochastic, continuous, and non-linear with distributed parameters, hence difficult to model properly. Therefore, simplifying assumptions are necessary to make the model more amenable to available mathematical techniques. Four different types of models are categorized depending on their purposes as descriptive, interpretive, predictive or explanatory models.

The modeling process is best summarized using Figure 2.20. Firstly, the model structure for the given unknown system is determined by applying relevant physical laws (mechanics, electromagnetism, optics, etc.) relating to the system as demonstrated in the example of the inverted pendulum. If that is not possible the input-output data of the given unknown system must be analyzed as a black-box as shown in the example of stock prices. Secondly, an experiment is designed to obtain further data which is then used to identify the unknown parameters in the system model. It was discussed in detail how this is mathematically reduced to an optimization problem and various numerical methods of parameter optimization were elucidated. Finally, in the validation phase, the model's appropriateness is examined by comparing the available data with the results obtained from simulations of the system with the parameters identified in the previous phase. Further experiments might be necessary to obtain additional data to improve parameter identification if the results are not satisfactory. If continued experiments do not yield a parameter set which results in a satisfactory model, it might be necessary to modify the model structure.

2.9 Exercises

A1: Explain in your own words what you understand when you hear the word "model".

A2: Explain how you would go about modeling an electro-mechanical system.

A3: Why is it difficult to develop a model to predict stock prices?

A4: What are the difficulties in designing experiments for validating a mathematical model of a process?

A5: What are the major difficulties in identifying the parameters of a model numerically and how can those difficulties be dealt with?

B1: A furniture manufacturer produces two sizes of boxes (large, small) that are used to make either a table or a chair. A table makes $3 profit and a chair makes $5 profit. If m small blocks and n large blocks are produced, how many tables and chairs should the manufacturer make in order to obtain the greatest profit? What else do you need to know?

B2: Investigate the population dynamics of whales and krill, where krill is assumed to be the main source of food for the whales. We will use the following model:

$$\dot{k} = (a - bw)\,k\,,$$
$$\dot{w} = (-m + nk)\,w\,.$$

k is the krill population (as a function of the time in years), w is the whale population, and a, b, m and n are positive constants. Firstly, interpret this system of differential equations and the meaning of the constants. Determine if there are any equilibrium points of this system. If so, can you say anything about the stability of these points? Write the equations for simulating this system. Consider the parameter values $a = 0.2, b = 0.0001, m = 0.5, n = 0.000001$. With the starting points $k(0) = 700000$ and $w(0) = 3000$, and time step $\Delta T = 0.3$ years, manually simulate $k(t)$ and $w(t)$ for $t = 0.3$ and $t = 0.6$. Using the parameter values above, use Mathematica (or any other simulation package) to simulate the model from several different starting points. To get interesting results do not just pick arbitrary starting points, but be systematic and think how they relate to the question about the stability of equilibrium points. Discuss your observations. Can you draw general conclusions not dependent on this particular choice of values? Investigate the effect of krill fishing on these populations. To model this, add a term $-rk$ to the equation for $\dot{k}$, where $r < a$. Interpret this term. Try out different values of r, simulate and discuss your observations. Could you suggest how the model could be refined or extended?

B3: Person A wants to avoid person B, who insists to casually meet A at lunchtime. At the university, there are two restaurants, one with a dining time of 20 min and one with a dining time of 50 min. For simplicity, we can assume that A and B have to go to lunch at 12.00, and that it is not an option not to have lunch. What strategy should A choose to minimize the time spent with B? Conversely, what is the most successful strategy for B? How much time will A have to spend with B? Hint: try first to intuitively understand and solve the problem.

B4: Use the golden section method to find the maximum of the function

$$f(x) = -x^2 + 21.6x + 3$$

in the interval $[0, \ 20]$ with a tolerance level of 0.001. Solve the problem with different starting points.

B5: The following linear control loop is given in the frequency domain:

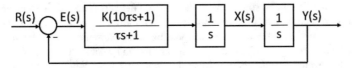

Determine the controller parameters K and τ such that $Z = \int_0^\infty e^2 dt$ is minimized with the constraint $J = \int_0^\infty x^2 dt = A$. Assume $r(t)$ to be a unit step. Discuss the case when $J < A$.

B6: Consider the following discrete time system

$$\mathbf{z}[k+1] = \mathbf{A}\mathbf{z}[k] + \mathbf{b}u[k], \qquad y[k+1] = \mathbf{c}\mathbf{z}[k+1],$$

where

$$\mathbf{z} = \begin{bmatrix} z_1 \\ z_2 \end{bmatrix}, \qquad \mathbf{A} = \begin{bmatrix} a_{11} & a_{12} \\ 0 & a_{22} \end{bmatrix}, \qquad \mathbf{b} = \begin{bmatrix} 0 \\ 1 \end{bmatrix}, \qquad \mathbf{c} = \begin{bmatrix} 1 & 0 \end{bmatrix}.$$

In this problem, we will explore some of the properties of this discrete time system as a function of the parameters, the initial conditions, and the inputs. (a) Assume $a_{12} = 0$ and that there is no input, $u = 0$. Write a closed form expression for the output of the system from a non-zero initial condition $\mathbf{z}[0] = (z_1[0], z_2[0])$ and give conditions on a_{11} and a_{22} under which the output gets smaller as k gets larger. (b) Now assume that $a_{12} \neq 0$ and write a closed form expression for the response of the system from a non-zero initial condition. Give conditions on the elements of $\mathbf{A}$ under which the output gets smaller as k gets larger. (c) Write a MATLAB program to plot the output of the system in response to a unit step input, $u[k] = 1, k \geq 0$. Plot the response of your system with $\mathbf{z}[0] = \mathbf{0}$ and $\mathbf{A}$ given by

$$\mathbf{A} = \begin{bmatrix} 0.5 & 1 \\ 0 & 0.25 \end{bmatrix}.$$

B7: Write an essay arguing for and against the statement in "A Philosophical Excursion" that search for reality is outside the scope of science.

C1: Historically, some concepts like submarines or space travel were first introduced in science fiction literature and became science fact with time. Others, like time travel or Startrek style "beaming" have not (yet?) made that transition. Consider Hari Seldon's "psychohistory". Do you think it might one day become a fact? Argue your case using concepts like gas dynamics, stochastic models and chaos.

C2: It is not obvious how to tell just looking at data whether a physical system is random or chaotic. Some have even argued philosophically that these concepts might be observationally equivalent. Develop criteria and methods for distinguishing the two.

C3: A doctor wants to prescribe a safe but effective dosage. How much should she prescribe, how often and when should the patient take the medicine?

C4: Coming back to the example of inverted pendulum in Section 2.3, we modify the problem the following way: let us assume that the cart moves on an elevated track and the pendulum is not a standing pole but a rope hanging down from the cart. The task is to move the cart horizontally in such way that the rope moves from an arbitrary swinging state to a desired position at complete stand still in the least possible time. Note that we are interested in *all* points on the rope being transformed from any position and velocity to a stand still at another position. How would you go about solving this problem? What kind of equations will you need to model the rope and the cart?

C5: In a futuristic bar you are being served your cocktails by robots. The robots are supposed to bring your cocktails as fast as possible without spilling them. How would you model the movement of the drinks in the cocktail glasses? What if the same robot is serving several different glasses at the same time? First, assume the robot moves on a straight line only. Secondly, assume it can move freely in a room. Thirdly, assume that the floor on which the robot moves is not smooth.

Chapter 3
Probability and Stochastic Processes

Probability is expectation founded upon partial knowledge. A perfect acquaintance with all the circumstances affecting the occurrence of an event would change expectation into certainty, and leave neither room nor demand for a theory of probabilities.

— George Boole

If we have an atom that is in an excited state and so is going to emit a photon, we cannot say when it will emit the photon. It has a certain amplitude to emit the photon at any time, and we can predict only a probability for emission; we cannot predict the future exactly.

— Richard Feynman

3.1 Introduction

This Chapter builds on Kolmogorov's axioms and reviews basic concepts of probability theory. Mathematical description and analysis of stochastic processes with emphasis on several relevant special classes like stationary and ergodic processes are explained. Some special processes such as the normal process, the Wiener process, the Markov process and white noise are then discussed.

Most people find concepts relating to probability difficult to understand and to employ correctly. An illustrative example is known as the Monty Hall puzzle, named after the Canadian-American TV-show host best known for his television game show "Let's Make a Deal" in the 1970's. The puzzle which became famous as a question from a reader's letter quoted in Marilyn vos Savant's column in *Parade* magazine in 1990 is stated as

Suppose you are on a game show, and you are given the choice of three doors: Behind one door is a car; behind the others, [toy] goats. You pick a door, say No. 1, and the host, who knows what is behind the doors, opens another door, say No. 3, which has a [toy] goat. He

© Springer International Publishing AG 2018
S. S. Hacısalihzade, *Control Engineering and Finance*, Lecture Notes in Control and Information Sciences 467, https://doi.org/10.1007/978-3-319-64492-9_3

then says to you, "Do you want to pick door No. 2?" Is it to your advantage to switch your choice?

An overwhelming majority of people will answer "It does not matter whether the contestant changes her choice or not, because there are two doors and the chances of the car being behind one or the other is fifty-fifty." Well, this is not true. Actually, by changing her original choice the contestant *doubles* her chance of winning the car! The reader will be able to ascertain the correctness of this statement by employing some very basic concepts which are presented in this Chapter.

3.2 History and Kolmogorov's Axioms

Probability theory and statistics deal with mathematical models of random events and with laws for observed quantities which are described with these models. This field of mathematics was not treated systematically until the 16th century. First known attempts to analyze games of chance mathematically are attributed to Cardano[1] in his book entitled *Liber de Ludo Aleae* (Book on Games of Chance) which was published posthumously in 1663. But also other well-known mathematicians like Galileo Galilei,[2] Pierre de Fermat[3] and Blaise Pascal[4] have contributed probability models to study the game of dice. Statistics (a word derived from Latin *Statisticum Collegium* meaning council of state) was seen for a long time as merely the art of collecting data, their visual representation and their interpretation especially in relation with population phenomena. Mathematical statistics came about with the application of probability models in the evaluation of observations of demographic data. Jakob Bernoulli[5] was the first to establish the relationship between descriptive statistics and probability theory in his work *Ars Conjectandi* (The Art of Conjecture), published posthumously in 1713.

Modern probability theory is surprisingly recent. It is built on the foundations laid by Andrey Nikolaevich Kolmogorov[6] in his manuscript *Grundbegriffe der Wahrscheinlichkeitstheorie* (Fundamental Concepts of Probability Theory) published in 1933. The axiomatic approach in this book is likened by some historians to Euclid's axiomatic approach to plane geometry. Kolmogorov formulated five axioms: Let E be a collection of elements $\xi, \eta, \zeta, \ldots$, which are called elementary

---

[1] Girolamo Cardano, Italian Renaissance mathematician (1501–1576); famous for his solution of the cubic equation and as the inventor of the universal joint.

[2] Galileo Galilei, Italian astronomer, physicist, engineer, mathematician and philosopher (1564–1642); universally considered to be the father of observational astronomy and more significantly of the scientific method.

[3] Pierre de Fermat, French lawyer and mathematician (1607–1665); probably best known for his "last theorem" which is very easy to state but which resisted proof until the end of the 20th century.

[4] Blaise Pascal, French mathematician, physicist and philosopher (1623–1662).

[5] Jakob Bernoulli, Swiss mathematician, (1654–1705).

[6] Andrey Nikolaevich Kolmogorov, Russian mathematician, (1903–1987).

events, and let $\mathfrak{F}$ be a set of subsets of E; the elements of the set $\mathfrak{F}$ are called random events:

I $\mathfrak{F}$ is a field of sets.
II $\mathfrak{F}$ contains the set E.
III A non-negative real number $P(A)$ is assigned to each set A in $\mathfrak{F}$; this number is called the probability of the event A.
IV $P(E)$ equals 1.
V   If A and B have no common elements, then $P(A \cup B) = P(A) + P(B)$.

A system of sets $\mathfrak{F}$, together with a definite assignment of numbers $P(A)$ satisfying the Axioms I - V, is called a *probability field*.

A field has the following properties with respect to the operations '+' and '·':

(i)	Closure	$\forall\, a, b \in \mathfrak{F}$	$a + b,\ a \cdot b \in \mathfrak{F}$.
(ii)	Associativity	$\forall\, a, b, c \in \mathfrak{F}$	$a + (b + c) = (a + b) + c,\ a \cdot (b \cdot c) = (a \cdot b) \cdot c$.
(iii)	Commutativity	$\forall\, a, b \in \mathfrak{F}$	$a + b = b + a,\ a \cdot b = b \cdot a$.
(iv)	Identity existence	$\forall\, a \in \mathfrak{F}$	$\exists\, 0, 1 \mid a + 0 = a,\ a \cdot 1 = a$.
(v)	Inverse existence	$\forall\, a \in \mathfrak{F}$	$\exists\, -a,\ a^{-1} \mid a + (-a) = 0,\ a \cdot a^{-1} = 1$.
(vi)	Distributivity	$\forall\, a, b, c \in \mathfrak{F}$	$a \cdot (b + c) = (a \cdot b) + (a \cdot c)$.

3.3 Random Variables and Probability Distributions

3.3.1 Random Variables

A probability space models an experiment consisting of states that occur randomly. Usually, a probability space is constructed with a specific kind of situation or experiment in mind with the proposition that each time a situation of that kind arises, the set of possible outcomes is the same and the probabilities are also the same. An experiment consists of a procedure and observations. There is uncertainty what will be observed; otherwise performing the experiment would be unnecessary [138]. Some examples of experiments include flipping a coin (will it land with heads or tails facing up?), blind dating (on a scale of one to ten, how hot will the date be?), giving a lecture (how many students will show up?).

A probability space consists of a sample space, Ω, which is the set of all possible outcomes; a set of events $\mathfrak{F}$, where each event is a set containing zero or more outcomes; and the assignment of probabilities to the events as a function P from events to probabilities.

Figure 3.1 illustrates a mapping X of the probability space Ω on the set of real numbers $\mathbb{R}$.

The random variable X is a mapping of Ω on $\mathbb{R}$, which assigns every point ω of the probability space Ω to a real number $X(\omega)$. In the first case, X can take a countable number of values (finite: $\{x_1, x_2, \dots, x_l\}$ or infinite: $\{x_1, x_2, \dots, x_n, \dots\}$), these are called discrete random variables. In the second case, X can take more than

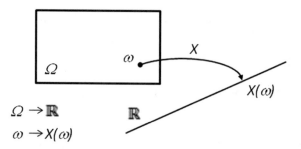

Fig. 3.1 Graphic representation of the mapping from the probability space Ω on real numbers $\mathbb{R}$

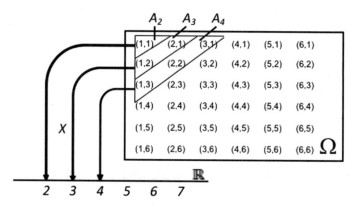

Fig. 3.2 Mapping of Ω into X and its inverse mapping in the case of rolling two dice

a countable number of values, *e.g.,* the interval $[0, 1]$, these are called non-discrete (or continuous) random variables.

Continuing the definition from above, (a) For discrete random variables X, the events $A_k = \{\omega | X(\omega) = x_k\}$ for $k = 1, 2, 3, \ldots$ must be events whose probability can be determined, (b) For continuous random variables X, the events $A_a = \{\omega | X(\omega) \leq a\}$ for any real number a, must be events whose probability can be determined.

In other words, $A_k = \{X = x_k\}$, $A_k = X^{-1}(x_k)$ meaning $\omega \in A_k$, exactly then, if $X(\omega) = \{x_k\}$. Or, in the continuous case $A_a = \{X \leq a\}$, $A_a = X^{-1}((-\infty, a])$.

Example: Let us look at the case of rolling two dice and define the random variable X as the sum of the rolled numbers after each roll. The inverse mapping X^{-1}, for instance for the number 4 is the set $A_4 = \{(1, 3), (2, 2), (3, 1)\}$. Here $x_k = k$. See Figure 3.2 for an illustration. □

Table 3.1 Sum of two rolled dice and their probabilities

Values of X	2	3	4	5	6	7	8	9	10	11	12
Corresponding probabilities	1/36	2/36	3/36	4/36	5/36	6/36	5/36	4/36	3/36	2/36	1/36

3.3.2 Probability Distribution of a Discrete Random Variable

A probability distribution is a mathematical description of a random phenomenon in terms of the probabilities of events. For instance, for a discrete random variable X on a probability space Ω, let the values of X be $x_1, x_2, x_3, \ldots, x_n, x_{n+1}, \ldots$ and the corresponding probabilities be $p_1, p_2, p_3, \ldots, p_n, p_{n+1}, \ldots$ where all $p_i \geq 0$ and $\sum_i p_i = 1$.

Example: Consider, as in the previous example, the sum of two rolled dice.

$$\sum_{i=1}^{11} p_i = 1 .$$

$\square$

3.3.3 Binomial Distribution

Let us now consider n experiments. Let E_i be the event "success in i-th experiment". The outcomes of the experiments (events) are assumed to beindependent of each other. It is also assumed that the probabilities are equal for all events: $P(E_i) = p \,\forall\, i$. Random variables $Y_i = I_{E_i}$ are considered. They are defined as $I_{E_i}(\omega) = 1$, if $\omega \in E_i$ and $I_{E_i}(\omega) = 0$ if $\omega \notin E_i$. I_{E_i} is called the indicator variable of E_i (it is customary to give the value 1 for success and 0 for failure). Random variables which can have only the values 0 and 1 are called Bernoulli variables. Let us assign $P(1) = p$, $P(0) = 1 - p = q$. Now, the random variable $X = \sum_{i=0}^{n} Y_i$ is called a binomial variable and often represented as $X \sim B(n, p)$.

Example: Let us consider an illustrative special case where $n = 3$, $p = 1/3$, $q = 2/3$. The probability for each possible outcome can be computed with the decision tree shown in Figure 3.3.

To be sure, checking the combined probability of all possible outcomes, $P(X = 3) + P(X = 2) + P(X = 1) + P(X = 0) = \frac{1}{27} + \frac{6}{27} + \frac{12}{27} + \frac{8}{27} = 1$, as it should be. $\square$

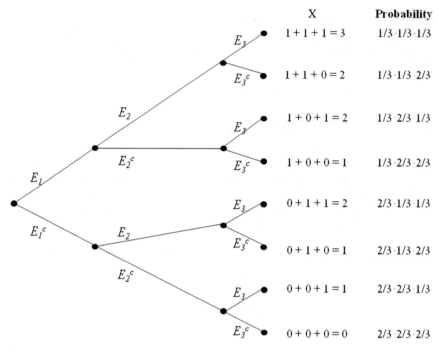

Fig. 3.3 E_i denotes a successful event. E_i^c denotes the complementary (failure) event

$$P(X = 3) = \left(\frac{1}{3}\right)^3 = \frac{1}{27}$$

$$P(X = 2) = 3\left(\frac{1}{3}\right)^2\left(\frac{2}{3}\right) = \frac{6}{27}$$

$$P(X = 1) = 3\left(\frac{1}{3}\right)\left(\frac{2}{3}\right)^2 = \frac{12}{27}$$

$$P(X = 0) = \left(\frac{2}{3}\right)^3 = \frac{8}{27}$$

In general, n experiments result in $\binom{n}{k} = \frac{n!}{k!(n-k)!}$ elementary events with k successes and $n - k$ failures. Hence,

$$P(X = k) = \binom{n}{k}p^k(1 - p)^{n-k}, \qquad k = 0, 1, 2, \ldots n. \qquad (3.1)$$

Let us note for now that for a random variable X with a binomial distribution, the expected value $E[X] = np$, and the variance $\text{Var}[X] = npq$. The meaning and the relevance of this statement will become evident very soon.

Example: Let us have a look at the annoying—at least for the passengers—overbooking issue in airlines. An airline knows that, on average, 4% of the passengers who have made reservations for a flight from London to Frankfurt do not show up at the gate. Therefore, it sells 129 tickets for 127 seats on an airplane. What is the probability that a passenger arrives at the gate and is told that there are no seats available even though her reservation status is "OK"?

Let E_i be the event the passenger T_i shows up at the gate for $i = 1, 2, \ldots, 129$. Let us further assume that all the events are independent and that the probability of showing up at the gate for all passengers is equal ($p = 0.96$). If X is the number of passengers at the gate, the probability that a passenger does not get a seat is

$$P(X \geq 128) = \binom{129}{128} \cdot 0.96^{128} \cdot 0.04^{(129-128)} + \binom{129}{129} \cdot 0.96^{129} \cdot 0.04^{(129-129)} \approx 3.29\% \,.$$

$\square$

3.3.4 Distribution Functions

The function $F_X(t) = P(X \leq t)$ is called the frequency distribution function of the random variable X, or simply the distribution function (some authors use the term cumulative distribution function). The distribution function in probability theory corresponds to the relative cumulative frequency distributions in descriptive statistics.

Example: Binomial variable X with $n = 8$, $p = 0.3$, $q = 0.7$. Let us calculate the probabilities for X taking the value k.

$$P(X = 0) = \binom{8}{0} \cdot 0.3^0 \cdot 0.7^8 = \frac{8!}{0!8!} \cdot 0.3^0 \cdot 0.7^8 \quad \approx 0.0576$$

$$P(X = 1) = \binom{8}{1} \cdot 0.3^1 \cdot 0.7^7 = \frac{8!}{1!7!} \cdot 0.3 \cdot 0.7^7 \quad \approx 0.1977$$

$$\vdots$$

$$P(X = 6) = \binom{8}{6} \cdot 0.3^6 \cdot 0.7^2 = \frac{8!}{6!2!} \cdot 0.3^6 \cdot 0.7^2 \quad \approx 0.0100$$

$$P(X = 7) = \binom{8}{7} \cdot 0.3^7 \cdot 0.7^1 = \frac{8!}{7!1!} \cdot 0.3^7 \cdot 0.7^1 \quad \approx 0.0012$$

$$P(X = 8) = \binom{8}{8} \cdot 0.3^8 \cdot 0.7^0 = \frac{8!}{8!0!} \cdot 0.3^8 \cdot 0.7^0 \quad \approx 0.0001 \,.$$

The results are shown in Table 3.2 and Figure 3.4. $\square$

Table 3.2 The distribution function shown in the example of a binomial variable X with $n = 8$; $p = 0.3$

k	$P(X = k)$	$P(X \leq k)$
0	0.0576	0.0576
1	0.1977	0.2553
2	0.2965	0.5518
3	0.2541	0.8059
4	0.1361	0.9420
5	0.0467	0.9887
6	0.0100	0.9987
7	0.0012	0.9999
8	0.0001	1.0000

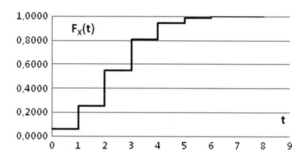

Fig. 3.4 Graphic representation of the distribution function calculated in Table 3.2

General properties of distribution functions:

1. $F(t)$ is monotonously increasing $(F(s) \geq F(t)$ if $s > t)$,
2. $F(-\infty) = 0$, $F(\infty) = 1$, $F(\pm\infty) = \lim_{t \to \pm\infty} F(t)$,
3. $F(t)$ is right continuous, $\lim_{t_n \to t+} F(t_n) = F(t)$.

These are also the necessary and sufficient conditions for a function to be a distribution function. Only continuous distribution functions $F_X(t)$ for which a so-called frequency density function $f_X(y)$ (or, again, simply the density function) exists such that

$$F_X(t) = \int_{-\infty}^{t} f_X(y)dy \tag{3.2}$$

are considered here. With some weak assumptions, at all points where $f(t)$ is continuous, $F'(t) = f(t)$. Loosely, it can be said that "the density is the derivative of the distribution function". The distribution function is defined here for continuous variables but this definition can easily be extended to discrete variables, in which case the distribution function is calculated with the formula

$$P(X \leq k) = \sum_{i=0}^{k} P(X \leq i). \tag{3.3}$$

3.3.5 Multidimensional Distribution Functions and Independence

Let X and Y be random variables on a probability space Ω. The function $F_{X,Y}(s, t) = P(X \leq s, \ Y \leq t)$ is called the (two dimensional) distribution function of X and Y. Correspondingly, $f(u, v)$ is a joint density function for the random variables X and Y if

$$F_{X,Y}(s, t) = \int_{v=-\infty}^{t} \int_{u=-\infty}^{s} f(u, v)du dv . \tag{3.4}$$

The properties of multidimensional distribution functions are similar to the ones described above for one dimensional random variables. Also, one can easily extend the two dimensional case to n dimensions. The multidimensional distribution of the random variables $X_1, X_2, \ldots, X_n$ on a probability space Ω is defined by the function $F_{X_1,X_2,\ldots,X_n}(t_1, t_2, \ldots, t_n) = P(X_1 \leq t_1, X_2 \leq t_2, \ldots, X_n \leq t_n)$.

The random variables $X_1, X_2, \ldots, X_n$ are said to be independent if $F_{X_1,X_2,\ldots,X_n}(t_1, t_2, \ldots, t_n) = F_{X_1}(t_1) \cdot F_{X_2}(t_2) \cdot \ldots \cdot F_{X_n}(t_n)$ or equivalently $f_{X_1,X_2,\ldots,X_n}(t_1, t_2, \ldots, t_n) = f_{X_1}(t_1) \cdot f_{X_2}(t_2) \cdot \ldots \cdot f_{X_n}(t_n)$ holds for all t_i.

In practice, however, for discrete random variables the condition

$$P(X_1 = t_1, X_2 = t_2, \ldots, X_n = t_n) = P(X_1 = t_1) \cdot P(X_2 = t_2) \cdot \ldots \cdot P(X_n = t_n),$$

and for continuous random variables the condition

$$P(X_1 < t_1, X_2 < t_2, \ldots, X_n < t_n) = P(X_1 < t_1) \cdot P(X_2 < t_2) \cdot \ldots \cdot P(X_n < t_n),$$

for all t_i is necessary and sufficient for the independence of random variables $X_1, X_2, \ldots, X_n$. In other words, in practice, looking at probabilities rather than distribution or density functions is enough to see whether random variables are independent.

A very useful theorem states that if the random variables $X_1, X_2, \ldots, X_n$ on a probability space Ω are independent and defining new random variables as a function of these variables such that $Y_k = g(X_k)$ for $k = 1, 2, \ldots, n$, then $Y_1, Y_2, \ldots, Y_n$ are also independent. (The proof of this theorem is left to the interested student.)

3.3.6 Expected Value and Further Moments

Let X be a discrete random variable which can take the values $\{x_1, x_2, \ldots, x_n, \ldots\}$ with corresponding probabilities $P(X = x_k) = p_k$. Further, let $g(\cdot)$ be a real function defined for all values of X. Then,

$$E[g(X)] = \sum_{k=1}^{\infty} g(x_k) p_k \tag{3.5}$$

is called the expected value of the function $g(X)$, assuming that the sequence above converges absolutely.

Similarly, let X be a continuous random variable with the density function $f_X(t)$ and let $g(\cdot)$ be a real function defined for all values of X. Then,

$$E[g(X)] = \int_{-\infty}^{\infty} g(t) f_X(t) dt \tag{3.6}$$

is called the expected value of the function $g(X)$, assuming that the integral converges absolutely.

Arguably, the most important special case is when $g(X) = X$. Then, for discrete and continuous cases the expected value of a random variable is calculated as

$$E[X] = \sum_{k=1}^{\infty} x_k p_k , \tag{3.7}$$

and

$$E[X] = \int_{-\infty}^{\infty} t f_X(t) dt , \tag{3.8}$$

respectively. Note the similarity to calculating the center of gravity of an object. Just as the center of gravity does not have to lie inside the object (think of concave objects like a rectangular metal "U" profile where the center of gravity is outside the profile), the expected value does not even have to be a possible random variable (think of rolling a fair die where the expected outcome is $\frac{1}{6}(1 + 2 + 3 + 4 + 5 + 6) = 3.5$).

Some properties of the expected value are:

1. For $X \geq 0$, $E[X] \geq 0$,
2. $E[X + Y] = E[X] + E[Y]$,
3. $E[cX] = c \cdot E[X]$,
4. $E[I_A] = P(A)$ with I_A being the indicator of the event A as defined above,
5. $E[X \cdot Y] = E[X] \cdot E[Y]$ if X and Y are independent random variables (note that, in general, the reversal of this statement is not true).

Let X be a random variable with the *expected value* $\mu = E[X]$. $E[X^k]$ is called the k-th moment of X, and $E[(X - \mu)^k]$ the k-th central moment of X. The second central moment of X, $E[(X - \mu)^2]$ is called the *variance* of X and written as $\text{Var}[X]$. The variance is a measure of deviation of the random variable from its expected value. $\sigma(X) = \sqrt{\text{Var}[X]}$ is called the *standard deviation* of X. The variance (or standard deviation) is a measure of dispersion of random variables from their expected values.

Fig. 3.5 Closing prices of two stocks normalized by their initial values during one year. IBM with a market capitalization of $150 billion has a volatility of 0,07. YY with a market capitalization of $3 billion has a volatility of 0,16

A useful relationship is

$$\text{Var}[X] = \text{E}[(X - \mu)^2] = \text{E}[X^2 - 2\mu X + \mu^2] = \text{E}[X^2] - 2\mu\text{E}[X] + \mu^2 = \text{E}[X^2] - \mu^2.$$
$$(3.9)$$

A popular mnemonic to remember this useful relationship is "Variance is the expected value of the square minus the square of the expected value".

Example: Looking at the closing prices of stocks, it is observed that some vary much more than others. This variation is called volatility and stocks commonly known as "blue chips" are less volatile than others. An important reason for blue chips to be less volatile lies in their high degrees of capitalization and broader basis of shareholders. Comparison of the closing prices of the quintessential blue chip IBM and a smaller Internet information provider YY during a year is depicted in Figure 3.5.   □

The *covariance* of the random variables X and Y is defined as

$$\text{Cov}[X, Y] = \text{E}[(X - \text{E}[X]) \cdot (Y - \text{E}[Y])].$$
$$(3.10)$$

The *correlation coefficient* of the random variables X and Y is defined as

$$\rho(X, Y) = \frac{\text{Cov}[X, Y]}{\sigma(X)\sigma(Y)},$$
$$(3.11)$$

and X and Y are said to be uncorrelated if $\rho(X, Y) = 0$. Note that, unlike variance or covariance, correlation is dimensionless. Also remember that independent random variables with finite first and second moments are un-correlated (but the reversal of this statement is true only if the random variables are normally distributed). A direct

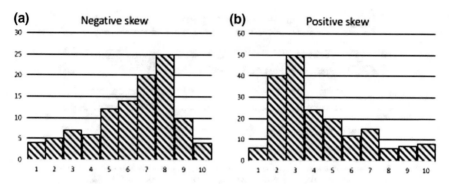

Fig. 3.6 a A distribution with negative skew, and **b** a distribution with positive skew. Distributions symmetrical with respect to their expected values have no skew

consequence of this fact is, for uncorrelated random variables X and Y $\mathrm{Var}[X + Y] = \mathrm{Var}[X] + \mathrm{Var}[Y]$. Further, $\mathrm{Var}[cX] = c^2 \mathrm{Var}[X]$ and $\sigma(cX) = c\sigma(X)$ for all real c.

Skewness is the normalized third central moment and defined as

$$\gamma_1 = E\left[\left(\frac{X - \mu}{\sigma}\right)^3\right] = \frac{E[(X - \mu)^3]}{(E[(X - \mu)^2])^{3/2}} = \frac{E[X^3] - 3\mu\sigma^2 - \mu^3}{\sigma^3}. \quad (3.12)$$

Skewness is a measure of the asymmetry of the probability distribution. As Figure 3.6 illustrates, a positive skewness indicates that the tail at the right hand side is longer.

Kurtosis is the normalized fourth central moment and defined as

$$\beta_2 = E\left[\left(\frac{X - \mu}{\sigma}\right)^4\right] = \frac{E[(X - \mu)^4]}{(E[(X - \mu)^2])^2}.$$

Kurtosis is a measure for "fat tails". In other words, a higher kurtosis means more of the variance is the result of infrequent extreme deviations, as opposed to frequent modestly sized deviations. Kurtosis provides a measurement about the extremities. Therefore, it provides an indication of the presence of outliers. Looking at investment returns as random variables with certain distributions, a risk averse investor prefers a distribution with a small kurtosis, because losses will be closer to the expected value rather than infrequent but huge, devastating losses (Nassim Taleb makes use of this concept in his best seller "The Black Swan" [129]). Kurtosis is sometimes referred to as the "volatility of volatility". As Figure 3.7 illustrates, distributions with narrower tails have smaller kurtosis and distributions with fatter tails have bigger kurtosis. The kurtosis for a standard normal distribution is three.[7] For this reason, some sources

---

[7] See Subsection 3.3.8 on Normal Distribution.

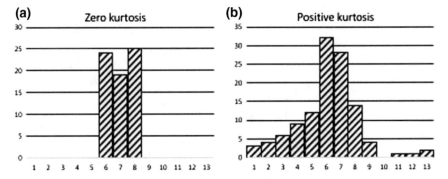

Fig. 3.7 Two distributions with the same variance but very different kurtosis

use the concept of "excess kurtosis" as the kurtosis above or below three having positive or negative excess kurtosis.

3.3.7 Correlation

Correlation is a statistical measure that indicates the extent to which two (or more) variables fluctuate together (see Figure 3.8). A positive correlation indicates the extent to which those variables increase or decrease in parallel; a negative correlation indicates the extent to which one variable increases as the other decreases.

The plot in Figure 3.8b might have the number of lung cancer deaths (Y) as the vertical axis and the number of cigarettes smoked per day (X) as the horizontal axis. What the plot says is that the probability of dying from lung cancer increases as the amount of cigarette smoking increases. Since X and Y are random variables both the proverbial "grand father who smoked two packages a day and still going on strong at 95" and the unfortunate "12 year old girl who never smoked a single cigarette but died of lung cancer" are covered by that probabilistic statement.

The plot in Figure 3.8c might have the percent of body fat as the vertical axis and the hours of weekly exercise as the horizontal axis. What this plot says is that the probability of having a high percentage of body fat decreases as the hours of weekly exercise increases.

It is important to remember that correlation does not mean causality. Figure 3.8b might very well be the plot of the relationship between the number of sunburn victims versus the amount of ice cream sold in a city. Although there is a clear correlation, it would be rather silly to claim that ice cream causes sun burns. In this particular case it is easy to see that both the number of sun burn victims and the amount of ice cream sales increases as the number of sunshine hours per day and the temperature

Fig. 3.8 Plots of random variable X versus random variable Y. **a** No correlation, **b** Positive correlation, **c** Negative correlation

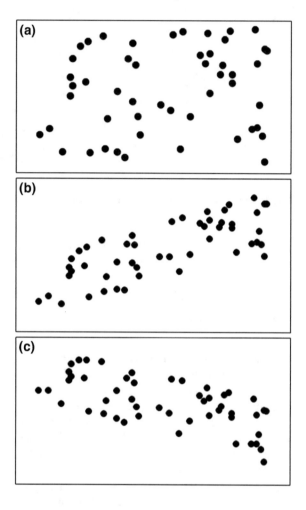

increase. However, it is sometimes difficult to see that causality. Famous examples in literature for faulty consequences drawn from statistical data include statements like "Students who smoke have lower grades at school. This means smoking reduces your academic performance" or "People living below power lines suffer from increased gastrointestinal problems. Therefore, high tension power lines have negative health effects". Both of these arguments were found to be wrong. In the first case common socioeconomic and familial reasons were found to cause both lower grades and a propensity to smoke. In the second case it was found that many of the slums with poor sanitation and open sewage were below high tension power lines and many diseases were caused by the germs spread through open sewage lines.

3.3.8 Normal Distribution

A random variable X is said to be normally distributed if it has the density function

$$f_X(t) = \frac{1}{\sqrt{2\pi}\,\sigma} e^{-\frac{(t-\mu)^2}{2\sigma^2}} . \tag{3.13}$$

μ (real) and σ (positive) are parameters. A special case of the normal distribution is obtained when $\mu = 0$ and $\sigma = 1$. This is called the *standard normal distribution* and has the density

$$\varphi(t) = \frac{1}{\sqrt{2\pi}} e^{-\frac{t^2}{2}} . \tag{3.14}$$

The distribution function

$$\Phi(t) = \frac{1}{\sqrt{2\pi}} \int_{-\infty}^{t} e^{-\frac{\tau^2}{2}} d\tau , \tag{3.15}$$

cannot be determined analytically. Therefore, both $\varphi(t)$ and $\Phi(t)$ are tabulated (See Appendix C). If the random variable X is normally distributed with the parameters μ and σ, then $Z = \frac{X-\mu}{\sigma}$ is a random variable with the standardized normal distribution. The shorthand description $X \sim N(\mu, \sigma)$ is used to say "X is normally distributed with the parameters μ and σ". One can show that for independent random variables X and Y such that $X \sim N(\mu_X, \sigma_X)$, $Y \sim N(\mu_Y, \sigma_Y)$

$$(cX + d) \sim N(c\mu_X + d, |c|\sigma_X) \text{ and}$$

$$(X + Y) \sim N\left(\mu_X + \mu_Y, \sqrt{\sigma_X^2 + \sigma_Y^2}\right) .$$

3.3.9 Central Limit Theorem

One of the most astounding and beautiful theorems in probability theory is the so-called central limit theorem. Let $\{X_1, X_2, \ldots, X_n\}$ be a set of independent real variables with identical distributions with finite moments $E[X_i] = \mu$ and $\mathrm{Var}[X_i] = \sigma^2$. Let a new random variable be defined as $S_n = \sum_{i=1}^{n} X_i$. The central limit theorem states that

$$\lim_{n \to \infty} P\left(\frac{S_n - n\mu}{\sqrt{n}\sigma} \le x\right) = \Phi(x) , \tag{3.16}$$

where $E[S_n] = \sum_{i=1}^{n} E[X_i] = n\mu$ and $Var[S_n] = \sum_{i=1}^{n} Var[X_i] = n\sigma^2$. The derivation of this theorem goes through the Chebyshev[8] inequality and the law of large numbers [120].

This theorem is so important, because, for the stated assumptions, it says that for large n, S_n has a normal distribution with mean μ and the variance σ^2. This is astounding, because it holds independent of the shapes of distributions of X_i. Lyapunov[9] has further shown that, with some additional weak assumptions, this holds even if the distributions of X_i are different. What makes the theorem even more astounding is, although "large" n is mentioned, in reality it might require only a few random variables as the illustrative example in Figure 3.9 shows.

Example: There are two florists called Azalea and Begonia in a small town. Years of experience has established that there will be 100 independent orders for long stemmed red roses on Valentine's Day. The probability that either Azalea or Begonia get an order is equal. How many roses do the florists each have to keep on stock so that they can fulfill all orders they receive with a probability of 99%?

Let us say X is the orders Azalea receives and has a binomial distribution with $n = 100$ and $p = 0.5$. Therefore, the probability that Azalea has enough roses for T orders can be computed as

$$P(X = k) = \binom{n}{k} p^k (1 - p)^{n-k}$$
$$= \binom{n}{k} \left(\frac{1}{2}\right)^k \left(\frac{1}{2}\right)^{n-k}$$
$$= \binom{n}{k} \left(\frac{1}{2}\right)^n .$$

In other words, the probability that Azalea can fulfill all orders is $P(X \leq T)$. The question is what should T be, such that $P(X \leq T) = 0.99$?

One could, of course, calculate the answer by computing all probabilities $P(1)$, $P(2)$, ... and sum them up until 0.99 is reached but this is rather tedious. A more elegant solution can be found using the central limit theorem. Let us just remember that the expected value and the variance of a binomial distribution are given as $E[X] = np$, and $Var[X] = npq$ respectively. In this case $E[X] = 50$, $Var[X] = 25$. Therefore, $\sigma = 5$. Plugging these numbers in the central limit theorem gives

$$P(X \leq T) = 0.99 \approx \Phi\left(\frac{T + \frac{1}{2} - 50}{5}\right),$$

---

[8]Pafnuty Lvovich Chebyshev, Russian mathematician (1821–1894); famous for his work on probability and number theory; also as the doctoral advisor of Alexandr Mikhailovich Lyapunov and Andrei Andreyevich Markov.

[9]Alexandr Mikhailovich Lyapunov, Russian mathematician (1857–1918).

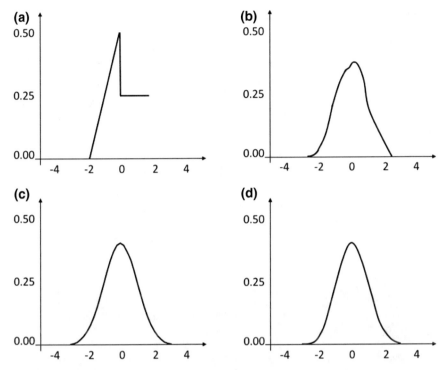

Fig. 3.9 **a** Begin with a rather unusual probabilitydensity function ($n = 1$). **b** Next, compute the density of the sum of two independent variables ($n = 2$), each having the density in (**a**). This density is already significantly smoother than the original. **c** Then, compute the density of the sum of three independent variables ($n = 3$), each having the density in (**a**). This density is even smoother than the preceding one. **d** Finally, compute the density of the sum of four independent variables ($n = 4$), each having the density in (**a**). This density appears qualitatively very similar to a normal density and any differences from it cannot be distinguished by the eye. (In all figures, the density shown has been rescaled by $\sqrt{n}$, so that its standard deviation is 1)

and looking at the standardized normal distribution table (see Appendix C) with the variable transformation

$$z = \frac{X - \mu}{\sigma}$$

yields

$$\left(\frac{T + \frac{1}{2} - 50}{5} \right) \approx 2.326 \, .$$

Solving this for T yields $T = 62$. This means the florist Azalea (or Begonia) needs to stock only 62 roses to satisfy 99% of its customers. But where does that $\frac{1}{2}$ come from? The answer is "histogram correction". $\qquad \square$

Fig. 3.10 Possible outcomes of rolling two dice and their probabilities

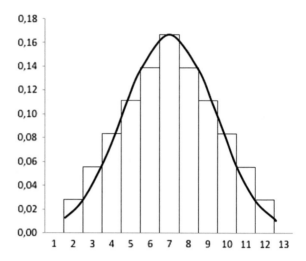

Histogram Correction

Consider rolling a fair die. Possible outcomes are $\{1, 2, 3, 4, 5, 6\}$. The mean is 3.5, standard deviation is 1.71.

With two rolls of the dice and adding the outcomes: expected value is $2 \times 3.5 = 7$, standard deviation is $\sqrt{2} \times 1.71 = 2.42$. Figure 3.10 shows the probability density function for that experiment.

One does not really need to use the normal approximation here, because the actual probabilities are easy to sort out. Let us say one wanted to anyway. How can one find P(roll 8 or 9)?

Actual probability: $\frac{5}{36} + \frac{4}{36} = 0.25$

Lower end: $z = (7.5 - 7)/2.42 = 0.20$

Upper end: $z = (9.5 - 7)/2.42 = 1.05$

Area: $P = (0.7063 - 0.1585)/2 = 0.2739$

Really precise area using computer normal tables is 0.2677 which is pretty close to 0.25.

Why was the interval 7.5 to 9.5 considered instead of the interval 8 to 9? The probability histogram has one bin for each possible value. The value itself is in the center of the bin. So, the bin for 8 is 1 wide and has height equal to $P(8)$. If only the interval from 8 to 9 in the normal approximation were considered, it would have missed the left half of the 8 bin, and the right half of the 9 bin. What is achieved by adding half bin widths is called histogram correction. It is especially important if one has relatively few possible values, or if a high level of precision is required.

Note that the normal distribution curve says that one can throw a 13 or a 1 or even 0 and −1 and so on. So, when using the central limit theorem, special care needs to be given to its range of applicability.

3.3.10 Log-Normal Distribution

A log-normal distribution is a continuous probability distribution of a random variable whose logarithm is normally distributed. Thus, if the random variable Y is log-normally distributed, then $X = \log(Y)$ has a normal distribution or conversely, if X has a normal distribution, then $Y = e^X$ has a log-normal distribution. The distribution of a variable might be modeled as log-normal if it can be thought of as the multiplicative product of several independent positive random variables. (Think of the central limit theorem being applied in the log-domain.) This distribution is widely used in finance as will become evident in the following Chapters.

If X is a log-normally distributed random variable, the parameters μ and σ are the mean and standard deviation of the variable's logarithm respectively (by definition, the variable's logarithm is normally distributed), therefore $X = e^{\mu + \sigma Z}$ with Z having a standard normal distribution. It follows that X has the probability density function

$$f_X(t) = \frac{1}{\sqrt{2\pi}\sigma t} e^{-\frac{(\log t - \mu)^2}{2\sigma^2}} \,,$$

and the distribution function

$$F_X(t) = \Phi\left(\frac{\log t - \mu}{\sigma}\right).$$

The first moments of X are given as

$$E[X] = e^{\mu + \frac{1}{2}\sigma^2} \,,$$

$$\text{Var}[X] = (e^{\sigma^2} - 1)e^{2\mu + \sigma^2} = (e^{\sigma^2} - 1)E[X]^2 \,,$$

or equivalently

$$\mu = \log(E[X]) - \frac{1}{2}\log\left(1 + \frac{\text{Var}[X]}{E[X]^2}\right) = \log(E[X]) - \frac{1}{2}\sigma^2 \,, \quad (3.17)$$

$$\sigma^2 = \log\left(1 + \frac{\text{Var}[X]}{E[X]^2}\right). \quad (3.18)$$

If $X \sim \mathcal{N}(\boldsymbol{\mu}, \boldsymbol{\Sigma})$ is an n dimensional multivariate normal distribution with $X = [x_i]^T$, $\boldsymbol{\mu} = [\mu_i]^T$ and $\boldsymbol{\Sigma} = [\sigma_{ij}]$ with $i, j = 1, \ldots, n$, then $Y = e^X$ has a multivariate log-normal distribution with the elements of the mean and covariance matrices given as

$$E[Y]_i = e^{\mu_i + \frac{1}{2}\sigma_{ii}} \,, \quad (3.19)$$

$$\text{Cov}[Y]_{ij} = e^{\mu_i + \mu_j + \frac{1}{2}(\sigma_{ii} + \sigma_{jj})}(e^{\sigma_{ij}} - 1). \quad (3.20)$$

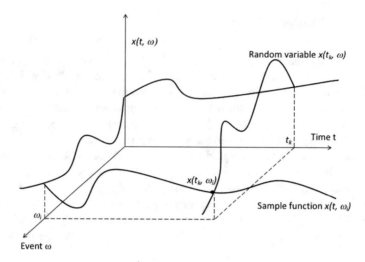

Fig. 3.11 Definition of a stochastic process $\{x(t, \omega), \ t \in T, \ \omega \in \Omega\}$ with T being the definition interval of t. Every sample function is a realization of the process and the set of all realizations is called the ensemble

3.4 Stochastic Processes

A statistical experiment E is defined by the triple $(\Omega, \mathfrak{F}, P)$ where Ω is the sample space consisting of all possible outcomes, $\mathfrak{F}$ is a set of events, where each event is a set containing zero or more outcomes and P is the probability assigned to an event. When each event ω of the experiment E is assigned a time function

$$x = x(t, \omega) \tag{3.21}$$

a family of functions called a stochastic process as shown in Figure 3.11 is obtained.

For a given event ω_i, $x(t, \omega_i)$ is a single time function. Each of these time functions are different realizations of the process. The set of all realizations is called the *ensemble*. At a given time t_k the function $x(t_k, \omega)$ is a random variable dependent solely on ω. For a given time t_k and event ω_i, $x(t_k, \omega_i)$ is just a number.

A stochastic process is a process which is non-deterministic. Its exact course is not determined in advance. The concept of a stochastic process needs to be defined more precisely, but before doing that let us examine several examples of stochastic processes.

Example: In technical applications, a time dependent process which is not a deterministic function is often called a stochastic process. This means that the process does not have a "regular" course. A typical application is the discovery made by Robert

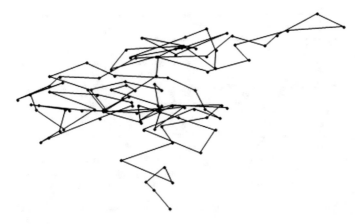

Fig. 3.12 A Brownian motion in two dimensions

Brown[10] in 1827 that small particles in a fluid have "very irregular" movements, the so-called Brownian motion as shown schematically in Figure 3.12.

Einstein could demonstrate in 1905 that the movements of the particles were due to their collision with liquid molecules. Actually, this was the first real confirmation of atomic theory. Only much later, in 1923, was Wiener[11] able to model these "irregular" movements mathematically by means of probability theory. It is remarkable that a totally unruly looking sample function like in Figure 3.12 can be described as a *Brownian motion* in a mathematically rigorous way. Figure 3.12 was generated by an Excel file

A1 = 0, B1 = 0,
A2 = A1+RANDBETWEEN(-100;100), B2 = B1+RANDBETWEEN(-100;100),

$\vdots$

A100 = A99+RANDBETWEEN(-100;100), B100= B99+RANDBETWEEN(-100;100)

$\square$

Example: The example above should not give the impression that the irregular course of a sample function is the only characteristic of a stochastic process. Let us now consider the output voltages $U(t)$ of a number of identical voltage generators. In a simplified fashion

$$U(t) = A \sin(\omega t + \theta).$$

The amplitude A, the frequency ω and the phase θ are all dependent on the generator. In other words, they are random variables dependent on the choice of the generator. Figure 3.13 shows two possible sample functions of the voltage $U(t)$.

---

[10]Robert Brown, Scottish botanist (1773 – 1858).

[11]Norbert Wiener, American mathematician, control engineer (1894–1964); famous as the "father of cybernetics".

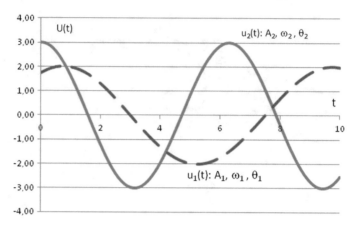

Fig. 3.13 Two sample functions of the stochastic process $\{U(t),\ t \in T\}$ spawned by the output voltages of the voltage generators

Obviously, once a specific generator is chosen, the course of $U(t)$ for $t > t_1$ is determined based on the course of $U(t)$ for $t < t_1$. This would not have been possible in the example of Brownian motion above. □

3.5 Mathematical Description of Stochastic Processes with Distribution Functions

A stochastic process is a function $x(t, \omega)$ of—at least— two independent variables $t \in T$ and $\omega \in \Omega$. As mentioned, the totality of all possible sample functions is often called the "ensemble". Of course, not all possible sample functions are equally probable. Chance chooses a sample function or $\omega \in \Omega$ according to a certain rule of probability. For a given $t \in T$ this rule can be defined by means of a distribution function of the random variable

$$F(X, t) = P[x(t, \omega) \le X].\qquad(3.22)$$

Since a single point in time is considered, $F(X, t)$ is called a first order distribution function. For any given t and ω the distribution function $F(X, t)$ is the probability that the random variable $x(t, \omega) \le X$. In the continuous case the first order density function is

$$f(x, t) = \frac{\partial F(x, t)}{\partial x}.\qquad(3.23)$$

If two points in time, t_1 and t_2, are now considered and the corresponding limits x_1 and x_2, the resulting second order distribution function is

$$F(x_1, x_2, t_1, t_2) = P[x(t_1, \omega) \le x_1 \cap x(t_2, \omega) \le x_2]. \tag{3.24}$$

The corresponding density function is thus

$$f(x_1, x_2, t_1, t_2) = \frac{\partial^2 F(x_1, x_2, t_1, t_2)}{\partial x_1 \partial x_2}. \tag{3.25}$$

In the general case, for the n-th order distribution function the definition is

$$F(x_1, x_2, \dots, x_n, t_1, t_2, \dots, t_n) = P[x(t_1, \omega) \le x_1 \cap \dots \cap x(t_n, \omega) \le x_n]. \tag{3.26}$$

This is the probability that the random vector $\mathbf{X} = [x(t_1, \omega), x(t_2, \omega), \dots, x(t_n, \omega)]^T$ is element-wise smaller than the limit $\mathbf{x} = [x_1, x_2, \dots, x_n]^T$ (the ^{T} in the superscript denotes the transpose).

The distribution functions in (3.26) cannot be chosen arbitrarily. Apart from the usual properties of distribution functions like

$$x_1 \le x_2 \Rightarrow F(x_1) \le F(x_2),$$
$$\lim_{x \to -\infty} F(x) = 0,$$
$$\lim_{x \to \infty} F(x) = 1,$$

these must be symmetrical in argument couples and fulfill the symmetry and compatibility conditions.[12]

Only well behaved distribution functions which have densities

$$F(x_1, x_2, \dots, x_n, t_1, t_2, \dots, t_n) = \int_{-\infty}^{x_1} \dots \int_{-\infty}^{x_n} f(\xi_1, \xi_2, \dots, \xi_n, t_1, t_2, \dots, t_n) d\xi_1 \dots d\xi_n \tag{3.27}$$

are considered in this Book.

In general, the calculation of the distribution function in (3.26) is a rather tedious task. The next example is meant as a simple illustration.

Example: Given is an oscillation

$$x(t) = a_1 \cos(t) + a_2 \sin(t),$$

where the amplitudes a_1 and a_2 are normally distributedrandom variables.

$$a_1 \sim N(\mu_1, 1),$$
$$a_2 \sim N(\mu_2, 1),$$

---

[12]Symmetry conditions: if $\{i_1, i_2, \dots, i_j\}$ is a permutation of $\{1, 2, \dots, j\}$, then $F(t_{i_1}, t_{i_2}, \dots, t_{i_j}, x_{i_1}, x_{i_2}, \dots, x_{i_j}) = F(t_1, t_2, \dots, t_j, x_1, x_2, \dots, x_j)$. Compatibility conditions are $F(t_1, \dots, t_i, t_{i+1}, \dots, t_j, x_1, \dots, x_i, \dots) = F(t_1, \dots, t_i, x_1, \dots, x_i)$.

or, with $\mathbf{a} = [a_1 \ a_2]^T$ and $\boldsymbol{\mu} = [\mu_1 \ \mu_2]^T$, more compactly

$$\mathbf{a} \sim N(\boldsymbol{\mu}, \mathbf{I}).$$

$\boldsymbol{\Sigma}$, the covariance matrix of $\mathbf{a}$, being the identity matrix which has only diagonal elements, tells us that a_1 and a_2 are uncorrelated.

What is the n-th order distribution function of $x(t)$?

To answer this question, define the random vector $\mathbf{X}$ as

$$\begin{pmatrix} x(t_1, \omega) \\ \vdots \\ x(t_n, \omega) \end{pmatrix} = \begin{pmatrix} \cos(t_1) & \sin(t_1) \\ \vdots & \vdots \\ \cos(t_n) & \sin(t_n) \end{pmatrix} \cdot \begin{pmatrix} a_1 \\ a_2 \end{pmatrix} = \mathbf{Ca}. \tag{3.28}$$

Because the random vector $\mathbf{a}$ is transformed linearly, the distribution function of $x(t)$, according to (3.26) and from what was stated at the end of Section 3.3.8, can be calculated as

$$F(x_1, x_2, \ldots, x_n, t_1, t_2, \ldots, t_n) \sim N(\mathbf{C}\boldsymbol{\mu}, \mathbf{CIC}^T).$$

□

Other than the distribution function, moments, especially the first two, play an important role in describing stochastic processes. Because of the density function in (3.23), the mean value of the process $\{x(t)| \ t \in T\}$ as a function of time can be defined as

$$m(t) = E[x(t)] = \int_{-\infty}^{\infty} x f(x, t) dx. \tag{3.29}$$

As Figure 3.14 shows, $m(t)$ represents the swing center of the ensemble made of all sample functions $x_i(t)$. Note that $m(t)$ is a *deterministic* function of time.

Variance of a random process is defined as

$$\text{Var}[x(t)] = E\{[x(t) - m(t)]^2\} = \int_{-\infty}^{\infty} x^2 f(x, t) dx - [m(t)]^2, \tag{3.30}$$

and it denotes the deviation of the process from its mean at time t. The correlation function $C(t, s)$ is the correlation between random variables at two different time points and is defined as the expected value of their products. The random variables x_1 and x_2 are viewed at instants t and s. Just like the mean function, the correlation function is a deterministic function. It is defined as

$$C(t, s) = E[x(t)x(s)] = \int_{-\infty}^{\infty} \int_{-\infty}^{\infty} x_1 x_2 f(x_1, x_2, t, s) dx_1 . dx_2 \tag{3.31}$$

Covariance is a measure of how much two variables change together and the covariance function $R(t, s)$ describes the variance of a random process. It is defined as

$$R(t, s) = \mathrm{E}[\{x(t) - m(t)\} \cdot \{x(s) - m(s)\}] \tag{3.32}$$

$$= \int_{-\infty}^{\infty} \int_{-\infty}^{\infty} [x(t) - m(t)][x(s) - m(s)] f(x_1, x_2, t, s) dx_1 dx_2 \tag{3.33}$$

$$= C(t, s) - m(t)m(s). \tag{3.34}$$

For n-dimensional processes $\{\mathbf{x}(t) |\ t \in T\}$, both parts of (3.32) can be defined separately for each component x_i. That way cross-covariance and cross-correlation functions can also be defined between two components of the process. For two components of the process vector $\{\mathbf{x}(t) |\ t \in T\}$ $x_i(t)$ and $x_j(t)$ the cross-covariance function is defined as

$$R_{x_i x_j}(t, s) = \mathrm{E}[\{x_i(t) - m_i(t)\} \cdot \{x_j(s) - m_j(s)\}] \tag{3.35}$$

$$= \int_{-\infty}^{\infty} \int_{-\infty}^{\infty} [x_i(t) - m_i(t)][x_j(s) - m_j(s)] f(x_i, x_j, t, s) dx_i dx_j.$$

Similarly, the cross-correlation function is

$$C_{x_i x_j}(t, s) = \mathrm{E}[x_i(t) \cdot x_j(s)] = \int_{-\infty}^{\infty} \int_{-\infty}^{\infty} x_i(t) x_j(s) f(x_i, x_j, t, s) dx_i dx_j. \tag{3.36}$$

Example: Making use of the linearity property of the expected value operator, the mean value and the covariance functions in the previous Example can be computed

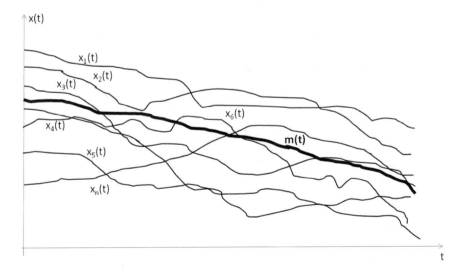

Fig. 3.14 Mean function $m(t)$ is deterministic

as

$$m(t) = E[a_1 \cos(t) + a_2 \sin(t)] \tag{3.37}$$
$$= E[a_1 \cos(t)] + E[a_2 \sin(t)]$$
$$= E[a_1] \cos(t) + E[a_2] \sin(t)$$
$$= \mu_1 \cos(t) + \mu_2 \sin(t),$$

where $\mu_1 = E[a_1]$, $\mu_2 = E[a_2]$.

$$R(t, s) = C(t, s) - m(t) \cdot m(s), \tag{3.38}$$

$$C(t, s) = E\{[x(t)] \cdot [x(s)]\} = E\{[a_1 \cos(t) + a_2 \sin(t)] \cdot [a_1 \cos(s) + a_2 \sin(s)]\}. \tag{3.39}$$

Since

$$\mathbf{a} \sim N(\boldsymbol{\mu}, \mathbf{I})$$

$$\Rightarrow \begin{pmatrix} E\{(a_1 - \mu_1)^2\} & E\{(a_1 - \mu_1) \cdot (a_2 - \mu_2)\} \\ E\{(a_1 - \mu_1) \cdot (a_2 - \mu_2)\} & E\{(a_2 - \mu_2)^2\} \end{pmatrix} = \begin{pmatrix} 1 & 0 \\ 0 & 1 \end{pmatrix} \tag{3.40}$$

$$\begin{pmatrix} E\{a_1^2\} - \mu_1^2 & E\{(a_1 a_2)\} - \mu_1 \mu_2 \\ E\{a_1 a_2\} - \mu_1 \mu_2 & E\{a_2^2\} - \mu_2^2 \end{pmatrix} = \begin{pmatrix} 1 & 0 \\ 0 & 1 \end{pmatrix} \tag{3.41}$$

$$E\{a_1^2\} = 1 + \mu_1^2$$

$$E\{a_2^2\} = 1 + \mu_2^2$$

$$E\{a_1 a_2\} = \mu_1 \mu_2$$

$$C(t, s) = \underbrace{E\{a_1^2\}}_{1+\mu_1^2} \cdot \cos(t) \cos(s) + \underbrace{E\{a_2^2\}}_{1+\mu_2^2} \cdot \sin(t) \sin(s)$$
$$+ \underbrace{E\{a_1 a_2\}}_{\mu_1 \mu_2} (\cos(t) \sin(s) + \sin(t) \cos(s)).$$

Similarly,

$$m(t) \cdot m(s) = [\mu_1 \cos(t) + \mu_2 \sin(t)] \cdot [\mu_1 \cos(s) + \mu_2 \sin(s)]$$
$$= \mu_1^2 \cos(t) \cos(s) + \mu_2^2 \sin(t) \sin(s)$$
$$+ \mu_1 \mu_2 [\cos(t) \sin(s) + \sin(t) \cos(s)].$$

Substituting all these results in (3.38) and remembering that since the *sine* and *cosine* functions are deterministic, they can be taken out of the E-operator which yields

$$R(t, s) = (1 + \mu_1^2) \cdot \cos(t) \cos(s) + (1 + \mu_2^2) \cdot \sin(t) \sin(s)$$
$$+ \mu_1 \mu_2 \cdot [\cos(t) \sin(s) + \sin(t) \cos(s)]$$
$$- \mu_1^2 \cos(t) \cos(s) - \mu_2^2 \sin(t) \sin(s)$$
$$- \mu_1 \mu_2 [\cos(t) \sin(s) + \sin(t) \cos(s)]$$
$$= \cos(t) \cos(s) + \sin(t) \sin(s)$$
$$= \cos(t - s)$$
$$= \cos(\epsilon) \text{ where } \epsilon = t - s.$$

This is an interesting result, because the covariance function $R(t, s)$ depends only on the difference of s and t and is independent of μ_1 and μ_2. □

3.6 Stationary and Ergodic Processes

A stochastic process is called *stationary* if its properties do not change with time. In other words, the statistical properties of the process do not depend on when one looks at the process. That, obviously, makes life a lot easier. Luckily enough, many technical processes can be modeled as stationary processes. Therefore, studying stationary processes has realistic significance.

The stochastic process $\{x(t), \ t \in T\}$ is called *strictly stationary* if the distribution functions in (3.26) are not affected by the choice of the time point when the process begins; or in other words, if the two processes $\{x(t)\}$ and $\{x(t + \tau)\}$ have the same distribution functions for all values of τ:

$$F(x_1, \ldots, x_n, t_1, \ldots, t_n) = F(x_1, \ldots, x_n, t_1 + \tau, \ldots, t_n + \tau) \qquad \forall \tau \quad (3.42)$$

for $t_1, \ldots, t_n \in T$. A special case is the stationary case when $n = 1$:

$$F(x, t) = F(x). \tag{3.43}$$

This means that the first order distribution is not a function of time. Consequently,

$$m(t) = E[x(t)] = \text{constant} . \tag{3.44}$$

Another special case is the second order distribution function

$$F(x_1, x_2, t_1, t_2) = F(x_1, x_2, t_1 + \tau, t_2 + \tau) . \tag{3.45}$$

This means that the correlation and covariance functions of a stationary process in (3.31) and (3.32) depend only on the difference $\epsilon = t - s$:

$$R(t, s) = R(\epsilon) , \tag{3.46}$$

$$C(t, s) = C(\epsilon) = E[x(t + \epsilon) \cdot x(t)] . \tag{3.47}$$

A weaker[13] form of stationarity is the so-called weak or *wide-sense stationarity*. This is the case if only the first two moments do not vary with time, meaning only (3.44), (3.46) and (3.47) hold but not (3.42) for $n > 2$.

A stationary process is called *ergodic* if the mean of all sample functions at any given time equals the mean of a sample function over time or, as the Science and Engineering Encyclopedia defines it "*An ergodic process is a random process that is stationary and of such a nature that all possible time averages performed on one signal are independent of the signal chosen and hence are representative of the time averages of each of the other signals of the entire random process*". Formally this means

$$E[x(t)] = \int_{-\infty}^{\infty} x f(x, t) dx = \lim_{T \to \infty} \frac{1}{2T} \int_{T_0}^{T_0 + 2T} x(t, \omega) dt \qquad \forall \omega, \forall T_0 . \tag{3.48}$$

Ergodicity implies that any realization of a random process goes through all possible states of the random process. Therefore, when calculating various statistical values such as mean value or autocorrelation function, observation for an infinite time is not needed. A single observation suffices to characterize the process fully. In other words, one can use (3.48) to substitute the time mean value of a single realization for the statistical mean value, thus simplifying the problem significantly. Also, any ergodic process must be a stationary process but the reversal of this statement does not hold in general.

---

[13]This is because the correlation function $R(t, s) = R(t + (-s), s + (-s)) = R(t - s, 0)$ and $C(t, s) = C(t + (-s), s + (-s)) = C(t - s, 0)$.

3.7 Spectral Density

In this Section another way of describing stochastic processes is shown. For this purpose the Fourier transform of the covariance function $R(\epsilon)$ is used (because covariance functions are non-negative definite). This way it is aimed to achieve the equivalence between the time and frequency domain descriptions which are familiar from deterministic systems.

A theorem by Bochner[14] states that a non-negative definite function $R(\epsilon)$ can always be represented as

$$R(\epsilon) = \frac{1}{2\pi} \int_{-\infty}^{\infty} e^{j\omega\epsilon} S(\omega)d\omega. \tag{3.49}$$

$S(\omega)$ is called the spectral density function and

$$F(\omega) = \int_{-\infty}^{\omega} S(\rho)d\rho \tag{3.50}$$

is called the spectral distribution function of the stochastic process.

From (3.49) follows that

$$S(\omega) = \int_{-\infty}^{\infty} e^{-j\omega\epsilon} R(\epsilon)d\epsilon. \tag{3.51}$$

Note that

$$\text{Var}[x(t)] = R(0) = \frac{1}{2\pi} \int_{-\infty}^{\infty} S(\omega)d\omega. \tag{3.52}$$

That is, the total variation of the spectral density function is equal to the variance of the process. In other words, the area under the curve defined by thespectral density $S(\omega)$ corresponds to the total variance (or the "power") of the process. Furthermore, $R(\epsilon)$ is an even function. Therefore, $S(\omega)$ is a real function.

Example: If the covariance function of a process is $R(\epsilon) = e^{-\alpha|\epsilon|}$ what is its power density function according to (3.51)?

---

[14]Salomon Bochner, American mathematician (1899–1982).

$$S(\omega) = \int_{-\infty}^{\infty} e^{-j\omega\epsilon} R(\epsilon) d\epsilon$$

$$= \int_{-\infty}^{\infty} e^{-j\omega\epsilon} e^{-\alpha|\epsilon|} d\epsilon$$

$$= \int_{-\infty}^{0} e^{(\alpha-j\omega)\epsilon} d\epsilon + \int_{0}^{\infty} e^{-(\alpha+j\omega)\epsilon} d\epsilon$$

$$= \left[\frac{1}{\alpha - j\omega} e^{(\alpha-j\omega)\epsilon} \right]_{-\infty}^{0} - \left[\frac{1}{\alpha + j\omega} e^{-(\alpha+j\omega)\epsilon} \right]_{0}^{\infty}$$

$$= \frac{1}{\alpha - j\omega} [1 - 0] - \frac{1}{\alpha + j\omega} [0 - 1]$$

$$= \frac{1}{\alpha - j\omega} + \frac{1}{\alpha + j\omega} = \frac{\alpha + j\omega}{\alpha^2 - \omega^2} + \frac{\alpha - j\omega}{\alpha^2 - \omega^2}$$

$$S(\omega) = \frac{2\alpha}{-\omega^2 + \alpha^2} \cdot$$

□

3.8 Some Special Processes

3.8.1 Normal (Gaussian) Process

A stochastic process is called *normal* if the n-th order distribution function of the random vector $\mathbf{x} = [x(t_1), \ldots, x(t_k)]^T$ is normal for all k and all $\{t_i \in T, i = 1, 2, \ldots, k\}$. Such processes are completely described by their mean functions

$$\mu_i = E[x(t_i)] \quad i = 1, 2, \ldots, k \tag{3.53}$$

and covariance functions

$$\sigma_{ij} = \text{Cov}[x(t_i), x(t_j)] \quad i, j = 1, 2, \ldots, k. \tag{3.54}$$

With

$$\boldsymbol{\mu} = \begin{pmatrix} \mu_1 \\ \vdots \\ \mu_n \end{pmatrix}, \quad \boldsymbol{\Sigma} = \begin{pmatrix} \sigma_{11} & \cdots & \sigma_{1n} \\ \vdots & \ddots & \vdots \\ \sigma_{n1} & \cdots & \sigma_{nn} \end{pmatrix}, \tag{3.55}$$

the n-th order density function can be given as an n-th order normal distribution:

$$f(x_1, x_2, \ldots, x_k, t_1, t_2, \ldots, t_k) \sim N(\boldsymbol{\mu}, \boldsymbol{\Sigma}). \tag{3.56}$$

A normal process is thus completely described by its mean vector and covariance matrix:

$$\boldsymbol{\mu}(t) = E[\mathbf{x}(t)]. \tag{3.57}$$

$$\begin{aligned}\boldsymbol{\Sigma}(s,t) = {} = {} & \text{Cov}[\mathbf{x}(s), \mathbf{x}(t)] \\ = {} & E[\{\mathbf{x}(s) - \boldsymbol{\mu}(s)\} \cdot \{\mathbf{x}(t) - \boldsymbol{\mu}(t)\}^T].\end{aligned} \tag{3.58}$$

If the value of a Gaussian process is independent of time, this means that for all $i \neq j$, $\sigma_{ij} = 0$. In that case the probability density function can be simplified as

$$f(x_1, x_2, ..., x_n, t_1, t_2, ..., t_n) = \prod_{j=1}^{n} \frac{1}{\sqrt{2\pi}\sigma_j} e^{-\frac{(x_j - \mu_j)^2}{2\sigma_j^2}} \tag{3.59}$$

$$= f(x_1, t_1) \cdot f(x_2, t_2) \cdots f(x_n, t_n). \tag{3.60}$$

In other words, as a product of first order density functions where σ_k is the variance of the k-th density function.

3.8.2 Markov Process

Let t_i and t be elements of T such that $t_1 < t_2 < \ldots < t_k < t$. A process $\{x(t), t \in T\}$ is called a *Markov process*[15] if

$$P[x(t) \leq \xi \mid x(t_1), x(t_2), \ldots, x(t_k)] = P[x(t) \leq \xi \mid x(t_k)] \tag{3.61}$$

In other words, a Markov process is a random process whose future probabilities are determined by its most recent values alone. If the distribution function of $x(t_1)$ is known (*i.e.,* the initial first order distribution)

$$F(x_1, t_1) = P[x(t) \leq x_1], \tag{3.62}$$

as well as the transition probability

$$F(x_t, t \mid x_s, s) = P[x(t) \leq x_t \mid x(s) = x_s], \tag{3.63}$$

---

[15] Andrey Andreyevich Markov, Russian mathematician (1856–1922); famous for his work on stochastic processes.

Bayes[16] theorem tells us that the distribution function of the random processes $\mathbf{x} = [x(t_1), x(t_2), \ldots, x(t_k)]^T$ can be calculated as

$$F(x_1, x_2, \ldots, x_k, t_1, t_2, \ldots, t_k) = F(x_k, t_k | x_{k-1}, t_{k-1}) \cdot F(x_{k-1}, t_{k-1} | x_{k-2}, t_{k-2}) \cdots$$
$$F(x_2, t_2 | x_1, t_1) \cdot F(x_1, t_1). \tag{3.64}$$

In other words, a Markov process is defined by two functions:
(a) The absolute probability function $F(y, s)$, and
(b) The transition probability distribution $F(x, t | y, s)$.

Example: Consider the bus ridership in a city. After examining several years of data, it was found that 30% of the people who regularly ride on buses in a given year do not regularly ride the bus in the next year. Also it was found that 20% of the people who do not regularly ride the bus in that year, begin to ride the bus regularly the next year. If 5'000 people ride the bus and 10'000 do not ride the bus in a given year, what is the distribution of riders/non-riders in the next year? In 2 years? In n years?

Firstly, determine how many people will ride the bus next year. Of the people who currently ride the bus, 70% of them will continue to do so. Of the people who do not ride the bus, 20% of them will begin to ride the bus. Thus: $5'000(0.7) + 10'000(0.2)$ = the number of people who ride bus next year = b_1. By the same argument as above: $5'000(0.3) + 10'000(0.8)$ = the number of people who do not ride the bus next year = b_2.

This system of equations is equivalent to the matrix equation: $\mathbf{Mx} = \mathbf{b}$ where

$$\mathbf{M} = \begin{pmatrix} 0.7 & 0.2 \\ 0.3 & 0.8 \end{pmatrix}, \quad \mathbf{x} = \begin{pmatrix} 5000 \\ 10000 \end{pmatrix} \quad \text{and} \quad \mathbf{b} = \begin{pmatrix} b_1 \\ b_2 \end{pmatrix}.$$

Note $\mathbf{b} = [5500 \quad 9500]^T$. For computing the result after 2 years, just use the same matrix $\mathbf{M}$, however use $\mathbf{b}$ in place of $\mathbf{x}$. Thus the distribution after 2 years is $\mathbf{Mb} = \mathbf{M}^2\mathbf{x}$. In fact, after n years, the distribution is given by $\mathbf{M}^n\mathbf{x}$. Incidentally, according to this model, the situation converges to a stationary situation. For instance

$$\mathbf{M}^{10} = \begin{pmatrix} 0.4006 & 0.3996 \\ 0.5994 & 0.6004 \end{pmatrix}, \quad \mathbf{M}^{11} = \begin{pmatrix} 0.4003 & 0.3998 \\ 0.5997 & 0.6002 \end{pmatrix},$$

$$\mathbf{M}^{12} = \begin{pmatrix} 0.4001 & 0.3999 \\ 0.5999 & 0.6001 \end{pmatrix}, \quad \mathbf{M}^{13} = \begin{pmatrix} 0.4 & 0.4 \\ 0.6 & 0.6 \end{pmatrix}.$$

□

Example: A Markov process with four states A, B, C, D is given. The transition probabilities from one state to the other shown in Figure 3.15 are known. This process

---

[16]Thomas Bayes, English mathematician and statistician, (1701–1761); famous for formulating the conditional probability theorem named after him.

generates strings like $ABCDABDABCA$.... If it runs long enough, all the states will be visited and visited in an unpredictable manner.   □

Example: Assuming now that the probabilities for the next choice depends not only on the last state as equation (3.61) states 'but on the last two states, what results is called a *second order Markov process* and (3.61) becomes

$$P[x(t) \leq \xi \,|\, x(t_1), x(t_2), \ldots, x(t_k)] = P[x(t) \leq \xi \,|\, x(t_k), x(t_{k-1})]. \qquad (3.65)$$

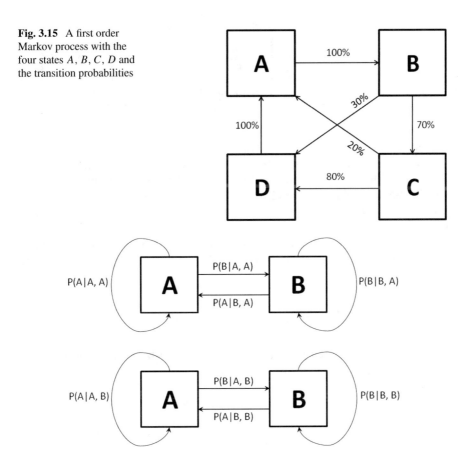

Fig. 3.15 A first order Markov process with the four states A, B, C, D and the transition probabilities

Fig. 3.16 A second order Markov process with two states A, B and the conditional transition probabilities between the states

Fig. 3.17 A second order Markov process with two states A, B is equivalent to a first order Markov process with four states

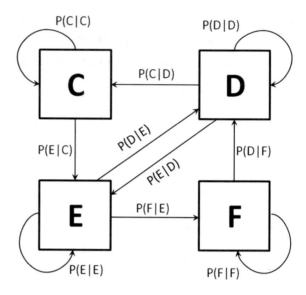

Let us look at an example with two states A and B as shown in Figure 3.16. The major difference to the previous figure is that the transition probabilities have now become conditional on two latest states. One can see that instead of needing just four probabilities, eight probabilities are now needed. Therefore, the situation is equivalent to a first order Markov process with four states C, D, E, F with C corresponding to A, A; D corresponding to A, B; E corresponding to B, A and F corresponding to B, B as shown in Figure 3.17.   □

3.8.3 Process with Independent Increments

The process $\{x(t),\ t \in T\}$ is called a *process with independent increments* if the random variables

$$\Delta_k = x(t_k) - x(t_{k-1}) \tag{3.66}$$
$$\Delta_{k-1} = x(t_{k-1}) - x(t_{k-2})$$
$$\cdots$$
$$\Delta_2 = x(t_2) - x(t_1)$$
$$\Delta_1 = x(t_1)$$

are all independent of each other for $t_i \in T$ for $i = 1, 2, \ldots, k$ and $t_1 < t_2 < \ldots < t_k$. If the increments in (3.66) are not correlated with each other, the process is called a process with uncorrelated increments.

A process with independent increments is completely determined by the distribution of the increments $\Delta(t, s) = x(t) - x(s)$ as well as the density function $f(x, t_1)$. If the density of $\Delta(t, s)$ depends only on the difference $t - s$ the process has stationary increments. If $\Delta(t, s)$ is distributed normally, then it is a process with independent normal increments. The covariance function of a process with independent normal increments is

$$R(s, t) = \text{Cov}[x(\min(s, t)), x(\min(s, t))] . \tag{3.67}$$

3.8.4 Wiener Process

The Wiener process, also known as Brownian motion which was introduced in Section 3.4, has the following properties:
(1) $x(0) = 0$,
(2) $x(t)$ is normal for $t \in T$,
(3) $E[x(t)] = 0$ for $t \geq 0$,
(4) The process has independent stationary increments.

Since the Wiener process is normal, it is completely determined by its mean value and covariance function:

$$m(t) = E[x(t)] = 0 , \tag{3.68}$$

$$\text{Var}[x(t)] = \sigma^2 t . \tag{3.69}$$

What is the covariance of the Wiener process? Let us remember the definition of covariance function:

$$R(t, s) = E[\{x(t) - m(t)\} \cdot \{x(s) - m(s)\}] = C(t, s) - m(t)m(s) . \tag{3.70}$$

With (3.68) equation (3.70) becomes for $0 \leq s < t$

$$
\begin{aligned}
R(s, t) &= E[x(s) \cdot x(t)] \\
&= E[x(s) \cdot \{x(t) - x(s) + x(s)\}] \\
&= E[x^2(s)] + E[x(s) \cdot \{x(t) - x(s)\}] ,
\end{aligned}
$$

and because of (4) above and because the product of expected values is the expected value of the products

$$
\begin{aligned}
&= E[x^2(s)] + E[x(s)] \cdot E[\{x(t) - x(s)\}] \\
&= \sigma^2 s + 0 = \sigma^2 s .
\end{aligned}
$$

Similarly, for $0 \leq t < s$, $R(s, t) = \sigma^2 t$. Remembering that $\Delta(t, s) = x(t) - x(s)$, for $s < t$, $\Delta(t, s) \sim N(0, t - s)$ and for $t < s$, $\Delta(t, s) \sim N(0, s - t)$. This leads to the formula for $s, t \geq 0$

$$R(s, t) = \sigma^2 \min(s, t) .\tag{3.71}$$

3.8.5   Gaussian White Noise

The n-dimensional process $\{\mathbf{x}(t), t \geq t_0\}$ is called *Gaussian white noise* if it can be determined completely by the following two properties:

$$E[\mathbf{x}(t)] = \mathbf{0} ,\tag{3.72}$$

$$E[\mathbf{x}(t) \cdot \mathbf{x}^T(s)] = \mathbf{Q}(t)\delta(t - s) .\tag{3.73}$$

$\mathbf{Q}$ is a positive semidefinite matrix and $\delta(\cdot)$ is the Dirac-pulse. Remember that the power spectral density of Gaussian white noise is flat.

3.9   Analysis of Stochastic Processes

So far, the study of random processes has focused on the distribution function. However, much can be done in terms of the first two moments: the mean and variance of the random variables. The elaborations here focus on the study of these two moments exclusively. This approach is especially convenient when dealing with random processes, because the entire probabilistic description of a random process would require the modeling of the distributions of all finite collections of random variables. Such a task is not tractable except in special cases. Therefore, so-called second-order theory is considered here. Essentially, this means focusing on the expected value and covariance functions of random processes.

3.9.1   Convergence

The sequence of random variables $\{X_n(\omega), n = 1, 2, \ldots\}$, where $X_n(\omega) = X(\omega, t_n)$, converges in the mean-square sense to $X(\omega)$ if

$$\lim_{n \to \infty} E[(X_n - X)^2] = 0 .\tag{3.74}$$

In this case the limit and the E-operator can be exchanged:

$$\lim_{n \to \infty} E[X_n] = E\left[\lim_{n \to \infty} X_n\right] = E[X] .\tag{3.75}$$

3.9.2 Continuity

Let $\{X_t, -\infty < t < \infty\}$ be a stochastic process. Hereafter, this notation is shortened to $\{X_t\}$ and it is assumed that t spans the entire real line. $\{X_t\}$ is said to be continuous in the mean-square sense at t if

$$\lim_{h \to 0} E[(X_{t+h} - X_t)^2] = 0. \qquad (3.76)$$

$\{X_t\}$ is continuous in the mean-square sense at t if and only if the mean function and the covariance function of $\{X_t\}$ are continuous in the mean-square sense.

3.9.3 Differentiability

$\{X_t\}$ is mean-square differentiable at t if $\lim_{h \to 0} \frac{X_{t+h} - X_t}{h}$ exists, in which case the limit is denoted as

$$\lim_{h \to 0} \frac{X_{t+h} - X_t}{h} = \dot{X}_t. \qquad (3.77)$$

$\{X_t\}$ is mean-square differentiable at t if and only if the mean function and the covariance function of $\{X_t\}$ are differentiable and if the generalized second derivative of the covariance function $\frac{\partial^2 R(s,t)}{\partial s \partial t}$ exists. Let us note that the Wiener process is continuous everywhere but not differentiable anywhere. This pathological property is revisited in Chapter 5.

3.9.4 Integrability

A second order stochastic process $\{x(t), t \in T\}$ is called mean-square Riemann integrable over the interval $[a, b]$ if the integrals $\int_a^b m(t)dt$ and $\int_a^b \int_a^b R(s, t)dsdt$ exist.

Note that the integral $\int_a^b x(t)dt$ is a random variable, since it is not integrated with respect to the sample points ω; rather, it is integrated with respect to time. It should also be emphasized that this integral is *not* a time average of a sample path. It is the integral of a stochastic process, and therefore randomness is preserved. The concept of integrability will play a central role in Chapter 5.

3.9.5   Brief Summary

Table 3.3 A summary of definitions for random variables X and Y. x^* denotes the complex conjugate of x

Concept	Discrete case	Continuous case
Probability of ...	$P(\cdot)$	$P(\cdot)$
Distribution function	$F_X(t) = P(X \le t)$	$F_X(t) = P(X \le t)$
Density function	$F_X(n) = \sum_{i=-\infty}^{n} f_X(i)$	$F_X(t) = \int_{-\infty}^{t} f_X(y)dy$
Expected value	$\mu_X = \mathrm{E}[X] = \sum_{k=1}^{\infty} x_k p_k$	$\mu_X = \mathrm{E}[X] = \int_{-\infty}^{\infty} t f_X(t)dt$
Variance	$\mathrm{Var}[X] = \mathrm{E}[(X - \mu_X)^2] = \mathrm{E}[X^2] - \mu_X^2$	$\mathrm{Var}[X] = \mathrm{E}[(X - \mu_X)^2] = \mathrm{E}[X^2] - \mu_X^2$
Standard deviation	$\sigma_X = \sigma(X) = \sqrt{\mathrm{Var}[X]}$	$\sigma_X = \sigma(X) = \sqrt{\mathrm{Var}[X]}$
Covariance	$\mathrm{Cov}[X, Y] = \mathrm{E}[(X - \mu_X) \cdot (Y - \mu_Y)]$	$\mathrm{Cov}[X, Y] = \mathrm{E}[(X - \mu_X) \cdot (Y - \mu_Y)]$
Correlation coefficient	$\rho_{X,Y} = \mathrm{Corr}(X, Y) = \frac{\mathrm{Cov}[X,Y]}{\sigma_X \sigma_Y}$	$\rho_{X,Y} = \mathrm{Corr}(X, Y) = \frac{\mathrm{Cov}[X,Y]}{\sigma_X \sigma_Y}$
Correlation function	$C(s, t) = \mathrm{Corr}(X(s), X(t))$	$C(s, t) = \mathrm{Corr}(X(s), X(t))$
Cross-correlation	$C_{xy}(n) = \sum_{m=-\infty}^{\infty} x^*(m)y(n + m)$	$C_{xy}(t) = \int_{-\infty}^{\infty} x^*(\tau)y(t + \tau)d\tau$
Autocorrelation	$C_{xx}(j) = \sum_{n=-\infty}^{\infty} x_n x_{n-j}^*$	$C_{xx}(\tau) = \int_{-\infty}^{\infty} x(t + \tau)x^*(t)dt$

3.10   Exercises

A1: What is a random variable? How can you describe random variables mathematically?

A2: What is a Bernoulli experiment? What are the properties of binomial distribution?

A3: How are distribution and density functions related?

A4: Explain the concepts of expected value, variance, skewness and kurtosis in your own words and mathematically.

A5: What is correlation? How is correlation different from cause–effect relationship?

A6: What is a random process? Explain the concepts ofensemble, realization and random variable in the context of random processes both in your own words and mathematically.

A7: Explain the central limit theorem in your own words. What is the significance of this theorem?

A8: What is an ergodic process? Give several examples of ergodic and non-ergodic processes and explain why they are or are not ergodic.

A9: What is a Markov process? What does the order of a Markov process mean?

A10: What is the difference between probability theory and statistics? Give several problems as examples of both cases.

B1: In a multiple choice test with 20 questions, each question has five possible answers. What is the probability of a student getting exactly half of the answers right simply by guessing? What is the numerical value of this probability correct to three digits?

B2: Let the random variable X be defined by its probability

$$P_X(x) = \binom{4}{x}(1/2)^4 .$$

(a) What is the standard deviation of the random variable X? b) What is the probability that X is within one standard deviation of the expected value?

B3: Two independent random variables X and Y are always uncorrelated but the converse is not always true. Provide an example of two random variables X and Y that are uncorrelated but NOT independent.

B4: X and Y are random variables on a probability space Ω. Their two dimensional distribution function is $F_{X,Y}(s, t) = (s + t)^2$. The distribution functions of X and Y are given as $F_X(s) = s^2$ and $F_Y(t) = t^2$. Are X and Y independent ? Why?

B5: Let us come back to the Monty Hall puzzle at the beginning of this Chapter. Suppose you are on a game show, and you're given the choice of three doors: behind one door is a car; behind the others, nothing. You pick a door, say the first one [but the door is not opened], and the host, who knows what is behind the doors, opens another door, say the third one, which has nothing. He then says to you, "Do you want to change your choice to the second door?" Is it to your advantage to switch your choice? Answer carefully and then conduct a simulation to verify your answer. Be prepared to be surprised! Find arguments to convince your suspecting friends and family of the correct solution. The result is so counterintuitive that many a famous mathematician got it wrong!

B6: Are the functions below possible probability density functions?

$$f_X(t) = \begin{cases} 1, & -1 \le t \le 1 \\ 0, & \text{else}. \end{cases}$$

$$f_Y(t) = \begin{cases} \sin(t), & 0 \le t \le 2\pi \\ 0, & \text{else}. \end{cases}$$

$$f_Z(t) = \begin{cases} |t|, & -1 \le t \le 1 \\ 0, & \text{else}. \end{cases}$$

B7: X and Y are random variables with the two dimensional density function

$$f_{X,Y}(u, v) = \begin{cases} \frac{c}{(1+u+v)^3)}, & \text{if } u > 0 \text{ and } v > 0 \\ 0, & \text{else}. \end{cases}$$

(a) what is the value of c? (b) What is the corresponding distribution function $F_{X,Y}(s, t)$?

B8: Given the function

$$f(x, y) = \begin{cases} 0.5(x + y) + kxy, & 0 \le x \le 1; \ 0 \le y \le 1 \\ 0, & \text{otherwise}. \end{cases}$$

(a) Find k such that $f(x, y)$ can be the joint probabilitydensity function of X and Y. (b) Are X and Y independent? (c) Are X and Y uncorrelated?

B9: X and Y are two randomly chosen points in the interval $[0, 1]$. The probability that the point is in a certain interval $[a, b]$ (with $0 \le a \le b \le 1$) is equal to the length of the interval $|X - Y|$. What is $E[|X - Y|]$?

B10: Prove the formulas in Section 3.3.10 for the first two moments of log-normal distribution including the multivariate case.

B11: A random process is given as

$$\xi(t) = A \cos(\omega_c t + \theta),$$

with A and ω_c as constants. θ is a random variable equally distributed in $(0, 2\pi)$. (a) Is $\xi(t)$ ergodic? (b) What is the autocorrelation function of $\xi(t)$? (c) What is the power spectral density of $\xi(t)$?

B12: What is the condition on the transition matrix for a Markov process to converge to a stationary situation?

B13: A sequence segment of bases adenine, cytosine, guanine and thymine (A, C, G, T) from the human preproglucagon gene *GTATTAAATCCGTAGTC*

TGAACTAACTA... shows that the bases appear in somehow regular basis. An empirical study shows that for the state space $S = A, C, G, T$ the transition probabilities are given by the first order Markov matrix

$$M = \begin{pmatrix} 0.359 & 0.384 & 0.143 & 0.156 \\ 0.305 & 0.284 & 0.199 & 0.182 \\ 0.167 & 0.023 & 0.331 & 0.437 \\ 0.151 & 0.177 & 0.345 & 0.357 \end{pmatrix}.$$

(a) Calculate the probability of encountering the sequence *CTGAC* in that gene, given the first base being *C*. (b) As the sequences grow longer, the probabilities of encountering them get smaller and smaller. Why? Propose a method to deal with that issue and demonstrate it on the sequence *CTGACCTGAC*. (Hint: Your solution should be very simple.)

B14: X is a binary random variable. It either takes the value "1" with probability p or "0". What is the maximum standard deviation of X?

B15: The voltage waveform $v(t)$ between the red terminal and ground of a power supply in your lab is equally likely to be either $+5$ V for all time t, or -5 V for all time t. For this random process $v(t)$, determine (a) the mean value of the process, (b) its autocorrelation function, (c) its autocovariance function, (d) whether it is strict-sense stationary and (e) whether it is ergodic.

B16: Suppose you have invented a pill to cure a certain disease. Sometimes the pill works and the patient is cured; sometimes it does not work. Let p be the fraction of the cases in which the pill works. To estimate p, we pick a random sample of n patients and give the pill to each. Then we count the number S of successful cures, and use S/n as our estimate of p. We want to say, with at least 95% certainty, that the observed value S/n is within 1% of the true value p. How big should the sample size be?

B17: Suppose the weights of children in a school are independent, normally distributed random variables. Girls have an average weight of $\mu_X = 40$ kg and the boys have an average weight of $\mu_Y = 50$ kg. The corresponding variances are $\sigma_X = 3$ kg and $\sigma_Y = 6$ kg. What is the probability that a boy and a girl have a combined weight between 90 kg and 100 kg? Write your answer in terms of the function $\Phi(z)$ and give a numerical value.

B18: Both figures below depict a deterministic decay process with additive noise $y(t) = Ae^{(-bt)} + n(t)$. Looking at the noise component of the figures below, which one is more likely to depict a Wiener process?

B19: Consider a homogeneous chain with state space $\{1, 2, 3\}$ and transition probability matrix $T = [t_{ij}]$ where the transition probabilities are given by $t_{ij} = \frac{i+j}{6+3i}$. (a) Write out the matrix T. (b) Can the chain produced by the matrix T be a Markov chain? (c) The initial probability distribution is given as $P(X_0 = 1) = 1/2$, $P(X_0 = 2) = 1/4$, $P(X_0 = 3) = 1/4$. Determine $P(X_1 = i)$ for each state $i = 1, 2, 3$.

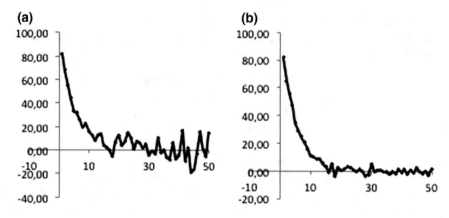

Fig. 3.18 Which one is more likely to depict a Wiener process?

B20: Consider modeling economic mobility in a hypothetical society as a Markov process. There are three states: upper, middle, and lower economic classes (indexed by 1, 2 and 3 respectively). Suppose that we conducted a study to discover the probabilities that a family will move up or down in economic class after a generation based on its current economic class. We can encode these probabilities into the following transition matrix:

$$\mathbf{T} = \begin{pmatrix} 0.6 & 0.1 & 0.1 \\ 0.3 & 0.8 & 0.2 \\ 0.1 & 0.1 & 0.7 \end{pmatrix}.$$

If a family starts in the lower economic class today ($n = 0$), what is the probability that it will have moved up to the upper economic class in (a) one generation ($n = 1$), (b) in ten generations ($n = 10$)? (Use two significant digits in your results.) (c) Comment your results.

B21: Show that for two independent random variables X and Y with identical distributions with finite variance for $Z_1 = X + Y$ and $Z_2 = X - Y$ to be independent X and Y must be normally distributed. Can you extend this result for three or more variables?

B22: Prove Stein's lemma, relevant for portfolio choice theory, which states that Z is a random variable with standard normal distribution if and only if $E[f'(Z)] = E[Zf(Z)]$ for all absolutely continuous functions $f(\cdot)$ with $E[|f'(Z)|] < \infty$.

C1: Some random processes (*e.g.,* stock prices over longer periods of time) are inherently time-varying. Develop several alternative algorithms for determining the momentary values of the process parameters.

C2: Continuity of stochastic processes can be tricky. A definition for continuity in the mean-square sense was given in Section 3.9. Analyze the problem and propose other definitions. Study the strengths and weaknesses of your proposals. Find extreme examples where all the definitions become problematic.

C3: In 1917 the Italian mathematician Francesco Paolo Cantelli made the seemingly simple conjecture that for a positive function $f(\cdot)$ and independent random variables X and Y with standard normal distributions $Z = X + f(X)Y$ is normal, then $f(\cdot)$ is a constant almost everywhere. This has not been proven yet. Find a proof or a counter example.

Chapter 4
Optimal Control

Engineers aren't boring people, we just get excited over boring things.

— Seen on a t-shirt

The scientific imagination always restrains itself within the limits of probability.

— Thomas Huxley

4.1 Introduction

Optimal Control is a very important and broad branch of control engineering. Clearly, no single chapter can possibly claim to cover all of this field. Therefore, this chapter, which is intended solely as an introduction or a refresher, begins with calculus of variations, goes through the fixed and variable endpoint problems as well as the variation problem with constraints. These results are then applied to dynamic systems, leading to the solution of the optimal control problem using the Hamilton-Jacobi and Pontryagin methods. The concept of dynamic programming is explained and the chapter ends with a brief introduction to differential games.

Most control engineering problems are underdetermined. Therefore, it makes sense to demand additional conditions for the solution of such problems.[1] Consider a linear system in the state space representation

$$\dot{\mathbf{x}} = \mathbf{A}\mathbf{x} + \mathbf{B}\mathbf{u} \qquad \mathbf{x}(0) = \mathbf{x}_0 , \qquad (4.1)$$

---

[1] This chapter which is an extended version of an Appendix in [51] is thought as an introduction to Optimal Control. For rigorous definitions, derivations and proofs, the reader is referred to standard textbooks *e.g.,* [4, 45, 75, 113].

© Springer International Publishing AG 2018
S. S. Hacısalihzade, *Control Engineering and Finance*, Lecture Notes in Control and Information Sciences 467, https://doi.org/10.1007/978-3-319-64492-9_4

with $\mathbf{x}(t)$ as the system state, $\mathbf{u}(t)$ as the input and $\mathbf{A}$ and $\mathbf{B}$ as the system matrix and the input matrix, respectively. The control problem is to find $\mathbf{u}(t)$, such that $\mathbf{x}(t_f) = \mathbf{0}$. A reasonable (and very common) approach is to define a cost functional and solve the control problem subject to the minimization of the cost functional (also called the performance index). It makes sense to include the system states' trajectories, inputs and final states in the cost functionals. Probably the most popular cost functional is the quadratic form

$$Z = \frac{1}{2}\mathbf{x}^T(t_f)\mathbf{S}\mathbf{x}(t_f) + \frac{1}{2}\int_{t_0}^{t_f} \mathbf{x}^T(t)\mathbf{Q}\mathbf{x}(t) + \mathbf{u}^T(t)\mathbf{P}\mathbf{u}(t)dt , \qquad (4.2)$$

where t_0 and t_f are the initial and final times of interest respectively. $\mathbf{Q}$ is a positive semidefinite matrix and $\mathbf{P}$ and $\mathbf{S}$ are positive definite matrices. Those weighting matrices are often chosen as diagonal matrices.

The first integrand $\mathbf{x}^T(t)\mathbf{Q}\mathbf{x}(t) \geq 0$ because $\mathbf{Q}$ is positive semidefinite. It represents the cost incurred at time t for the state trajectories which deviate from 0. The second integrand $\mathbf{u}^T(t)\mathbf{P}\mathbf{u}(t) > 0$ for $\mathbf{u}(t) \neq \mathbf{0}$, because $\mathbf{P}$ is positive definite. It represents the cost at time t incurred by the control effort to get $\mathbf{x}(t)$ to $\mathbf{0}$. The total cost is the cumulative cost incurred during the time interval of interest. The choice of the weighting matrices $\mathbf{Q}$, $\mathbf{P}$ and $\mathbf{S}$ affects the trade-off between the requirements of controlling the state and the effort (energy) to do so as well as how closely a desired final state is reached.

Some special cases are of particular interest.

$$Z = \int_{t_0}^{t_f} dt = t_f - t_0 , \qquad (4.3)$$

for a given t_0 optimizing Z results in the shortest time.

$$Z = \int_{t_0}^{t_f} \mathbf{u}^T(t)\mathbf{R}\mathbf{u}(t)dt , \qquad (4.4)$$

for a given t_0 and t_f as well as $\mathbf{R}$ this corresponds to the minimal energy solution.

$$Z = \mathbf{x}^T(t_f)\mathbf{S}\mathbf{x}(t_f) , \qquad (4.5)$$

for a given t_f and $\mathbf{S}$ this cost functional corresponds to attaining the final state $\mathbf{x}(t_f) = \mathbf{0}$ no matter how much the states deviate from zero during the process and how much energy it requires to do so.

4.2 Calculus of Variations

4.2.1 Subject Matter

Example: Let us begin with an illustrative case. Assume that the functional V represents the length of a curve $x(t)$.

$$V(x) = \int_a^b \sqrt{1 + \dot{x}(t)^2}\,dt \;. \tag{4.6}$$

An "interesting" problem is to find $x(t)$ that minimizes V, whereas, in general

$$V(x) = \int_a^b L[x(t), \dot{x}(t), t]\,dt \;. \tag{4.7}$$

□

Definition: A functional V has a relative minimum at x_0 if a neighborhood Ω of x_0 exists such that $V(x) \geq V(x_0)\; \forall x \in \Omega$ for small values of $|x - x_0|$ and $|\dot{x} - \dot{x}_0|$. •

Theorem: If the variation of the functional V exists and has a relative minimum at x_0, then $\delta V = 0$ at x_0.

$$\delta V = \left. \frac{\partial V}{\partial x} \right|_{x_0} \delta x \quad \text{is the first variation of } V\,, \tag{4.8}$$

where $\delta x = x_1 - x_0$, $x_1 \in \Omega$. The total variation $\Delta V = V(x_1) - V(x_0) = V(x_0 + \delta x) - V(x_0)$ is non-negative, because $V(x_1) \geq V(x_0)$. ◇

The Taylor series expansion of the total variation around x_0 is

$$\Delta V = \left. \frac{\partial V}{\partial x} \right|_{x_0} \delta x + \frac{1}{2!} \cdot \left. \frac{\partial^2 V}{\partial x^2} \right|_{x_0} (\delta x)^2 + \frac{1}{3!} \cdot \left. \frac{\partial^3 V}{\partial x^3} \right|_{x_0} (\delta x)^3 + \ldots \tag{4.9}$$

$$= \delta V + \frac{1}{2!} \delta^2 V + \frac{1}{3!} \delta^3 V + \ldots \tag{4.10}$$

where

$$\delta V = \left. \frac{\partial V}{\partial x} \right|_{x_0} \delta x \quad \text{is the first variation of } V\,,$$

$$\delta^2 V = \left. \frac{\partial^2 V}{\partial x^2} \right|_{x_0} (\delta x)^2 \quad \text{is the second variation of } V\,,$$

$$\delta^3 V = \ldots \;.$$

4.2.2   Fixed Endpoint Problem

Given

$$V(x) = \int_a^b L[x(t), \dot{x}(t), t]dt \,,$$

$$x(a) = A \,,$$

$$x(b) = B \,,$$

where L has continuous first and second derivatives. V has a minimum at x. $x_1 = x + \delta x$ and $\delta x(a) = \delta x(b) = 0$ (fixed endpoint). Defining ΔV as the difference between the functional values with x_1 and x as

$$\Delta V = V(x + \delta x) - V(x) \tag{4.11}$$

$$= \int_a^b L[x + \delta x, \dot{x} + \delta \dot{x}, t]dt - \int_a^b L[x, \dot{x}, t]dt \,. \tag{4.12}$$

Firstly, writing the Taylor expansion of this difference in functionals as

$$\Delta V = \int_a^b \left[\frac{\partial L(x, \dot{x}, t)}{\partial x} \delta x + \frac{\partial L(x, \dot{x}, t)}{\partial \dot{x}} \delta \dot{x} + \frac{1}{2!} \frac{\partial^2 L(x, \dot{x}, t)}{\partial x^2} \delta x^2 \right.$$

$$\left. + \frac{1}{2!} \frac{\partial^2 L(x, \dot{x}, t)}{\partial x \partial \dot{x}} \delta x \delta \dot{x} + \frac{1}{2!} \frac{\partial^2 L(x, \dot{x}, t)}{\partial \dot{x}^2} \delta \dot{x}^2 + \dots \right] dt \,. \tag{4.13}$$

Taking the first variation of the functional difference as

$$\delta V = \int_a^b \left[\frac{\partial L(x, \dot{x}, t)}{\partial x} \delta x + \frac{\partial L(x, \dot{x}, t)}{\partial \dot{x}} \delta \dot{x} \right] dt \,, \tag{4.14}$$

and with the help of partial integration rule

$$\int_a^b u dv = uv \Big|_a^b - \int_a^b v du \,,$$

the first variation of the functional difference becomes

$$\delta V = \int_a^b \left[\frac{\partial L(x, \dot{x}, t)}{\partial x} - \frac{d}{dt} \frac{\partial L(x, \dot{x}, t)}{\partial \dot{x}} \right] \delta x dt + \frac{\partial L(x, \dot{x}, t)}{\partial \dot{x}} \delta x \Big|_a^b \,. \tag{4.15}$$

Also, because the initial value is known and the endpoint is fixed

$$\delta x(a) = 0 \,,$$

$$\delta x(b) = 0 \,,$$

(4.15) becomes

$$\delta V = \int_a^b \left[\frac{\partial L(x, \dot{x}, t)}{\partial x} - \frac{d}{dt} \frac{\partial L(x, \dot{x}, t)}{\partial \dot{x}} \right] \delta x \, dt . \tag{4.16}$$

Finally, remembering that $\delta V = 0$ is necessary for V to be optimal for any variation in x leads to the so-called *Euler equation*

$$\frac{\partial L(x, \dot{x}, t)}{\partial x} - \frac{d}{dt} \frac{\partial L(x, \dot{x}, t)}{\partial \dot{x}} = 0 . \tag{4.17}$$

Theorem: For a continuous and differentiable function x and the functional $V(x) = \int_a^b L(x, \dot{x}, t) dt$ which is minimized by x where L has continuous first and second partial derivatives and the beginning and end points are fixed, x satisfies the Euler equation. ◇

Example: Getting back to the example at the beginning of this section, where $x(a) = A$, $x(b) = B$ and $V(x) = \int_a^b \sqrt{1 + \dot{x}(t)^2} dt$ the Euler equation becomes (because L is not a function of x the first term vanishes)

$$\frac{d}{dt} \frac{\partial L(x, \dot{x}, t)}{\partial \dot{x}} = 0 .$$

Integrating this equation with respect to time yields

$$\frac{\partial L(x, \dot{x}, t)}{\partial \dot{x}} = \text{constant} ,$$

$$\frac{\partial L(x, \dot{x}, t)}{\partial \dot{x}} = \frac{\dot{x}}{\sqrt{1 + \dot{x}^2}} = \text{constant} .$$

Therefore, $\dot{x} = K_1$ hence $x(t) = K_1 t + K_2$. As expected, the answer is a straight line[2] and the boundary conditions can be used to calculate its slope K_1 and its y-intercept K_2. ☐

For multivariable systems with the state vector **x** comprising n continuous and differentiable components, the Euler equation becomes n equations

$$\frac{\partial L(\mathbf{x}, \dot{\mathbf{x}}, t)}{\partial x_i} - \frac{d}{dt} \frac{\partial L(\mathbf{x}, \dot{\mathbf{x}}, t)}{\partial \dot{x}_i} = 0 \quad i = 1, \ldots, n . \tag{4.18}$$

---

[2] A more interesting problem is to find the curve with the fastest trajectory of a ball rolling only due to gravity from a high point A to a lower point B. This is known as the brachistochrone curve and its solution is left to the interested reader.

At this point, it might be interesting to note that algebraic equations have numbers as solutions (*e.g.*, $x = 3$). Differential equations have functions as solutions (*e.g.*, $f(x) = e^{3x} - 3e^{-2x} + 1$). Calculus of variations problems have differential equations as solutions (*e.g.*, $2\ddot{x} - \dot{x} + 3x = 0$).

4.2.3 Variable Endpoint Problem

Variable endpoint problems in calculus of variations occur when *e.g.*, looking for the shortest distance from a given point to a given curve. Consider the situation in Figure 4.1.

Again with $V(x) = \int_a^b L[x(t), \dot{x}(t), t]dt$ the Euler equation satisfies

$$\Delta V = V(x + h) - V(x) \tag{4.19}$$

$$= \int_{t_0}^{t_1 + \delta t_1} L(x + h, \dot{x} + \dot{h}, t)dt - \int_{t_0}^{t_1} L(x, \dot{x}, t)dt \tag{4.20}$$

$$= \int_{t_1}^{t_1 + \delta t_1} L(x + h, \dot{x} + \dot{h}, t)dt + \int_{t_0}^{t_1} \left[L(x + h, \dot{x} + \dot{h}, t) - L(x(t), \dot{x}(t), t) \right]dt . \tag{4.21}$$

Let us rewrite the first term of the last equation using the mean value theorem as

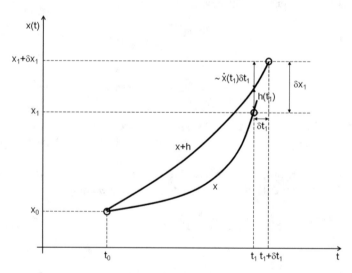

Fig. 4.1 The variable endpoint problem. x_0 is the fixed starting point. The endpoint is not known. x represents the unknown function which optimizes the given functional

$$\int_{t_1}^{t_1+\delta t_1} L(x+h, \dot{x}+\dot{h}, t)dt \approx L(x, \dot{x}, t)\Big|_{t_1} \delta t_1,$$

and develop the Taylor series for the second part as

$$\int_{t_0}^{t_1} \left[L(x+h, \dot{x}+\dot{h}, t) - L(x, \dot{x}, t)\right] dt \approx \int_{t_0}^{t_1} \left[\frac{\partial L(x, \dot{x}, t)}{\partial x}h + \frac{\partial L(x, \dot{x}, t)}{\partial \dot{x}}\dot{h}\right] dt.$$

The last expression can be calculated with partial integration as

$$\int_{t_0}^{t_1} \left[\frac{\partial L(x, \dot{x}, t)}{\partial x} - \frac{d}{dt}\frac{\partial L(x, \dot{x}, t)}{\partial \dot{x}}\right] hdt + \frac{\partial L(x, \dot{x}, t)}{\partial \dot{x}}h\Big|_{t_0}^{t_1}. \qquad (4.22)$$

Euler equation tells us that the value of the integrand in the last equation vanishes. Therefore,

$$\Delta V \approx \frac{\partial L(x, \dot{x}, t)}{\partial \dot{x}}h\Big|_{t_0}^{t_1} + L(x, \dot{x}, t)\Big|_{t_1} \delta t_1. \qquad (4.23)$$

With the initial and endpoint variations

$$h(t_0) = 0,$$
$$h(t_1) \approx \delta x_1 - \dot{x}(t_1)\delta t_1,$$

one can now calculate the first variation of the functional V as

$$\delta V = \frac{\partial L(x, \dot{x}, t)}{\partial \dot{x}}\Big|_{t_1} [\delta x_1 - \dot{x}(t_1)\delta t_1] + L(x, \dot{x}, t)\Big|_{t_1} \delta t_1. \qquad (4.24)$$

Equating the first variation of the functional V to 0 to calculate the optimum finally gives us the generalized boundary condition

$$\frac{\partial L(x, \dot{x}, t)}{\partial \dot{x}}\Big|_{t_1} \delta x_1 + \left[L(x, \dot{x}, t) - \dot{x}\frac{\partial L(x, \dot{x}, t)}{\partial \dot{x}}\right]\Big|_{t_1} \delta t_1 = 0. \qquad (4.25)$$

Several special cases of the generalized boundary condition are of particular interest.

When δt_1 and δx_1 are independent (4.25) becomes the two equations

$$\frac{\partial L(x, \dot{x}, t)}{\partial \dot{x}}\Big|_{t_1} = 0, \qquad (4.26)$$

$$\left[L(x, \dot{x}, t) - \dot{x}\frac{\partial L(x, \dot{x}, t)}{\partial \dot{x}}\right]\Big|_{t_1} = 0. \qquad (4.27)$$

If t_1 is given (*i.e.*, $\delta t_1 = 0$) (4.25) becomes

$$\frac{\partial L(x, \dot{x}, t)}{\partial \dot{x}}\bigg|_{t_1} = 0. \tag{4.28}$$

If x_1 is given (*i.e.*, $\delta x_1 = 0$) (4.25) becomes

$$\left[L(x, \dot{x}, t) - \dot{x}\frac{\partial L(x, \dot{x}, t)}{\partial \dot{x}} \right]\bigg|_{t_1} = 0. \tag{4.29}$$

If the endpoint lies on a curve $y(t)$ (*i.e.*, $x(t_1) = y(t_1)$ and $\delta x_1 \approx \dot{y}(t_1)\delta t_1$) (4.25) becomes the transversality condition

$$\left[L(x, \dot{x}, t) + (\dot{y} - \dot{x})\frac{\partial L(x, \dot{x}, t)}{\partial \dot{x}} \right]\bigg|_{t_1} = 0. \tag{4.30}$$

Example: An illustrative case is looking for the minimal length curve between the origin $(0, 0)$ and the line $y(t) = 2 - t$. The Euler equation states that the curve has the form $x(t) = K_1 t + K_2$. Since $x(0) = 0$, $K_2 = 0$. Transversality condition (4.30) can thus be written as

$$\left[\sqrt{1 + \dot{x}^2} + (\dot{y} - \dot{x})\frac{\dot{x}}{\sqrt{1 + \dot{x}^2}} \right]\bigg|_{t_1} = 0,$$

and because $\dot{x}(t_1) = K_1$ and $\dot{y}(t_1) = -1$

$$\sqrt{1 + K_1^2} + (-1 - K_1)\frac{K_1}{\sqrt{1 + K_1^2}} = 0 \quad \Rightarrow \quad K_1 = 1.$$

As expected, the answer is the line $x(t) = t$ which is perpendicular to the line $y(t) = 2 - t$. □

4.2.4 Variation Problem with Constraints

Consider the problem

$$\min_x V(x) = \int_{t_0}^{t_f} L[x(t), \dot{x}(t), t]dt, \tag{4.31}$$

subject to the constraint

$$g(x, \dot{x}, t) = 0. \tag{4.32}$$

The solution can be obtained by using Lagrange's method which was discussed in Subsection 2.5.2 and defining an auxiliary function $F = L + \lambda g$. This method can then be applied to F as described above. Note that here λ is no longer a constant but a function of time.

The generalization and solution of (4.31) for m constraints $g_1 = 0, g_2 = 0, \ldots g_m = 0$ is straight forward. First define an auxiliary function $F = L + \lambda_1 g_1 + \lambda_2 g_2 + \cdots + \lambda_m g_m$ with the Lagrange multipliers $\lambda_1, \lambda_2, \ldots, \lambda_m$ (again remembering that λ_i are functions) and apply Lagrange's method.

4.3 Optimal Dynamic Systems

4.3.1 Fixed Endpoint Problem

Consider the system given by its state space description as $\dot{\mathbf{x}} = \mathbf{f}(\mathbf{x}, \mathbf{u}, t)$ with $\mathbf{x}(t)$ as its state vector and $\mathbf{u}(t)$ as its input vector. Also given is an objective function $Z = \int_{t_0}^{t_1} L(\mathbf{x}, \mathbf{u}, t) dt$ (with L positive definite) as well as the initial and final states $\mathbf{x}(t_0)$ and $\mathbf{x}(t_1)$ respectively. The interesting question is which $\mathbf{u}(t)$ minimizes Z?

That seemingly intractable question can be answered by applying the calculus of variations methods described in the previous sections. The constraints are given by the system description

$$\mathbf{g}(\mathbf{x}, \dot{\mathbf{x}}, \mathbf{u}, \dot{\mathbf{u}}, t) = \mathbf{f}(\mathbf{x}, \mathbf{u}, t) - \dot{\mathbf{x}} = \mathbf{0}. \tag{4.33}$$

Let us now define a new objective function

$$\Gamma = \int_{t_0}^{t_1} \Lambda(\mathbf{x}, \dot{\mathbf{x}}, \mathbf{u}, \dot{\mathbf{u}}, t) dt, \tag{4.34}$$

with

$$\Lambda(\mathbf{x}, \dot{\mathbf{x}}, \mathbf{u}, \dot{\mathbf{u}}, t) = L(\mathbf{x}, \mathbf{u}, t) + \sum_{j=1}^{n} \lambda_j \left[f_j(\mathbf{x}, \mathbf{u}, t) - \dot{x}_j \right]. \tag{4.35}$$

The resulting Euler equations are

$$\frac{\partial}{\partial x_i} \Lambda - \frac{d}{dt} \frac{\partial}{\partial \dot{x}_i} \Lambda = 0 \qquad i = 1, 2, \ldots, n, \tag{4.36}$$

$$\frac{\partial}{\partial u_k} \Lambda - \frac{d}{dt} \frac{\partial}{\partial \dot{u}_k} \Lambda = 0 \qquad k = 1, 2, \ldots, r. \tag{4.37}$$

The first part of the Euler equations can be written more explicitly as

$$
\frac{\partial}{\partial x_i} \left\{ L(\mathbf{x}, \mathbf{u}, t) + \sum_{j=1}^{n} \lambda_j \left[f_j(\mathbf{x}, \mathbf{u}, t) - \dot{x}_j \right] \right\}
$$

$$
- \frac{d}{dt} \frac{\partial}{\partial \dot{x}_i} \left\{ L(\mathbf{x}, \mathbf{u}, t) + \sum_{j=1}^{n} \lambda_j \left[f_j(\mathbf{x}, \mathbf{u}, t) - \dot{x}_j \right] \right\} = 0.
\tag{4.38}
$$

But $\dot{x}_j$ is not a function of x_i. Also $L(\mathbf{x}, \mathbf{u}, t)$ and $f_j(\mathbf{x}, \mathbf{u}, t)$ are not functions of $\dot{x}_i$. Therefore, (4.38) becomes

$$
\frac{\partial}{\partial x_i} \left[L(\mathbf{x}, \mathbf{u}, t) + \sum_{j=1}^{n} \lambda_j f_j(\mathbf{x}, \mathbf{u}, t) \right] + \dot{\lambda}_i = 0,
\tag{4.39}
$$

or

$$
\dot{\lambda}_i = -\frac{\partial}{\partial x_i} \left[L(\mathbf{x}, \mathbf{u}, t) + \sum_{j=1}^{n} \lambda_j f_j(\mathbf{x}, \mathbf{u}, t) \right].
\tag{4.40}
$$

The second part of the Euler equation can be written more explicitly as

$$
\frac{\partial}{\partial u_k} \left\{ L(\mathbf{x}, \mathbf{u}, t) + \sum_{j=1}^{n} \lambda_j \left[f_j(\mathbf{x}, \mathbf{u}, t) - \dot{x}_j \right] \right\}
$$

$$
- \frac{d}{dt} \frac{\partial}{\partial \dot{u}_k} \left\{ L(\mathbf{x}, \mathbf{u}, t) + \sum_{j=1}^{n} \lambda_j \left[f_j(\mathbf{x}, \mathbf{u}, t) - \dot{x}_j \right] \right\} = 0.
\tag{4.41}
$$

But the second large braces in (4.41) is not a function of $\dot{\mathbf{u}}$. Therefore, (4.41) becomes

$$
\frac{\partial}{\partial u_k} \left[L(\mathbf{x}, \mathbf{u}, t) + \sum_{j=1}^{n} \lambda_j f_j(\mathbf{x}, \mathbf{u}, t) \right] = 0.
\tag{4.42}
$$

The large brackets in (4.40) and (4.42) are called Hamiltonians[3] and can be written as

$$
H(\mathbf{x}, \mathbf{u}, \boldsymbol{\lambda}, t) = L(\mathbf{x}, \mathbf{u}, t) + \boldsymbol{\lambda}^T \mathbf{f}(\mathbf{x}, \mathbf{u}, t).
\tag{4.43}
$$

---

[3] William Rowan Hamilton, Irish physicist, astronomer and mathematician (1805–1865).

Thus, the answer of the interesting question posed at the beginning of this section can be calculated as the solution of the set of following partial differential equations:

$$\dot{\lambda} = -\frac{\partial H}{\partial \mathbf{x}}, \qquad (4.44)$$

$$\mathbf{0} = \frac{\partial H}{\partial \mathbf{u}}, \qquad (4.45)$$

$$\dot{\mathbf{x}} = \frac{\partial H}{\partial \lambda}. \qquad (4.46)$$

The system of n equations in (4.44) describes an adjoint system and derives from (4.39). The system of n equations in (4.46) are the actual system description and constitute the constraints in the mathematical problem. Particular solutions can be found to these $2n$ equations making use of the n initial conditions and n final conditions of the system's states. The equation (4.45) derives from (4.42) and delivers the optimal $\mathbf{u}(t)$.

To solve the problem of finding the input of a system which optimizes a given objective function it is enough to follow the recipe below:

1: Build the Hamiltonian $H(\mathbf{x}, \mathbf{u}, \lambda, t) = L(\mathbf{x}, \mathbf{u}, t) + \lambda^T \mathbf{f}(\mathbf{x}, \mathbf{u}, t)$.

2: Solve $\frac{\partial H}{\partial \mathbf{u}} = \mathbf{0}$ to find $\mathbf{u}^* = \mathbf{u}^*(\mathbf{x}, \lambda, t)$.

3: Insert $\mathbf{u}^*$ in the Hamiltonian to get $H^*(\mathbf{x}, \lambda, t) = H(\mathbf{x}, \mathbf{u}^*, \lambda, t)$

4: Solve the $2n$ equations $\dot{\mathbf{x}} = \frac{\partial H^*}{\partial \lambda}$ and $\dot{\lambda} = -\frac{\partial H^*}{\partial \mathbf{x}}$ using the initial and final conditions $\mathbf{x}(t_0)$ and $\mathbf{x}(t_1)$.

5: Insert $\mathbf{x}(t)$ and $\lambda(t)$ from the previous step in the solution of the second step to get $\mathbf{u}^* = \mathbf{u}^*(\mathbf{x}, \lambda, t) = \mathbf{u}^*(t)$.

Example: Given the system
$$\dot{x} = -x + u,$$

and the objective function
$$Z = \int_0^1 (x^2 + u^2) dt,$$

with the initial and final conditions $x(0) = 1$ and $x(1) = 0$.

Begin by building the Hamiltonian with $L = x^2 + u^2$ and $f(x, u) = -x + u$ as

$$H(x, u, \lambda, t) = x^2 + u^2 + \lambda(-x + u).$$

Equate the derivative of the Hamiltonian with respect to u to zero to find u^*

$$\frac{\partial H}{\partial u} = 2u + \lambda = 0 \quad \Rightarrow \quad u^* = -\lambda/2.$$

Build the Hamiltonian with this u^* as

$$H^* = x^2 + \lambda^2/4 - \lambda x - \lambda^2/2 = x^2 - \lambda^2/4 - \lambda x.$$

Hence the system equation

$$\frac{\partial H^*}{\partial \lambda} = \dot{x} = -x - \lambda/2,$$

and the equation of the adjoint system

$$-\frac{\partial H^*}{\partial x} = \dot{\lambda} = \lambda - 2x.$$

Solving the last two equations yields

$$x(t) = K_1 e^{-\sqrt{2}t} + K_2 e^{\sqrt{2}t},$$
$$\lambda(t) = -2K_1(1 - \sqrt{2})e^{-\sqrt{2}t} - 2K_2(1 + \sqrt{2})e^{\sqrt{2}t}.$$

$x(0) = 1$ and $x(1) = 0$ yields $1 = K_1 + K_2$ and $0 = K_1 e^{-\sqrt{2}} + K_2 e^{\sqrt{2}}$. Solving these for K_1 and K_2 results in

$$K_1 = \frac{1}{1 - e^{-2\sqrt{2}}},$$
$$K_2 = \frac{1}{1 - e^{2\sqrt{2}}}.$$

Finally, the optimal control variable is given as

$$u^* = \frac{1 - \sqrt{2}}{1 - e^{-2\sqrt{2}}} e^{-\sqrt{2}t} + \frac{1 + \sqrt{2}}{1 - e^{2\sqrt{2}}} e^{\sqrt{2}t}.$$

□

4.3.2 Variable Endpoint Problem

For the variable endpoint problem the generalized boundary condition with constraints becomes

$$\frac{\partial \Lambda(\mathbf{x}, \dot{\mathbf{x}}, \mathbf{u}, \dot{\mathbf{u}}, t)}{\partial \dot{\mathbf{x}}}\bigg|_{t_1} \delta \mathbf{x}_1 + \left[\Lambda(\mathbf{x}, \dot{\mathbf{x}}, \mathbf{u}, \dot{\mathbf{u}}, t) - \dot{\mathbf{x}}\frac{\partial \Lambda(\mathbf{x}, \dot{\mathbf{x}}, \mathbf{u}, \dot{\mathbf{u}}, t)}{\partial \dot{\mathbf{x}}}\right]\bigg|_{t_1} \delta t_1 = 0,$$

$$\tag{4.47}$$

with

$$\Lambda = L(\mathbf{x}, \mathbf{u}, t) + \boldsymbol{\lambda}^T [\mathbf{f}(\mathbf{x}, \mathbf{u}, t) - \dot{\mathbf{x}}] .$$

Since neither $L(\cdot)$ nor $\mathbf{f}(\cdot)$ are functions of $\dot{\mathbf{x}}$, (4.47) simplifies to

$$- \boldsymbol{\lambda}^T\bigg|_{t_1} \delta \mathbf{x}_1 + \left\{L(\mathbf{x}, \mathbf{u}, t) + \boldsymbol{\lambda}^T [\mathbf{f}(\mathbf{x}, \mathbf{u}, t) - \dot{\mathbf{x}}] + \boldsymbol{\lambda}^T \dot{\mathbf{x}}\right\}\bigg|_{t_1} \delta t_1 = 0, \tag{4.48}$$

or with the Hamiltonian

$$H^* = L(\mathbf{x}, \mathbf{u}, t) + \boldsymbol{\lambda}^T [\mathbf{f}(\mathbf{x}, \mathbf{u}, t) - \dot{\mathbf{x}}] + \boldsymbol{\lambda}^T \dot{\mathbf{x}},$$

even simpler to

$$- \boldsymbol{\lambda}^T\bigg|_{t_1} \delta \mathbf{x}_1 + H^*\bigg|_{t_1} \delta t_1 = 0. \tag{4.49}$$

To solve the problem of finding the input of a system which optimizes a given objective function where the final state is not given it is enough to follow the modified recipe below:

1: Build the Hamiltonian $H(\mathbf{x}, \mathbf{u}, \boldsymbol{\lambda}, t) = L(\mathbf{x}, \mathbf{u}, t) + \boldsymbol{\lambda}^T \mathbf{f}(\mathbf{x}, \mathbf{u}, t)$.

2: Solve $\frac{\partial H(\mathbf{x}, \mathbf{u}, \boldsymbol{\lambda}, t)}{\partial \mathbf{u}} = \mathbf{0}$ to find $\mathbf{u}^* = \mathbf{u}^*(\mathbf{x}, \boldsymbol{\lambda}, t)$.

3: Insert $\mathbf{u}^*$ in the Hamiltonian to get $H^*(\mathbf{x}, \boldsymbol{\lambda}, t) = H(\mathbf{x}, \mathbf{u}^*, \boldsymbol{\lambda}, t)$

4: Solve the $2n$ equations $\dot{\mathbf{x}} = \frac{\partial H^*(\mathbf{x}, \boldsymbol{\lambda}, t)}{\partial \boldsymbol{\lambda}}$ and $\dot{\boldsymbol{\lambda}} = -\frac{\partial H^*(\mathbf{x}, \boldsymbol{\lambda}, t)}{\partial \mathbf{x}}$ using the initial conditions $\mathbf{x}(t_0)$ and the generalized boundary conditions

$$- \boldsymbol{\lambda}^T \delta \mathbf{x}\bigg|_{t_1} + H^*(\mathbf{x}, \boldsymbol{\lambda}, t)\delta t\bigg|_{t_1} = 0 \tag{4.50}$$

5: Insert $\mathbf{x}(t)$ and $\boldsymbol{\lambda}(t)$ from the previous step in the solution of the second step to get $\mathbf{u}^* = \mathbf{u}^*(\mathbf{x}, \boldsymbol{\lambda}, t) = \mathbf{u}^*(t)$.

Following special cases are of particular interest:

If $\delta\mathbf{x}(t_1)$ and δt_1 are independent

$$\boldsymbol{\lambda}^T\Big|_{t_1} = \mathbf{0}, \tag{4.51}$$

$$H^*(\mathbf{x}, \boldsymbol{\lambda}, t)\Big|_{t_1} = 0. \tag{4.52}$$

If t_1 is given

$$\boldsymbol{\lambda}^T\Big|_{t_1} = \mathbf{0}.$$

If $\mathbf{x}(t_1)$ is given

$$H^*(\mathbf{x}, \boldsymbol{\lambda}, t)\Big|_{t_1} = 0.$$

If the endpoint lies on a curve (so-called rendez-vous problem), then $\delta\mathbf{x} = \mathbf{y}(t_1)dt$

$$\boldsymbol{\lambda}^T\dot{\mathbf{y}}\Big|_{t_1} + H^*(\mathbf{x}, \boldsymbol{\lambda}, t)\Big|_{t_1} = 0, \tag{4.53}$$

$$x(t_1) = y(t_1). \tag{4.54}$$

Example: Given the system

$$\dot{x} = -2\sqrt{2}x + u,$$

$$Z = \int_0^1 (x^2 + u^2)dt.$$

with $x(0) = 2$ and $x(1)$ free

$$H = x^2 + u^2 + \lambda(-2\sqrt{2}x + u),$$

$$\frac{\partial H}{\partial u} = 2u + \lambda = 0 \quad \Rightarrow \quad u^* = -\lambda/2,$$

$$\dot{x} = -2\sqrt{2}x - \lambda/2,$$
$$\dot{\lambda} = -2x + 2\sqrt{2}\lambda,$$

$$H^* = x^2 + \lambda^2/4 - 2\sqrt{2}\lambda x - \lambda^2/2.$$

With the initial condition $x(0) = 2$ and (because $t_1 = 1$ is given) with $\lambda(1) = 0$

$$\lambda(t) = 0.687(e^{-3t} - e^{-6}e^{3t}),$$

$$u^*(t) = -\lambda(t)/2 = -0.343(e^{-3t} - e^{-6}e^{3t}).$$

□

4.3.3 Generalized Objective Function

Let us now consider a new objective function where the endpoint (the final state) is variable and is penalized

$$Z = S(\mathbf{x}(t_1), t_1) + \int_{t_0}^{t_1} L(\mathbf{x}, \mathbf{u}, t)dt. \tag{4.55}$$

The recipe for the solution changes only slightly:

1: Build the Hamiltonian $H(\mathbf{x}, \mathbf{u}, \lambda, t) = L(\mathbf{x}, \mathbf{u}, t) + \lambda^T \mathbf{f}(\mathbf{x}, \mathbf{u}, t)$.

2: Solve $\frac{\partial H(\mathbf{x}, \mathbf{u}, \lambda, t)}{\partial \mathbf{u}} = \mathbf{0}$ to find $\mathbf{u}^* = \mathbf{u}^*(\mathbf{x}, \lambda, t)$.

3: Insert $\mathbf{u}^*$ in the Hamiltonian to get $H^*(\mathbf{x}, \lambda, t) = H(\mathbf{x}, \mathbf{u}^*, \lambda, t)$.

4: Solve the $2n$ equations $\dot{\mathbf{x}} = \frac{\partial H^*(\mathbf{x}, \lambda, t)}{\partial \lambda}$ and $\dot{\lambda} = -\frac{\partial H^*(\mathbf{x}, \lambda, t)}{\partial \mathbf{x}}$ using the initial conditions $\mathbf{x}(t_0)$ and the generalized boundary conditions

$$\left(\frac{\partial S}{\partial \mathbf{x}} - \lambda\right)^T \delta \mathbf{x}\bigg|_{t_1} + \left(H^*(\mathbf{x}, \lambda, t) + \frac{\partial S}{\partial t}\right) \delta t\bigg|_{t_1} = 0. \tag{4.56}$$

5: Insert $\mathbf{x}(t)$ and $\lambda(t)$ from the previous step in the solution of the second step to get $\mathbf{u}^* = \mathbf{u}^*(\mathbf{x}, \lambda, t) = \mathbf{u}^*(t)$.

4.4 Optimization with Limited Control Variables

The previous section has shown that finding the optimal $\mathbf{u}(t)$ that minimizes a given penalty function is achieved firstly by building a Hamiltonian and equating the partial derivative of the Hamiltonian with respect to $\mathbf{u}(t)$ to zero. What happens if the resulting control variable gets too big? One way of dealing with this problem is penalizing $\mathbf{u}(t)$ in the objective function more heavily. A more general solution for the case that $\mathbf{u}$ belongs to a set U was given by Pontryagin[4] [104] who has shown that the solution can be reformulated as

$$\min_{\mathbf{u} \in U} H \rightarrow \mathbf{u}^* . \tag{4.57}$$

Example: A system is given by its state equations as

$$\dot{x}_1 = x_2 ,$$
$$\dot{x}_2 = u ,$$

where the control variable is limited to $|u| \leq 1$. Further, the initial and final states are given as $x_1(0) = x_{10}$, $x_2(0) = x_{20}$, $x_1(T) = 0$, $x_2(T) = 0$. The objective is to bring the system in the shortest possible time to the given endpoint $(0, 0)$. In other words, the objective function to minimize is

$$Z = \int_0^T 1 \, dt .$$

Again begin by building the Hamiltonian

$$H = 1 + \lambda_1 x_2 + \lambda_2 u ,$$

and find the $u^*(t)$ which minimizes the Hamiltonian

$$\min_{u \in U} H \rightarrow u^* = -\text{signum}(\lambda_2) = \pm 1 .$$

Therefore,

$$H^* = 1 + \lambda_1 x_2 - \lambda_2 \text{signum}(\lambda_2) ,$$

$$\dot{x}_1 = \frac{\partial H^*}{\partial \lambda_1} = x_2 ,$$

$$\dot{x}_2 = \frac{\partial H^*}{\partial \lambda_2} = -\text{signum}(\lambda_2) ,$$

$$\dot{\lambda}_1 = -\frac{\partial H^*}{\partial x_1} = 0 ,$$

$$\dot{\lambda}_2 = -\frac{\partial H^*}{\partial x_2} = -\lambda_1 .$$

---

[4]Lev Semyonovich Pontryagin, Russian mathematician (1908–1988).

Fig. 4.2 Optimal $u(t)$ is shown together with the adjoint variables

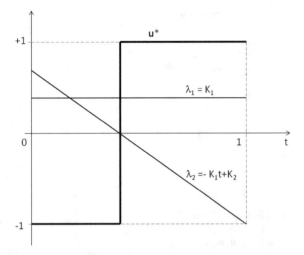

With the given initial and final conditions the generalized boundary conditions become (because the endpoint is given)

$$H^*\big|_T = 0 \quad \Rightarrow \quad |\lambda_2(T)| = 1,$$

$$\lambda_1 = K_1, \qquad \lambda_2 = -K_1 t + K_2.$$

u^* has the value either $+1$ or -1 as Figure 4.2 shows.

Let us now look at the phase portrait of the system as shown in Figure 4.3. When $u = +1$

$$\frac{dx_2}{dx_1} = \frac{1}{x_2},$$

$$\Rightarrow \frac{1}{2}x_2^2 = x_1 + C.$$

When $u = -1$

$$\frac{dx_2}{dx_1} = -\frac{1}{x_2},$$

$$\Rightarrow \frac{1}{2}x_2^2 = -x_1 + C.$$

Figure 4.4 illustrates how the optimal controller works. If the initial condition is below the switching curve the controller works "full gas" with $u = +1$ until the system reaches the so-called switching curve. At that point the controller switches to "full brake" with $u = -1$ until the origin is reached. If the initial condition is above

Fig. 4.3 Phase portrait of the system **a** when $u = +1$ and **b** when $u = -1$. The system should reach the origin in the shortest possible time

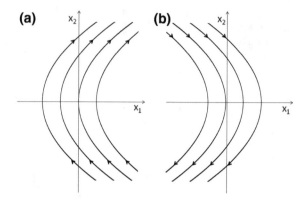

Fig. 4.4 Phase portrait of the optimal bang-bang controller with the switching curve

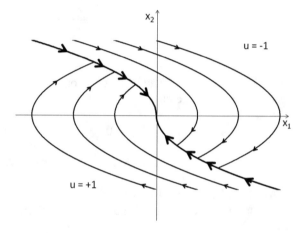

the switching curve the controller works "full brake" with $u = -1$ until the system reaches the so-called switching curve. At that point the controller switches to "full gas" with $u = +1$ until the origin is reached. Such a controller is called a bang-bang controller. □

Important results in Optimal Control Theory were developed during the space race between the Soviet Union and the USA which was a part of the Cold War (1945–1990). The example above indicates the solution of the problem of a space ship starting at point A in space and coming to a stand still at point B in the shortest possible time with no concern for fuel. It has also been noticed that nature makes use of bang-bang controllers. For instance, human eye movements called saccades which move the eyes from one looking direction to another while not smoothly pursuing an object occur in a time optimal fashion [29].

4.5 Optimal Closed-Loop Control

A method for finding the control variable $\mathbf{u}^*(t)$ which optimizes a given objective function (open-loop control) was shown in the previous section. Consider now the problem of finding the control variable $\mathbf{u}^*(\mathbf{x}(t), t)$ as a function of the system states which optimizes a given objective function $\Lambda = \int_{t_0}^{t_1} L(\mathbf{x}, \mathbf{u}, t)dt$ (*i.e.*, closed-loop control as illustrated in Figure 4.5).

Let us again begin by defining a scalar function

$$V(\mathbf{x}, t) = \int_t^{t_1} L[\mathbf{x}(\tau), \mathbf{u}^*(\mathbf{x}, \tau), \tau]d\tau \tag{4.58}$$

as the minimal value of the performance index for a starting state $\mathbf{x}$ and the time t. In other words, $V(\mathbf{x}, t)$ is the objective function for the optimal trajectory which starts at $\mathbf{x}(t)$ and ends at $\mathbf{x}(t_1)$. Assuming that $V(\mathbf{x}, t)$ has continuous second derivatives, it follows

$$\dot{V}(\mathbf{x}, t) = \frac{\partial V(\mathbf{x}, t)}{\partial \mathbf{x}} \cdot \dot{\mathbf{x}} + \frac{\partial V(\mathbf{x}, t)}{\partial t}, \tag{4.59}$$

or, with $\nabla V(\mathbf{x}, t)$ as the gradient vector of $V(\mathbf{x}, t)$

$$\dot{V}(\mathbf{x}, t) = \nabla V^T(\mathbf{x}, t) \cdot \dot{\mathbf{x}} + \frac{\partial V(\mathbf{x}, t)}{\partial t}. \tag{4.60}$$

Since the system is optimal along the trajectory

$$\dot{\mathbf{x}} = \mathbf{f}(\mathbf{x}, \mathbf{u}^*(\mathbf{x}, t), t), \tag{4.61}$$

(4.60) becomes

$$\dot{V}(\mathbf{x}, t) = \nabla V^T(\mathbf{x}, t) \cdot \mathbf{f}(\mathbf{x}, \mathbf{u}^*(\mathbf{x}, t), t) + \frac{\partial V(\mathbf{x}, t)}{\partial t}. \tag{4.62}$$

Fig. 4.5 Structure of the optimal closed-loop controller

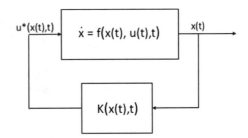

From the definition of $V(\mathbf{x}, t)$ follows

$$\dot{V}(\mathbf{x}, t) = -L(\mathbf{x}, \mathbf{u}^*(\mathbf{x}, t), t). \tag{4.63}$$

Finally, from (4.62) and (4.63) follows

$$\nabla \mathbf{V}^T(\mathbf{x}, t) \cdot \mathbf{f}(\mathbf{x}, \mathbf{u}^*(\mathbf{x}, t), t) + \frac{\partial V(\mathbf{x}, t)}{\partial t} + L(\mathbf{x}, \mathbf{u}^*(\mathbf{x}, t), t) = 0. \tag{4.64}$$

The partial derivative of (4.64) with respect to $\mathbf{x}$ gives

$$\frac{\partial}{\partial \mathbf{x}}(\nabla \mathbf{V}^T \cdot \mathbf{f}) + \frac{\partial^2 V}{\partial \mathbf{x} \partial t} + \frac{\partial L}{\partial \mathbf{x}} = 0. \tag{4.65}$$

Rewriting (4.65) as

$$\nabla \mathbf{V}^T \cdot \frac{\partial \mathbf{f}}{\partial \mathbf{x}} + \mathbf{f}^T \cdot \frac{\partial \nabla \mathbf{V}}{\partial \mathbf{x}} + \frac{\partial^2 V}{\partial \mathbf{x} \partial t} + \frac{\partial L}{\partial \mathbf{x}} = 0. \tag{4.66}$$

Finally,

$$\frac{d}{dt} \nabla \mathbf{V} = \mathbf{f}^T \cdot \frac{\partial \nabla \mathbf{V}}{\partial \mathbf{x}} + \frac{\partial^2 V}{\partial \mathbf{x} \partial t}, \tag{4.67}$$

results in

$$\frac{d}{dt} \nabla \mathbf{V} = -\left(\nabla \mathbf{V}^T \cdot \frac{\partial \mathbf{f}}{\partial \mathbf{x}} + \frac{\partial L}{\partial \mathbf{x}} \right). \tag{4.68}$$

Building the Pontryagin function according to the Pontryagin minimum principle and (4.43) as

$$H(\mathbf{x}, \boldsymbol{\lambda}, t) = H(\mathbf{x}, \mathbf{u}^*(\mathbf{x}, t), \boldsymbol{\lambda}, t) = \boldsymbol{\lambda}^T \cdot \boldsymbol{f}(\mathbf{x}, \mathbf{u}^*(\mathbf{x}, t)) + L(\mathbf{x}, \mathbf{u}^*(\mathbf{x}, t)), \tag{4.69}$$

and the Euler equations for the adjoint system in (4.44) as

$$\dot{\boldsymbol{\lambda}} = -\frac{\partial H(\mathbf{x}, \boldsymbol{\lambda}, t)}{\partial \mathbf{x}} \quad \text{or} \tag{4.70}$$

$$\frac{d}{dt} \boldsymbol{\lambda} = -\left(\boldsymbol{\lambda}^T \cdot \frac{\partial \mathbf{f}}{\partial \mathbf{x}} + \frac{\partial L}{\partial \mathbf{x}} \right). \tag{4.71}$$

Looking at equations (4.68), (4.70) and (4.71) it can be seen that $\nabla \mathbf{V} = \boldsymbol{\lambda}$. Since $\mathbf{u}^*$ is calculated according to the Pontryagin principle from

$$\min_{\mathbf{u} \in U} H(\mathbf{x}, \mathbf{u}, \nabla \mathbf{V}, t),$$

with (4.69) it follows that

$$\mathbf{u}^* = \mathbf{u}^*(\mathbf{x}, \nabla V, t) \text{ and}$$
$$H^*(\mathbf{x}, \nabla V, t) = \nabla V^T \cdot \mathbf{f}(\mathbf{x}, \mathbf{u}^*(\mathbf{x}, \nabla V, t), t) + L(\mathbf{x}, \mathbf{u}^*(\mathbf{x}, \nabla V, t), t) \,.$$

Comparing this equation with equation (4.64) finally results in the Hamilton-Jacobi equation[5]:

$$H^*(\mathbf{x}, \nabla V(\mathbf{x}, t), t) + \frac{\partial V(\mathbf{x}, t)}{\partial t} = 0 \,. \tag{4.72}$$

Hence, the optimal closed-loop control problem is reduced to the solution of this non-linear first order partial differential equation for $V(\mathbf{x}, t)$ with the boundary conditions $V(\mathbf{x}_1, t_1) = 0$ for a fixed final state problem and $V(\mathbf{x}, t_1) = 0$ for a variable final state problem.

Let us again write down the recipe for the solution of finding the closed-loop optimal control problem.

1. For the dynamic system given in its state representation $\dot{\mathbf{x}} = \mathbf{f}(\mathbf{x}, \mathbf{u}, t)$ and the performance index in the form $\Lambda = \int_{t_0}^{t_1} L(\mathbf{x}, \mathbf{u}, t) dt$ build the Pontryagin function

$$H(\mathbf{x}, \mathbf{u}, \nabla V, t) = \nabla V^T \cdot \mathbf{f}(\mathbf{x}, \mathbf{u}, t) + L(\mathbf{x}, \mathbf{u}, t) \,.$$

2. Minimize H with respect to permissible control variables

$$\min_{\mathbf{u} \in U} H(\mathbf{x}, \mathbf{u}, \nabla V, t) \quad \rightarrow \quad \mathbf{u}^* = \mathbf{u}^*(\mathbf{x}, \nabla V, t) \,.$$

3. Build the optimal Pontryagin function with that optimal $\mathbf{u}^*$

$$H^*(\mathbf{x}, \nabla V, t) = H(\mathbf{x}, \mathbf{u}^*(\mathbf{x}, \nabla V, t), \nabla V, t) \,.$$

4. Solve the Hamilton-Jacobi equation

$$H^*(\mathbf{x}, \nabla V, t) + \frac{\partial V}{\partial t} = 0 \,,$$

with the boundary conditions $V(\mathbf{x}_1, t_1) = 0$ for a fixed final state problem and $V(\mathbf{x}, t_1) = 0$ for a variable final state problem resulting in $V(\mathbf{x}, t)$ and $\nabla V(\mathbf{x}, t)$.

5. Substitute the $V(\mathbf{x}, t)$ and $\nabla V(\mathbf{x}, t)$ from the previous step in $\mathbf{u}^*$ resulting from the second step to get

$$\mathbf{u}^*(\mathbf{x}, t) = \mathbf{u}^*(\mathbf{x}, \nabla V, t) \,.$$

---

[5]Carl Gustav Jacob Jacobi, Prussian mathematician (1804–1851).

Example: A controller is sought for the system $\dot{x} = -2x + u$, which optimizes the performance index $\Lambda = \int_0^\infty (x^2 + u^2)dt$. The control variable u is unbounded.

Begin by building the Pontryagin function

$$H = \nabla V \cdot (-2x + u) + x^2 + u^2 .$$

To optimize H with respect to u the derivative of H can simply be equated to zero because u is unbounded.

$$\frac{\partial H}{\partial u} = 0 \quad \Rightarrow \quad \nabla V + 2u = 0 \quad \Rightarrow \quad u^* = -\nabla V/2 .$$

Insert this optimal u^* in the Pontryagin function

$$H^* = -2\nabla V x - \frac{(\nabla V)^2}{2} + x^2 + \frac{(\nabla V)^2}{4}$$

$$H^* = -2\nabla V x - \frac{(\nabla V)^2}{4} + x^2 .$$

Using the optimal Pontryagin function build the Hamilton-Jacobi equation

$$-2\nabla V x - \frac{(\nabla V)^2}{4} + x^2 + \frac{\partial V}{\partial t} = 0 .$$

One can solve this equation with the *Ansatz* $V = \alpha x^2$. Consequently $\nabla V = 2\alpha x$, and because $\frac{\partial V}{\partial t} = 0$

$$-4\alpha x^2 - \alpha^2 x^2 + x^2 = x^2(1 - 4\alpha - \alpha^2) = 0 .$$

α must be positive so that V is always positive. Hence, $\alpha = \sqrt{5} - 2$ and $\nabla V = 2(\sqrt{5} - 2)x$. This gives us the optimal controller as

$$u^*(x, t) = -(\sqrt{5} - 2)x .$$

The block diagram of this optimally controlled system is shown in Figure 4.6. $\square$

Some remarks are in order here.

(1) The optimal control variable $\mathbf{u}^*(\mathbf{x}, t)$ as a function of the state vector $\mathbf{x}(t)$ is optimal for *all* initial conditions $\mathbf{x}(0) = \mathbf{x}_0$ and represents the solution of the optimal closed-loop control problem.

(2) The control variable can be bounded or unbounded.

(3) No two point boundary condition problem needs to be solved. Rather, a partial differential equation has to be solved. The Hamilton-Jacobi method is especially well suited for linear or continuously linearized systems with a quadratic performance

Fig. 4.6 Optimal
closed-loop controller

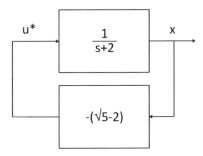

index. In that case, the partial differential equation can be solved analytically and
the solution is given by the matrix Riccati equation as shown in the next section.

(4) The Hamilton-Jacobi equation can also be derived using the optimality principle [11] instead of using calculus of variations as shown here (see Section 4.8).

(5) To achieve an optimal control *all* system states have to be fed back.

4.6 A Simple Cash Balance Problem

This section is an application of the methods presented in this chapter to a real world
financial problem. An important area of finance is called corporate finance. This also
involves making decisions about investment and dividend policies of a company over
time and how to finance these outlays.

Let us consider a problem formulation which goes back to [118] in which a
company has a known demand for cash over time for certain outlays. To satisfy this
cash demand, the company must keep some cash on hand. If the cash is too much,
the company "loses" money in terms of opportunity cost, in that it can earn higher
returns by buying securities like government bonds. On the other hand, if the cash
balance is too small, the company has to sell securities to meet the cash demand and
thus pay a broker's commission. The problem is to find the optimal trade-off between
the cash and security balances.

The state equations are

$$\dot{x}_1(t) = r_1(t)x_1(t) - d(t) + u(t) - \alpha|u(t)|, \quad x_1(0) = x_{10},$$
$$\dot{x}_2(t) = r_2(t)x_2(t) - u(t), \quad x_2(0) = x_{20}.$$

The control constraints are
$$-U_2 \le u(t) \le U_1,$$

and the objective function to maximize is

$$Z = x_1(T) + x_2(T).$$

$x_1(t)$ denotes the cash balance at time t, $x_2(t)$ denotes the security balance at time t, $d(t)$ is the instantaneous rate of demand for cash at time t (note that it can be both positive and negative). The control variable $u(t)$ is the rate of sale of securities in dollars (a negative sales rate means a rate of purchase). $r_1(t)$ is the interest rate earned on the cash balance and $r_2(t)$ is the interest rate earned on the security balance. α denotes the broker's commission in dollars per dollar's worth of securities bought or sold ($0 < \alpha < 1$). T denotes the time horizon.

To solve the problem, begin by defining the Hamiltonian with the help of the adjoint variables λ_1, λ_2 and leaving out the explicit time dependencies as

$$H = \lambda_1(r_1 x_1 - d + u - \alpha|u|) + \lambda_2(r_2 x_2 - u).$$

The adjoint system's differential equations are

$$\dot{\lambda}_1 = -\frac{\partial H}{\partial x_1} = -\lambda_1 r_1, \qquad \lambda_1(T) = 1,$$

$$\dot{\lambda}_2 = -\frac{\partial H}{\partial x_2} = -\lambda_2 r_2, \qquad \lambda_2(T) = 1.$$

The solution of these differential equations can readily be found as

$$\lambda_1(t) = e^{\int_t^T r_1(\tau)d\tau},$$

$$\lambda_2(t) = e^{\int_t^T r_2(\tau)d\tau}.$$

It is not always possible to find a "physical" explanation for the adjoint variables but here one can interpret λ_1 and λ_2 as the future value (at time T) of one dollar held in the cash account or the future value of one dollar invested in securities from time t to T.[6]

To find the optimal control variable $u(t)$ which maximizes the Hamiltonian above employ a little trick[7] which gets rid of the absolute value function and say

$$u(t) = u_1(t) - u_2(t), \qquad u_1(t) \geq 0, \ u_2(t) \geq 0 \ \forall t \in (0, T).$$

In order to make $u = u_1$ when u_1 is strictly positive and $u = -u_2$ when u_2 is strictly positive, the quadratic constraint $u_1 u_2 = 0$ is also imposed so that at most one of u_1 and u_2 can be non-zero. Actually, the optimal properties of the solution will automatically cause this constraint to be satisfied. The reason is that the

---

[6]See Chapter 6 for a detailed explanation of this interpretation.

[7]This nifty trick is very useful when one has to deal with the absolute value of a variable which is rather nasty in analytical solutions.

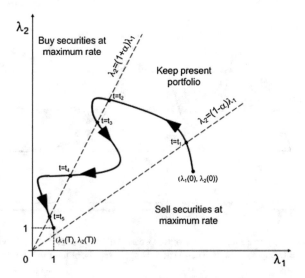

Fig. 4.7 Optimal cash balance policy illustrated in the adjoint system space. The $\lambda_1 - \lambda_2$ plane has three regions which correspond to three possible regimes: sell securities at maximum rate, keep present portfolio, and buy securities at maximum rate. The trajectory of the adjoint system, parametrized in time, starts at $(\lambda_1(0), \lambda_2(0))$ and ends at $(\lambda_1(T), \lambda_2(T))$ which is given as $(1, 1)$. The black dots indicate points where a regime change occurs. The dashed lines separating the regimes are called lines of singular control (after [118])

broker's commission must be paid on *every* transaction, which makes it nonsensical to simultaneously buy and sell securities. Thus, because $u_1 u_2 = 0$, $|u|^2 = |u_1 + u_2|^2 = |u_1^2 + u_2^2| = |u_1 - u_2|^2$. Therefore, $|u|$ can be substituted by $u_1 + u_2$ and the part of the Hamiltonian which depends on the control variable rewritten as

$$\tilde{H} = u_1[(1 - \alpha)\lambda_1 - \lambda_2] - u_2[(1 + \alpha)\lambda_1 - \lambda_2].$$

Maximizing H with respect to u is the same as maximizing $\tilde{H}$ with respect to u_1 and u_2. Since $\tilde{H}$ is linear in u_1 and u_2 the optimal strategy is, as shown in Section 4.4, bang-bang as follows

$$u^* = u_1^* - u_2^*,$$

where u_1^* is either 0 or U_1 and the switch between these values occur when signum$[(1 - \alpha)\lambda_1 - \lambda_2]$ changes. Similarly, u_2^* is either 0 or U_2 and the switch between these values occur when signum$[-(1 + \alpha)\lambda_1 + \lambda_2]$ changes. Figures 4.7 and 4.8 illustrate the optimal control policy.

In this example, it was assumed that over-drafts and short selling is allowed. In case over-drafts and short selling is forbidden, the additional constraints

$$x_1(t) \geq 0, \qquad x_2(t) \geq 0, \qquad \forall t \in (0, T)$$

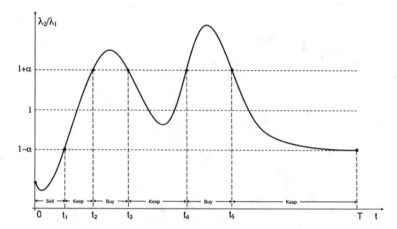

Fig. 4.8 Another way of looking at the optimal cash balance policy in the previous Figure is represented in the $(t, \lambda_2/\lambda_1)$ graph. The black dots again represent the time points where regime changes occur

have to be taken into account. The solution of the problem with this extension is left for the interested reader as an exercise.

4.7   Optimal Control of Linear Systems

4.7.1   Riccati Equation

A general solution for the optimal feedback control problem was shown in the previous section. In this section a special case which is widely used is studied: find the optimal closed-loop control variable $\mathbf{u}^*(\mathbf{x}, t)$ for the linear system given by its state space equation

$$\dot{\mathbf{x}} = \mathbf{A}\mathbf{x} + \mathbf{B}\mathbf{u},\tag{4.73}$$

which optimizes the quadratic performance index

$$\Lambda = \int_0^T (\mathbf{x}^T \mathbf{Q}\mathbf{x} + \mathbf{u}^T \mathbf{P}\mathbf{u})dt.\tag{4.74}$$

Let us remember several points before applying the Hamilton-Jacobi method: there are n states and r inputs. The final state $\mathbf{x}(T)$ is not fixed. $\mathbf{P}$ is a symmetrical, positive definite matrix. $\mathbf{Q}$ is a symmetrical, positive semidefinite matrix. The control variable $\mathbf{u}$ is unbounded but by choosing a large matrix $\mathbf{P}$ it can be made arbitrarily small.

1. The Pontryagin function for the given system and the objective function is

$$H(\mathbf{x}, \mathbf{u}, \nabla \mathbf{V}, t) = \nabla \mathbf{V}^T \cdot (\mathbf{A}\mathbf{x} + \mathbf{B}\mathbf{u}) + \mathbf{x}^T \mathbf{Q}\mathbf{x} + \mathbf{u}^T \mathbf{P}\mathbf{u} \,. \qquad (4.75)$$

2. Let us minimize the Pontryagin function by equating its derivative with respect to **u** to zero. (This can be done, because **u** is unbounded.)

$$\mathbf{B}^T \cdot \nabla \mathbf{V} + 2 \cdot \mathbf{P} \cdot \mathbf{u} = \mathbf{0} \quad \Rightarrow \quad \mathbf{u}^*(\mathbf{x}, t) = -\frac{1}{2}\mathbf{P}^{-1} \cdot \mathbf{B}^T \cdot \nabla \mathbf{V}(\mathbf{x}, t) \,. \quad (4.76)$$

$\mathbf{P}^{-1}$ exists always, because **P** is positive definite.

3. The optimal Pontryagin function is therefore

$$H^*(\mathbf{x}, \nabla \mathbf{V}, t) = \nabla \mathbf{V}^T \cdot \mathbf{A}\mathbf{x} - \frac{1}{2}\nabla \mathbf{V}^T \cdot \mathbf{B} \cdot \mathbf{P}^{-1} \cdot \mathbf{B}^T \cdot \nabla \mathbf{V} + \mathbf{x}^T \mathbf{Q}\mathbf{x} \qquad (4.77)$$

$$+ \frac{1}{4}\nabla \mathbf{V}^T \cdot \mathbf{B} \cdot \mathbf{P}^{-1} \cdot \mathbf{P} \cdot \mathbf{P}^{-1} \cdot \mathbf{B}^T \cdot \nabla \mathbf{V}$$

$$= \nabla \mathbf{V}^T \cdot \mathbf{A}\mathbf{x} - \frac{1}{4}\nabla \mathbf{V}^T \cdot \mathbf{B} \cdot \mathbf{P}^{-1} \cdot \mathbf{B}^T \cdot \nabla \mathbf{V} + \mathbf{x}^T \mathbf{Q}\mathbf{x} \,.$$

4. To solve the Hamilton-Jacobi equation

$$\nabla \mathbf{V}^T \cdot \mathbf{A} \cdot \mathbf{x} - \frac{1}{4}\nabla \mathbf{V}^T \cdot \mathbf{B} \cdot \mathbf{P}^{-1} \cdot \mathbf{B}^T \cdot \nabla \mathbf{V} + \mathbf{x}^T \mathbf{Q}\mathbf{x} + \frac{\partial V}{\partial t} = 0 \,, \qquad (4.78)$$

make use of the *Ansatz* $V(\mathbf{x}, t) = \mathbf{x}^T \mathbf{R}(t)\mathbf{x}$ which is a quadratic form and always non-negative. The integrand in the performance index

$$V(\mathbf{x}, t) = \int_t^T (\mathbf{x}^T \mathbf{Q}\mathbf{x} + \mathbf{u}^T \mathbf{P}\mathbf{u})d\tau \,, \qquad (4.79)$$

is a positive definite function. Hence the matrix $\mathbf{R}(t)$ is a positive definite symmetrical matrix for $t < T$ and for $t = T$ $\mathbf{R}(T) = \mathbf{0}$. Substituting the partial derivatives of the *Ansatz*

$$\nabla \mathbf{V} = 2\mathbf{R}(t)\mathbf{x} \quad \text{and} \quad \frac{\partial V}{\partial t} = \mathbf{x}^T \dot{\mathbf{R}}(t)\mathbf{x} \,, \qquad (4.80)$$

in the Hamilton-Jacobi equation yields

$$2\mathbf{x}^T \mathbf{R}(t)\mathbf{A}\mathbf{x} - \mathbf{x}^T \mathbf{R}(t)\mathbf{B}\mathbf{P}^{-1}\mathbf{B}^T \mathbf{R}(t)\mathbf{x} + \mathbf{x}^T \mathbf{Q}\mathbf{x} + \mathbf{x}^T \dot{\mathbf{R}}(t)\mathbf{x} = \mathbf{0} \,. \qquad (4.81)$$

Equation (4.81) can now be written as a quadratic form

$$\mathbf{x}^T \left[2\mathbf{R}(t)\mathbf{A} - \mathbf{R}(t)\mathbf{B}\mathbf{P}^{-1}\mathbf{B}^T \mathbf{R}(t) + \mathbf{Q} + \dot{\mathbf{R}}(t) \right] \mathbf{x} = \mathbf{0} \,. \qquad (4.82)$$

For a quadratic form,

$$\mathbf{x}^T \mathbf{M} \mathbf{x} = \mathbf{x}^T \mathbf{M}_s \mathbf{x}, \tag{4.83}$$

where $\mathbf{M}_s = (\frac{\mathbf{M}+\mathbf{M}^T}{2})$ is the symmetrical part of $\mathbf{M}$. Therefore, the Hamilton-Jacobi equation is fulfilled for all states $\mathbf{x}$ if only the symmetrical part of the matrix in the large brackets in (4.82) is zero. All terms of that bracket is symmetrical except for the first one. The symmetrical part of the first term can be calculated as

$$\text{sym}(2\mathbf{R}(t)\mathbf{A}) = 2 \cdot \frac{\mathbf{R}(t)\mathbf{A} + \mathbf{A}^T \mathbf{R}^T(t)}{2} = \mathbf{R}(t)\mathbf{A} + \mathbf{A}^T \mathbf{R}^T(t). \tag{4.84}$$

Since $\mathbf{R}(t)$ is symmetrical, the matrix Riccati equation follows from (4.82)

$$\dot{\mathbf{R}}(t) + \mathbf{Q} - \mathbf{R}(t)\mathbf{B}\mathbf{P}^{-1}\mathbf{B}^T \mathbf{R}(t) + \mathbf{R}(t)\mathbf{A} + \mathbf{A}^T \mathbf{R}^T(t) = \mathbf{0}, \tag{4.85}$$

with the boundary condition $\mathbf{R}(T) = \mathbf{0}$.

This first order ordinary non-linear differential equation can be solved for $\mathbf{R}(t)$ numerically, whereas the integration must be performed backwards in time, starting with the final state.

5. The optimal closed-loop control variable is hence

$$\mathbf{u}^*(\mathbf{x}, t) = -\mathbf{P}^{-1}\mathbf{B}^T \mathbf{R}(t)\mathbf{x}. \tag{4.86}$$

This equation is of the form

$$\mathbf{u}^*(\mathbf{x}, t) = -\mathbf{K}^T(t)\mathbf{x}, \tag{4.87}$$

with $\mathbf{K}(t) = \mathbf{R}(t)\mathbf{B}\mathbf{P}^{-1}$. The elements of the matrix $\mathbf{K}(t)$ are the variable feedback gains of the optimal system. Equation (4.87) gives the structure of the optimal controller as shown in Figure 4.9. The components of the control vector $\mathbf{u}^*$ in an optimally closed-loop controlled system are linear combinations of the state vector $\mathbf{x}$. This is an important insight, because for the controller to be optimal, *all* states have to be fed back and not only the output as is the case in classical control theory.

Example: Which controller minimizes

$$\Lambda = \int_0^T (x^2 + u^2)dt \,,$$

for the system

$$\dot{x} = -x + u \,?$$

In this problem, all the matrices are reduced to scalars such that $\mathbf{A} = a = -1$, $\mathbf{B} = b = 1$, $\mathbf{Q} = q = 1$, $\mathbf{P} = p = 1$. Consequently, the matrix $\mathbf{R}(t)$ is reduced to a scalar time function $r(t)$. The matrix Riccati equation becomes a first order scalar non-linear differential equation.

$$\dot{r}(t) + 1 - r^2(t) - 2r(t) = 0 \,,$$

with the boundary condition $r(T) = 0$. Actually, in this special case, there is an analytical solution of the Riccati equation which is given by

$$r(t) = \frac{e^{-\sqrt{2}(t-T)} - e^{\sqrt{2}(t-T)}}{(\sqrt{2}+1)e^{-\sqrt{2}(t-T)} + (\sqrt{2}-1)e^{\sqrt{2}(t-T)}} \,.$$

Using this $r(t)$ the optimal controller is

$$u^*(x, t) = -r(t)x \,.$$

□

An interesting special case of the matrix Riccati equation (4.85) is obtained when the final time in the performance index (4.74) $T = \infty$. In that case the scalar function $V(\mathbf{x}, t)$ must be independent of t. In other words

$$V(\mathbf{x}, t_1) = V(\mathbf{x}, t_2) \qquad \forall \, t_1 \text{ and } t_2 \,. \tag{4.88}$$

Fig. 4.9 Optimal closed-loop controller requires all of the system states to be fed back and not just the output

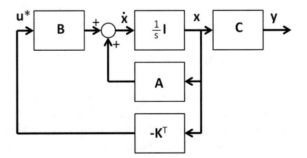

Consequently, the Riccati matrix $\mathbf{R}(t)$ becomes a constant matrix $\mathbf{R}_0$ and the feedback gains become constant and much easier to implement. Also, the Riccati (differential) equation becomes an algebraic equation, because $\mathbf{R}_0$ is constant, hence $\dot{\mathbf{R}}_0 = \mathbf{0}$. In other words (4.85) becomes

$$\mathbf{A}^T \mathbf{R}_0^T + \mathbf{R}_0 \mathbf{A} - \mathbf{R}_0 \mathbf{B} \mathbf{P}^{-1} \mathbf{B}^T \mathbf{R}_0 + \mathbf{Q} = \mathbf{0}, \tag{4.89}$$

which is a system of $n(n+1)/2$ non-linear algebraic equations. There are, in general, several solutions for $\mathbf{R}_0$. However, only one of the solutions is a positive definite matrix $\mathbf{R}_0$. This solution is actually the same one as one would get by solving (4.85) and letting $T \to \infty$.

Let us now revisit the previous example, this time taking the limit of the integral in the performance index as $T = \infty$. The scalar function $r(t)$ becomes a constant r_0 and the optimal state feedback control becomes $u^*(x, t) = -r_0 x$. r_0 can be computed by calculating $r(t \to \infty)$.

$$
\begin{aligned}
r_0 &= \lim_{T \to \infty} r(t) \\
&= \lim_{T \to \infty} \frac{e^{\sqrt{2}T} - e^{-\sqrt{2}T}}{(\sqrt{2}+1)e^{\sqrt{2}T} + (\sqrt{2}-1)e^{-\sqrt{2}T}} \\
&= \lim_{T \to \infty} \frac{1 - e^{-2\sqrt{2}T}}{(\sqrt{2}+1) + (\sqrt{2}-1)e^{-2\sqrt{2}T}} \\
&= \frac{1}{\sqrt{2}+1} \\
&= \sqrt{2} - 1.
\end{aligned}
$$

To be sure, one can solve the algebraic Riccati equation:

$$-2r_0 - r_0^2 + 1 = 0 \Rightarrow r_{0_{1,2}} = -1 \pm \sqrt{2},$$

and taking the positive solution as $r_0 = \sqrt{2} - 1$ gives the same result as above.

$\square$

4.7.2 Optimal Control When Not All States Are Measurable

As pointed out in the previous Subsection, optimal closed-loop control requires *all* of the system states to be fed back with appropriate gains. However, in practical applications not all states are always measurable. How does one proceed in such cases? The solution to this problem lies in *estimating* the unmeasurable states with an observer and using those estimates as if they were true measurements of the states as illustrated in Figure 4.10. However, a necessary condition for this to be possible,

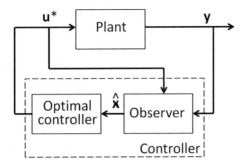

Fig. 4.10 When not all system states of the plant are available they must be estimated and used as true states for optimal closed-loop control. The observer is essentially a mathematical model of the plant. It makes use of the plant's input vector $\mathbf{u}^*$ and its output vector $\mathbf{y}$ to compute the estimated system states $\hat{\mathbf{x}}$. However, the initial values of the unmeasurable states are not known. This constitutes an additional problem

is that the plant has to be observable. In other words, it must be possible to determine $\mathbf{x}$ from given $\mathbf{u}$ and measured $\mathbf{y}$. For a linear system this condition is met if the observability matrix $\mathbf{Q}_o$

$$\mathbf{Q}_o = \begin{bmatrix} \mathbf{C}^T \\ \mathbf{C}^T \mathbf{A} \\ \mathbf{C}^T \mathbf{A}^2 \\ \vdots \\ \mathbf{C}^T \mathbf{A}^{n-1} \end{bmatrix}, \tag{4.90}$$

has the rank n.

In cases where the plant can be modeled as a linear system, its matrices $\mathbf{A}, \mathbf{B}, \mathbf{C}$ must be known. However, the initial conditions of the plant and the model are not necessarily the same, because the initial conditions of the plant states that cannot be measured are not known. Again, using a quadratic performance index for the error between the measured and modeled system outputs leads to a Riccati equation for the calculation of the observer gains if one assumes the structure shown in Figure 4.11.

Hence, there are two optimization problems: (1) Optimize the observer where the observer gains $\mathbf{K}_1$ are computed as the solution of a Riccati equation, (2) Optimize the controller gains $\mathbf{K}_2$ also through the solution of a Riccati equation. *Separation principle* says that these two problems can be solved independent of each other.

4.8 Dynamic Programming

There are many applications where a system given by its state space description

$$\dot{\mathbf{x}} = \mathbf{f}(\mathbf{x}, \mathbf{u}), \tag{4.91}$$

Fig. 4.11 The observer
which is not privy to all
initial states of the plant
computes an estimated
output vector $\hat{\mathbf{y}}$ which is
compared with the measured
system output $\mathbf{y}$. The
difference $\tilde{\mathbf{y}} = \mathbf{y} - \hat{\mathbf{y}}$ is then
fed back to the model after
being amplified by $\mathbf{K}_1$. The
eigenvalues of this inner
observer loop must be far in
the left half plane, so that the
estimation error $\tilde{\mathbf{y}} \to \mathbf{0}$
quickly

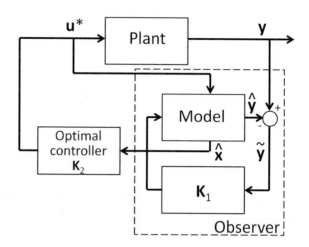

with $\mathbf{u}$ being the decision vector and $\mathbf{x}$ the state vector, is controlled by a controller
which generates piece-wise constant decision values as shown in Figure 4.12. Corre-
spondingly, $\mathbf{x}_1$ is the initial state, $\mathbf{u}_1$ is the first decision, $\mathbf{x}_2 = \psi(\mathbf{x}_1, \mathbf{u}_1)$ is the state
after the first decision, $\mathbf{u}_2$ is the second decision, $\mathbf{x}_3 = \psi(\mathbf{x}_2, \mathbf{u}_2)$ is the state after
the second decision, ..., $\mathbf{x}_N = \psi(\mathbf{x}_{N-1}, \mathbf{u}_{N-1})$ is the final state in a process with N
steps. What are the decision values which minimize the objective function

$$Z = Z(\mathbf{x}_1, \mathbf{x}_2, \ldots, \mathbf{x}_N, \mathbf{u}_1, \mathbf{u}_2, \ldots, \mathbf{u}_N)? \tag{4.92}$$

Assuming $\mathbf{x}_1$ is known and because all the following states $\mathbf{x}_i, i > 1$ depend on
$\mathbf{x}_1$, (4.92) becomes

$$Z = Z(\mathbf{u}_1, \mathbf{u}_2, \ldots, \mathbf{u}_N). \tag{4.93}$$

Let us further assume that in a process with N steps, the contribution of the
remaining $N - k$ steps to the objective function after k decisions depend only on the

Fig. 4.12 Piece-wise
constant decision values

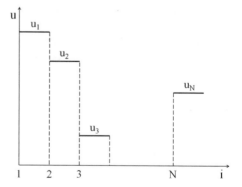

state of the system after the k-th decision and the decisions after the k-th decision (this is known as the Markov property as discussed in Subsection 3.8.2). With that assumption a central result in Optimal Control Theory can now be derived, namely the celebrated *optimality principle*:

$$Z = L(\mathbf{x}_1, \mathbf{u}_1) + L(\mathbf{x}_2, \mathbf{u}_2) + \cdots + L(\mathbf{x}_N, \mathbf{u}_N).$$ (4.94)

$$\mathbf{x}_i = \psi(\mathbf{x}_{i-1}, \mathbf{u}_{i-1}) \quad i = 2, 3, \ldots, N.$$ (4.95)

$$Z_1(\mathbf{x}_1) = \min_{\mathbf{u}_1} L(\mathbf{x}_1, \mathbf{u}_1)$$

$$Z_2(\mathbf{x}_1) = \min_{\mathbf{u}_1, \mathbf{u}_2} [L(\mathbf{x}_1, \mathbf{u}_1) + L(\mathbf{x}_2, \mathbf{u}_2)]$$

$$\vdots$$

$$Z_N(\mathbf{x}_1) = \min_{\mathbf{u}_i} [L(\mathbf{x}_1, \mathbf{u}_1) + L(\mathbf{x}_2, \mathbf{u}_2) + \ldots + L(\mathbf{x}_N, \mathbf{u}_N)]$$

$$= \min_{\mathbf{u}_1} \min_{\mathbf{u}_2} \cdots \min_{\mathbf{u}_N} [L(\mathbf{x}_1, \mathbf{u}_1) + L(\mathbf{x}_2, \mathbf{u}_2) + \ldots + L(\mathbf{x}_N, \mathbf{u}_N)]$$

$$= \min_{\mathbf{u}_1} [L(\mathbf{x}_1, \mathbf{u}_1) + \min_{\mathbf{u}_2} \min_{\mathbf{u}_3} \cdots \min_{\mathbf{u}_N} \{L(\mathbf{x}_2, \mathbf{u}_2) + \cdots + L(\mathbf{x}_N, \mathbf{u}_N)\}]$$

$$= \min_{\mathbf{u}_1} [L(\mathbf{x}_1, \mathbf{u}_1) + Z_{N-1}(\mathbf{x}_2)]$$

$$= \min_{\mathbf{u}_1} [L(\mathbf{x}_1, \mathbf{u}_1) + Z_{N-1}(\psi(\mathbf{x}_1, \mathbf{u}_1))] \quad N = 2, 3, \ldots.$$ (4.96)

This means that an optimal strategy has the property that, independent of the initial state and the first decision, the remaining control decisions build an optimal strategy starting with the state resulting from the first control decision. *In other words, each control decision of an optimal strategy is itself optimal.*

The main advantages of dynamic programming are

- The absolute minimum can be found (compare that with the other numerical optimization methods discussed in Section 2.5).
- No two point boundary condition problem has to be solved.
- Limitations of the control variable simplify the solution of the problem, because the condition $|u| \le m$ limits the number of possible control decisions at each step.
- The functions $f(\cdot)$ and $L(\cdot)$ do not need to be differentiable (actually, these functions can even be given as tables).
- Optimality principle excludes a large number of unnecessary strategies automatically.

The major disadvantage of dynamic programming lies in the number of calculations that are necessary for higher order systems (the so-called "curse of dimensionality"), because the number of operations increase exponentially with the number of variables.

Fig. 4.13 Salesman's routes
and the distance between the
cities

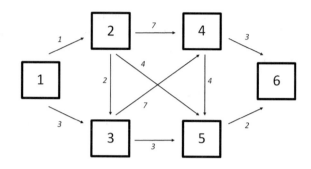

Example: Let us now consider a simplified version of the celebrated traveling salesman problem as an example to see how dynamic programming actually works [132].

There are six cities 1–6 with the topology and the distances between them shown
in Figure 4.13. The salesman's problem is to find the shortest route from 1 to 6.

Let d_{ij} be the distance between city i and j, both indices varying from 1 to 6.
Let further V_i be the shortest path from city i to the destination 6 along permissible
paths. Applying optimality principle yields

$$V_i = \min(d_{i6})$$
$$= \min(d_{ij} + d_{j6})$$
$$= \min(d_{ij} + V_j) \text{ where } V_6 = 0.$$

Starting from the destination and calculating backwards yields

$V_6 = 0$
$V_5 = \min(d_{56} + V_6) = 2$
$V_4 = \min(d_{45} + V_5, \ d_{46} + V_6) = \min(4 + 2, \ 3 + 0) = 3$
$V_3 = \min(d_{34} + V_4, \ d_{35} + V_5) = \min(7 + 3, \ 3 + 2) = 5$
$V_2 = \min(d_{23} + V_3, \ d_{24} + V_4, \ d_{25} + V_5,) = \min(2 + 5, \ 7 + 3, \ 4 + 2) = 6$
$V_1 = \min(d_{12} + V_2, \ d_{13} + V_3) = \min(1 + 6, \ 3 + 5) = 7.$

The solution is therefore $1 \rightarrow 2 \rightarrow 5 \rightarrow 6$. One can see that *not* all possible
routes have to be calculated to find the optimal one. The savings in computational
would be even more significant if there were more cities and more routes.   □

4.9   Discrete Time Systems

4.9.1   Remembering the Basics

So far in this chapter it has been assumed that the systems of interest are continuous in time and can be appropriately represented by differential equations. However, there are cases where the system time is discrete rather than continuous. Such systems are represented by difference equations. Let us now take a path there through familiar terrain.

Time continuous linear systems can be represented in the state space by the linear equation system

$$\dot{\mathbf{x}}(t) = \mathbf{A}\mathbf{x}(t) + \mathbf{B}\mathbf{u}(t) , \qquad (4.97)$$

where $\mathbf{x}(t)$ represents the system state, $\mathbf{u}(t)$ the input vector, $\mathbf{A}$ the system matrix and $\mathbf{B}$ the input matrix. Transforming this system of equations in the Laplace domain where s denotes the complex Laplace operator complex results in

$$s\mathbf{X}(s) = \mathbf{A}\mathbf{X}(s) + \mathbf{B}\mathbf{U}(s) + \mathbf{x}(0) , \qquad (4.98)$$

which can be rewritten as

$$\mathbf{X}(s) = (s\mathbf{I} - \mathbf{A})^{-1}\mathbf{x}(0) + (s\mathbf{I} - \mathbf{A})^{-1}\mathbf{B}\mathbf{U}(s) . \qquad (4.99)$$

Defining the fundamental matrix as the inverse Laplace transform of the resolvent matrix $(s\mathbf{I} - \mathbf{A})^{-1}$ as

$$\mathscr{L}^{-1}\{(s\mathbf{I} - \mathbf{A})^{-1}\} = \boldsymbol{\Phi}(t) = e^{\mathbf{A}t} , \qquad (4.100)$$

the solution of the differential equation system in (4.97) can be written as

$$\mathbf{x}(t) = \boldsymbol{\Phi}(t)\mathbf{x}(0) + \int_0^t \boldsymbol{\Phi}(t - \tau)\mathbf{B}\mathbf{u}(\tau)d\tau . \qquad (4.101)$$

Note that this solution is the superposition of a part dependent on the initial system state but independent of the input vector and a part dependent on the input vector but independent of the initial system state (see Appendix A for a derivation of this important result).

Example: Find the step response of the second order linear system given by its differential equation $\ddot{y} + \dot{y} = u(t)$ where $u(t)$ is the Heaviside function.

Firstly, define the system states as $y(t) = x_1(t)$, $\dot{y}(t) = x_2(t)$. Therefore, $\dot{x}_1(t) = x_2(t)$, $\dot{x}_2(t) = -x_2(t) + u(t)$. The system and input matrices are thus

$$\mathbf{A} = \begin{bmatrix} 0 & 1 \\ 0 & -1 \end{bmatrix}, \qquad \mathbf{b} = \begin{bmatrix} 0 \\ 1 \end{bmatrix} .$$

Therefore, the resolvent matrix can be calculated as

$$(s\mathbf{I} - \mathbf{A}) = \begin{bmatrix} s & -1 \\ 0 & s+1 \end{bmatrix} \quad \Rightarrow \quad (s\mathbf{I} - \mathbf{A})^{-1} = \begin{bmatrix} \frac{1}{s} & \frac{1}{s(s+1)} \\ 0 & \frac{1}{s+1} \end{bmatrix},$$

and the fundamental matrix as

$$\boldsymbol{\Phi}(t) = \mathcal{L}^{-1}\{(s\mathbf{I} - \mathbf{A})^{-1}\} = \begin{bmatrix} 1 & 1 - e^{-t} \\ 0 & e^{-t} \end{bmatrix}.$$

Remembering that the Laplace transform of the Heaviside function is $U(s) = \frac{1}{s}$,

$$(s\mathbf{I} - \mathbf{A})^{-1}\mathbf{b}U(s) = \begin{bmatrix} \frac{1}{s^2(s+1)} \\ \frac{1}{s(s+1)} \end{bmatrix}.$$

Taking the inverse Laplace transform as

$$\mathcal{L}^{-1}\{(s\mathbf{I} - \mathbf{A})^{-1}\mathbf{b}U(s)\} = \begin{bmatrix} t - 1 + e^{-t} \\ 1 - e^{-t} \end{bmatrix},$$

finally leads to the answer

$$\begin{bmatrix} x_1(t) \\ x_2(t) \end{bmatrix} = \begin{bmatrix} 1 & 1 - e^{-t} \\ 0 & e^{-t} \end{bmatrix} \cdot \begin{bmatrix} x_1(0) \\ x_2(0) \end{bmatrix} + \begin{bmatrix} t - 1 + e^{-t} \\ 1 - e^{-t} \end{bmatrix}.$$

$\square$

A difference equation[8] needs to be employed for the analyzing a discrete time linear system.

$$\mathbf{x}(n+1) = \mathbf{A}\mathbf{x}(n) + \mathbf{B}\mathbf{u}(n). \tag{4.102}$$

Transforming this system of equations with the z-transform to solve it yields

$$z\mathbf{X}(z) = \mathbf{A}\mathbf{x}(z) + \mathbf{B}\mathbf{U}(z) + z\mathbf{x}(0), \tag{4.103}$$

or rewriting it

$$\mathbf{X}(z) = (z\mathbf{I} - \mathbf{A})^{-1}z\mathbf{x}(0) + (z\mathbf{I} - \mathbf{A})^{-1}\mathbf{B}\mathbf{U}(z). \tag{4.104}$$

Defining the fundamental matrix (transition matrix) now as the inverse z-transform of $(z\mathbf{I} - \mathbf{A})^{-1}z$ as

$$\mathcal{Z}\{(z\mathbf{I} - \mathbf{A})^{-1}z\} = \boldsymbol{\Phi}(n) = \mathbf{A}^n, \tag{4.105}$$

the solution of the system of difference equations in (4.102) can be written as

---

[8]Some authors use the term recurrence relation for difference equations.

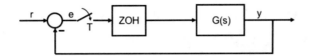

Fig. 4.14 Example of a sampled system. The linear system given by its transfer function is sampled at multiples of the sample time T and the result of the sampling is kept constant with a zero order hold element (ZOH)

$$\mathbf{x}(n) = \boldsymbol{\Phi}(n)\mathbf{x}(0) + \sum_{k=0}^{n-1} \boldsymbol{\Phi}(n-1-k)\mathbf{B}\mathbf{u}(k). \qquad (4.106)$$

As in the continuous case, this solution is the superposition of a part dependent on the initial system state but independent of the input vector and a part dependent on the input vector but independent of the initial system state. It helps to remember that the role played by an integrator in continuous time systems is played by a flip-flop (delay element) in discrete time systems.

Many technical systems are a mixture of time continuous and time discrete systems. These are often called sampled systems.

Example: Consider the system in Figure 4.14.

For $G(s) = \frac{1}{s(s+1)}$ the state space equations can be written as

$$\begin{bmatrix} x_1(n+1) \\ x_2(n+1) \end{bmatrix} = \begin{bmatrix} 1 & 1-e^{-T} \\ 0 & e^{-T} \end{bmatrix} \cdot \begin{bmatrix} x_1(n) \\ x_2(n) \end{bmatrix} + \begin{bmatrix} T-1+e^{-T} \\ 1-e^{-T} \end{bmatrix} u(n).$$

Begin with $t = 0$ corresponding to $n = 0$. The next time point is $t = T$ which corresponds to $n = 1$, and so on. With $y(n) = x_1(n)$, $u(n) = r(n) - y(n)$ the last set of equations become

$$\begin{bmatrix} x_1(n+1) \\ x_2(n+1) \end{bmatrix} = \begin{bmatrix} 2-T-e^{-T} & 1-e^{-T} \\ -1+e^{-T} & e^{-T} \end{bmatrix} \cdot \begin{bmatrix} x_1(n) \\ x_2(n) \end{bmatrix} + \begin{bmatrix} T-1+e^{-T} \\ 1-e^{-T} \end{bmatrix} r(n).$$

$\square$

In general, for the system $\dot{\mathbf{x}}(t) = \mathbf{A}\mathbf{x}(t) + \mathbf{B}\mathbf{u}(t)$ where $\mathbf{u}(t)$ is sampled and held, hence piece-wise constant,

$$\mathbf{x}(k+1) = \boldsymbol{\Phi}(T)\mathbf{x}(k) + \mathbf{D}(T)\mathbf{u}(k), \qquad (4.107)$$

where

$$\boldsymbol{\Phi}(T) = e^{\mathbf{A}T} \,, \tag{4.108}$$

$$\mathbf{D}(T) = \int_0^T \boldsymbol{\Phi}(T - \tau)\mathbf{B}d\tau \,, \tag{4.109}$$

$$\mathbf{x}(NT) = \boldsymbol{\Phi}(NT)\mathbf{x}(0) + \sum_{k=0}^{N-1} \boldsymbol{\Phi}([N - k - 1]T)\mathbf{D}(T)\mathbf{u}(kT) \,. \tag{4.110}$$

Note that $\mathbf{u}$ does not appear in the calculation of $\mathbf{D}(T)$, because it is constant during a sampling period.

4.9.2   Time Optimization

Let us consider a second order linear system with an unbounded scalar control variable u. The system is governed by the set of difference equations

$$\mathbf{x}(k + 1) = \boldsymbol{\Phi}(T)\mathbf{x}(k) + \mathbf{d}(T)u(k) \,.$$

How does one bring the system from any initial condition to a complete stand still at the origin of the state space in shortest possible time? Let us begin by answering another question: from which initial conditions can one bring the system to the zero point in a single sampling period ($k = 1$)? Or, formally, what is $\mathbf{x}(0)$ such that $\mathbf{x}(T) = 0$? To answer that question one needs to solve

$$\mathbf{x}(T) = \boldsymbol{\Phi}(T)\mathbf{x}(0) + \mathbf{d}(T)u(0) = \mathbf{0} \,.$$

The solution is simply a straight line going through the origin given by:

$$\mathbf{x}(0) = -\mathbf{s}_0 u(0) \,,$$

where $\mathbf{s}_0 = \boldsymbol{\Phi}(-T)\mathbf{d}(T)$ (note that $\{\boldsymbol{\Phi}(T)\}^{-1} = \boldsymbol{\Phi}(-T)$). In other words, if the system is already moving on the direction of $\mathbf{s}_0$, it can be brought to the zero point in a single sampling period as Figure 4.15 illustrates.

The next question is, naturally, from which initial conditions can one bring the system to the zero point in two sampling periods ($k = 2$)? Or, formally, what is $\mathbf{x}(0)$ such that $\mathbf{x}(2T) = 0$? In the first sampling period the system comes from the point $\mathbf{x}(0)$ to the point $\mathbf{x}(T)$. In the second sampling period it comes from $\mathbf{x}(T)$ to $\mathbf{0}$. Formally,

$$\mathbf{x}(T) = \boldsymbol{\Phi}(T)\mathbf{x}(0) + \mathbf{d}(T)u(0) := \mathbf{0} \,,$$
$$\mathbf{x}(T) = -\mathbf{s}_0 u(T) \,, \Rightarrow$$
$$\mathbf{x}(0) = -\mathbf{s}_0 u(0) - \mathbf{s}_1 u(T) \,,$$

where $s_0 = \boldsymbol{\Phi}(-T)\mathbf{d}(T)$ and $s_1 = \boldsymbol{\Phi}(-T)s_0 \,(= \boldsymbol{\Phi}(-2T)\mathbf{d}(T))$. Looking at Figure 4.16 one can see that the solution set spans the whole $(x_1 - x_2)$ plane. In other words, one can bring the system from any initial state to the origin in two sampling periods, provided that s_0 and s_1 are linearly independent. This means that the "controllability matrix" $[\mathbf{d}(T) \quad \boldsymbol{\Phi}(-T)\mathbf{d}(T)]$ must be a regular matrix.

One can thus compute the *optimal open-loop control sequence* which is needed to bring a second order linear sampled system from any initial state to the zero point in the shortest possible time.

System equations:

$$\mathbf{x}(T) = \boldsymbol{\Phi}\mathbf{x}(0) + \mathbf{d}u(0),$$
$$\mathbf{x}(2T) = \boldsymbol{\Phi}^2\mathbf{x}(0) + \boldsymbol{\Phi}\mathbf{d}u(0) + \mathbf{d}u(T) := \mathbf{0}.$$

Initial state:

$$\mathbf{x}(0) = -s_0 u(0) - s_1 u(T)$$
$$= -\mathbf{S}\mathbf{u},$$

where $\mathbf{S} = [s_0 \quad s_1]$ and $\mathbf{u} = [u(0) \quad u(T)]^T$.

Therefore, the optimal control vector can be calculated as

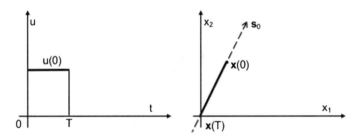

Fig. 4.15 For the system to be brought from any point in one step to the zero point, the initial state must lie on the vector s_0

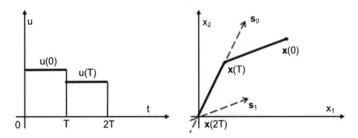

Fig. 4.16 The second order system can be brought from any point in the state space to the zero point in two sampling periods if the vectors s_0 and s_1 are linearly independent

$$\mathbf{u} = -\mathbf{S}^{-1}\mathbf{x}(0)$$
$$= \alpha \mathbf{x}(0),$$

where

$$\mathbf{u} = \begin{bmatrix} u(0) \\ u(T) \end{bmatrix}, \qquad \mathbf{x}(0) = \begin{bmatrix} x_1(0) \\ x_2(0) \end{bmatrix}, \qquad \alpha = \begin{bmatrix} \alpha_1' \\ \alpha_2' \end{bmatrix}.$$

Finally, the optimal control sequence is, hence,

$$u(0) = \alpha_1' \mathbf{x}(0),$$
$$u(T) = \alpha_2' \mathbf{x}(0).$$

The *optimal closed-loop control sequence* can be calculated such that $u(T) = f(\mathbf{x}(T))$ in the following way:

$$\mathbf{x}(T) = -s_0 u(T) - s_1 u(2T) = -\mathbf{S}\begin{bmatrix} u(T) \\ u(2T) \end{bmatrix},$$

where $u(2T) = 0$. Therefore, the optimal control vector is

$$\begin{bmatrix} u(T) \\ u(2T) \end{bmatrix} = -\mathbf{S}^{-1}\mathbf{x}(T).$$

This means the optimal closed-loop sequence is

$$u(T) = \alpha_1' \mathbf{x}(T),$$
$$u(0) = \alpha_1' \mathbf{x}(0).$$

Or more concisely

$$u(kT) = \alpha_1' \mathbf{x}(kT), \qquad i = 0, 1.$$

Example: Which controller can bring the sampled system given by its transfer function $G(s) = \frac{1}{s(s+1)}$ and held with a ZOH, in the shortest possible time to the zero point?

The transition matrix is

$$\boldsymbol{\Phi}(T) = \begin{bmatrix} 1 & 1 - e^{-T} \\ 0 & e^{-T} \end{bmatrix},$$

d becomes

$$\mathbf{d}(T) = \begin{bmatrix} T - 1 + e^{-T} \\ 1 - e^{-T} \end{bmatrix},$$

The base vectors $\mathbf{s}_0 = \boldsymbol{\Phi}(-T)\mathbf{d}(T)$ and $\mathbf{s}_1 = \boldsymbol{\Phi}(-2T)\mathbf{d}(T)$ are

$$\mathbf{s}_0 = \begin{bmatrix} T + 1 - e^{-T} \\ e^{-T} - 1 \end{bmatrix}, \quad \mathbf{s}_1 = \begin{bmatrix} T - e^{-2T} + e^{T} \\ e^{2T} - e^{T} \end{bmatrix}.$$

This makes the $\mathbf{S}$ matrix $\mathbf{S} = [\mathbf{s}_0 \ \ \mathbf{s}_1]$

$$\mathbf{S} = \begin{bmatrix} T + 1 - e^{-T} & T - e^{-2T} + e^{T} \\ e^{-T} - 1 & e^{2T} - e^{T} \end{bmatrix}.$$

Thus, the feedback matrix $\boldsymbol{\alpha}$ becomes

$$\boldsymbol{\alpha} = -\mathbf{S}^{-1} = \begin{bmatrix} \boldsymbol{\alpha}'_1 \\ \boldsymbol{\alpha}'_2 \end{bmatrix},$$

with

$$\boldsymbol{\alpha}'_1 = \begin{bmatrix} \frac{e^{T}}{T(1-e^{-T})} & \frac{T+e^{T}-e^{2T}}{T(1-e^{-T})^2} \end{bmatrix}.$$

Finally, the optimal closed-loop control is

$$u(kT) = \boldsymbol{\alpha}'_1 \mathbf{x}(kT), \quad k = 0, 1.$$

□

It is now not difficult to see that an n-th order system can be brought from any initial point in the state space to the zero point in at most n sampling periods when the control variable is unbounded. Derivation of the optimal controller is left to the reader as an exercise.

4.9.3 Optimization with a Quadratic Objective Function

In analogy to continuous time systems, one begins with the discrete time system given by its state equations

$$\mathbf{x}(k + 1) = \boldsymbol{\Phi}\mathbf{x}(k) + \mathbf{D}\mathbf{u}(k), \tag{4.111}$$

and defines the objective function

$$Z_N = \sum_{k=1}^{N} \left[\mathbf{x}^T(k)\mathbf{Q}\mathbf{x}(k) + \mathbf{u}^T(k-1)\mathbf{P}\mathbf{u}(k-1) \right], \tag{4.112}$$

with the symmetrical, at least positive semidefinite matrices $\mathbf{Q}$ and $\mathbf{P}$.

Let us consider the case where the control variable $\mathbf{u}$ is unbounded and define $f_N(\cdot)$ as

$$f_N(\mathbf{x}(0)) := \min Z_N = \min_{\mathbf{u}(0),\mathbf{u}(1),\dots,\mathbf{u}(N-1)} \sum_{k=1}^{N} [\mathbf{x}^T(k)\mathbf{Q}\mathbf{x}(k) + \mathbf{u}^T(k-1)\mathbf{P}\mathbf{u}(k-1)].$$

(4.113)

Thus for the last $N - j$ steps

$$f_{N-j}(\mathbf{x}(j)) = \min_{\mathbf{u}(j),\mathbf{u}(j+1),\dots,\mathbf{u}(N-1)} \sum_{k=j+1}^{N} [\mathbf{x}^T(k)\mathbf{Q}\mathbf{x}(k) + \mathbf{u}^T(k-1)\mathbf{P}\mathbf{u}(k-1)].$$

(4.114)

Making use of the optimality principle one can write

$$f_{N-j}(\mathbf{x}(j)) = \min_{\mathbf{u}(j)}[\mathbf{x}^T(j+1)\mathbf{Q}\mathbf{x}(j+1) + \mathbf{u}^T(j)\mathbf{P}\mathbf{u}(j) + f_{N-j-1}(\mathbf{x}(j+1))], \quad (4.115)$$

$$f_0(\mathbf{x}(N)) = 0,$$

$$f_1(\mathbf{x}(N-1)) = \min_{\mathbf{u}(N-1)} \left[\mathbf{x}^T(N)\mathbf{Q}\mathbf{x}(N) + \mathbf{u}^T(N-1)\mathbf{P}\mathbf{u}(N-1)\right],$$

$$f_2(\mathbf{x}(N-2)) = \min_{\mathbf{u}(N-2)} \left[\mathbf{x}^T(N-1)\mathbf{Q}\mathbf{x}(N-1) + \mathbf{u}^T(N-2)\mathbf{P}\mathbf{u}(N-2) + f_1(\mathbf{x}(N-1))\right],$$

$$\vdots$$

$$f_N(\mathbf{x}(0)) = \min_{\mathbf{u}(0)} \left[\mathbf{x}^T(1)\mathbf{Q}\mathbf{x}(1) + \mathbf{u}^T(0)\mathbf{P}\mathbf{u}(0) + f_{N-1}(\mathbf{x}(1))\right].$$

It can be proven using mathematical induction that $f_{N-j}(\mathbf{x}(j))$ is a quadratic form and can be written as

$$f_{N-j}(\mathbf{x}(j)) = \mathbf{x}^T(j)\mathbf{R}(N-j)\mathbf{x}(j),$$

(4.116)

and

$$f_{N-j-1}(\mathbf{x}(j+1)) = \mathbf{x}^T(j+1)\mathbf{R}(N-j-1)\mathbf{x}(j+1).$$

(4.117)

$\mathbf{R}(N-j)$ is an $n \times n$ symmetrical, at least positive semidefinite matrix with $\mathbf{R}(0) = \mathbf{0}$. Substituting this in (4.115) yields

$$\mathbf{x}^T(j)\mathbf{R}(N-j)\mathbf{x}(j) = \min_{\mathbf{u}(j)}[\mathbf{x}^T(j+1)\mathbf{S}(N-j-1)\mathbf{x}(j+1) + \mathbf{u}^T(j)\mathbf{P}\mathbf{u}(j)], \quad (4.118)$$

where $\mathbf{S}(N-j-1) = \mathbf{Q} + \mathbf{R}(N-j-1)$.

Substituting the system equation (4.111) in (4.118) results in

$$\mathbf{x}^T(j)\mathbf{R}(N-j)\mathbf{x}(j) = \min_{\mathbf{u}(j)} \left[[\boldsymbol{\Phi}\mathbf{x}(j) + \mathbf{D}\mathbf{u}(j)]^T \mathbf{S}(N-j-1)[\boldsymbol{\Phi}\mathbf{x}(j) + \mathbf{D}\mathbf{u}(j)] + \mathbf{u}^T(j)\mathbf{P}\mathbf{u}(j) \right].$$
(4.119)

The value of the large brackets on the right hand side is minimal when its derivative with respect to $\mathbf{u}(j)$ vanishes. Thus,

$$2[\mathbf{D}^T\mathbf{S}(N-j-1)\mathbf{D} + \mathbf{P}]\mathbf{u}(j) + 2\mathbf{D}^T\mathbf{S}(N-j-1)\boldsymbol{\Phi}\mathbf{x}(j) := \mathbf{0},$$

which finally results in the optimal control law as

$$\mathbf{u}(j) = -\left(\mathbf{D}^T[\mathbf{Q} + \mathbf{R}(N-j-1)]\mathbf{D} + \mathbf{P} \right)^{-1} \left(\mathbf{D}^T[\mathbf{Q} + \mathbf{R}(N-j-1)]\boldsymbol{\Phi} \right)\mathbf{x}(j).$$
(4.120)

In other words

$$\mathbf{u}(j) = \boldsymbol{\alpha}(N-j)\mathbf{x}(j)$$

where $\boldsymbol{\alpha}(N-j)$ is a time variable feedback matrix. With $\mathbf{R} = \mathbf{0}$,

$$\boldsymbol{\alpha}(1) = -\left(\mathbf{D}^T\mathbf{Q}\mathbf{D} + \mathbf{P} \right)^{-1} \left(\mathbf{D}^T\mathbf{Q}\boldsymbol{\Phi} \right),$$

$$\boldsymbol{\alpha}(2) = -\left(\mathbf{D}^T[\mathbf{Q} + \mathbf{R}(1)]\mathbf{D} + \mathbf{P} \right)^{-1} \left(\mathbf{D}^T[\mathbf{Q} + \mathbf{R}(1)]\boldsymbol{\Phi} \right),$$

$$\vdots$$

$$\boldsymbol{\alpha}(N) = -\left(\mathbf{D}^T[\mathbf{Q} + \mathbf{R}(N-1)]\mathbf{D} + \mathbf{P} \right)^{-1} \left(\mathbf{D}^T[\mathbf{Q} + \mathbf{R}(N-1)]\boldsymbol{\Phi} \right).$$

When one now substitutes this result in (4.119) it results in

$$\mathbf{x}^T(j)\mathbf{R}(N-j)\mathbf{x}(j) = \mathbf{x}^T(j)[\boldsymbol{\Phi} + \mathbf{D}\boldsymbol{\alpha}(N-j)]^T \mathbf{S}(N-j-1)[\boldsymbol{\Phi} + \mathbf{D}\boldsymbol{\alpha}(N-j)]\mathbf{x}(j)$$
$$+ \mathbf{x}^T(j)\boldsymbol{\alpha}^T(N-j)\mathbf{P}\boldsymbol{\alpha}(N-j)\mathbf{x}(j),$$

which means

$$\mathbf{R}(N-j) = [\boldsymbol{\Phi} + \mathbf{D}\boldsymbol{\alpha}(N-j)]^T [\mathbf{Q} + \mathbf{R}(N-j-1)][\boldsymbol{\Phi} + \mathbf{D}\boldsymbol{\alpha}(N-j)] + \boldsymbol{\alpha}^T(N-j)\mathbf{P}\boldsymbol{\alpha}(N-j) \quad (4.121)$$
$$= \boldsymbol{\Phi}^T [\mathbf{Q} + \mathbf{R}(N-j-1)][\boldsymbol{\Phi} + \mathbf{D}\boldsymbol{\alpha}(N-j)]. \quad (4.122)$$

Equations (4.120) and (4.121) can be used to compute $\alpha(1)$, $\mathbf{R}(1)\alpha(2)$, $\mathbf{R}(2)$, ... A notable special case is when $N = \infty$. In this case α and $\mathbf{R}$ have the constant values:

$$\alpha = -\left(\mathbf{D}^T[\mathbf{Q} + \mathbf{R}]\mathbf{D} + \mathbf{P}\right)^{-1}\left(\mathbf{D}^T[\mathbf{Q} + \mathbf{R}]\mathbf{\Phi}\right),$$
$$\mathbf{R} = \mathbf{\Phi}^T[\mathbf{Q} + \mathbf{R}][\mathbf{\Phi} + \mathbf{D}\alpha].$$

4.10 Differential Games

4.10.1 Introduction

Encyclopedia of Complexity and Systems Science [91] gives the definition of differential games as

> Differential games is a mathematical theory which is concerned with problems of conflicts [modeled] as game problems in which the state of the players depends on time in a continuous way. The positions of the players are the solution [of] differential equations. Differential games can be described from two different points of view, depending mainly on the field of application. Firstly, they can be considered as games where time is continuous. This aspect is often considered for applications in economics or management sciences. Secondly, they also can be viewed as control problems with several controllers having different objectives. In this way, differential games are a part of control theory with conflicts between the players. The second aspect often concerns classical applications of control theories...

The beginnings of differential games can be traced back to Isaacs [61] during the Cold War. After Pontryagin published his maximum principle, it became apparent that differential games and optimal control were related. Actually, one can see differential games as the generalization of optimal control problems where there are more than one controller. Not surprisingly, the problems become significantly more complex in this generalization and it is no longer evident what is meant by an "optimal solution". Therefore, there are a number of different optimal solutions with fancy names like minimax, Pareto or Nash optimal.[9, 10] Examples of applications of differential games in finance can be found, for instance, in [24].

For illustration purposes, let us begin with two players A and B with the m dimensional control variables $\mathbf{u}_A(t) \in U_A$ and $\mathbf{u}_B(t) \in U_B$ available to them. The state space representation of the n dimensional system is given by

$$\dot{\mathbf{x}}(t) = \mathbf{f}(\mathbf{x}(t), \mathbf{u}_A(t), \mathbf{u}_B(t), t), \qquad \mathbf{x}(0) = \mathbf{x}_0, \qquad t \in [0, T]. \tag{4.123}$$

---

[9]John Forbes Nash, American mathematician (1928–2015).

[10]This section follows the structure of the lecture notes [18] and some parts are borrowed from it.

The goal of the i-th player (here $i = A, B$) is to maximize its objective function

$$J_i = \Psi_i(\mathbf{x}(T)) - \int_0^T L_i(\mathbf{x}(t), \mathbf{u}_A(t), \mathbf{u}_B(t), t)dt \,. \tag{4.124}$$

Ψ_i are terminal pay-offs, while L_i account for running costs. As discussed earlier in this chapter one can consider open-loop $u_i = u_i(t)$ and closed-loop $u_i = u_i(\mathbf{x}(t), t)$ strategies. Such dynamic systems are studied later in the section. However, it helps to first look at static games and introduce some useful concepts.

4.10.2 Static Games

A game for two players A and B is given by their respective sets of possible strategies A and B which are assumed to be compact metric spaces and respective objective functions Φ^A and Φ^B which are assumed to be continuous. If the first player chooses a strategy $a \in A$ and the second player chooses $b \in B$ their respective pay-offs are $\Phi^A(a, b)$ and $\Phi^B(a, b)$. The goal of each player is to maximize his own pay-off. It is further assumed that each player knows both pay-off functions Φ^A and Φ^B but not the strategy of the other player.

The simplest class of games consists of so-called *bi-matrix games*, where each player has a finite set of strategies to choose from. Say

$$A = \{a_1, a_2, \ldots, a_m\}, \qquad B = \{b_1, b_2, \ldots, b_n\}. \tag{4.125}$$

Each pay-off function is thus determined by its $m \times n$ values $\Phi_{ij}^A = \Phi^A(a_i, b_j)$ and $\Phi_{ij}^B = \Phi^B(a_i, b_j)$. In other words, the game can be completely represented by an $m \times n$ bi-matrix composed of these pay-off functions.

As mentioned above, in general, one cannot simply speak of an optimal solution of the game. An outcome that is optimal for one player is not optimal for the other. Moreover, different concepts of optimality can be defined based on the information available to the players and on their willingness to cooperate.

The strategy pair (a^*, b^*) is said to be *Pareto optimal* if there exists no other pair $(a, b) \in A \times B$ such that $\Phi^A(a, b) > \Phi^A(a^*, b^*)$ and $\Phi^B(a, b) \geq \Phi^B(a^*, b^*)$ or $\Phi^B(a, b) > \Phi^B(a^*, b^*)$ and $\Phi^A(a, b) \geq \Phi^A(a^*, b^*)$. In other words, it is not possible to increase the pay-off of one player without decreasing the pay-off of the other. To find strategies which are Pareto optimal (there can be more than one) one has to solve the optimization problem

$$\max_{(a,b)\in A \times B} \lambda \Phi^A(a, b) + (1 - \lambda)\Phi^B(a, b), \tag{4.126}$$

where $\lambda \in [0, 1]$. Any pair (a^*, b^*) satisfying (4.126) yields a Pareto optimum. In Chapter 8 this approach is used to determine a portfolio of stocks which at the

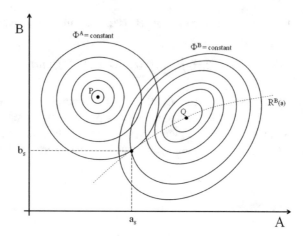

Fig. 4.17 The figure shows the contour lines of the two pay-off functions Φ^A and Φ^B. Here the leader (player A) chooses the horizontal coordinate, the follower (player B) chooses the vertical coordinate. The pay-off function Φ^A attains its global maximum at P. Φ^B attains its maximum at Q. If the leader chooses a strategy $a \in A$, then $R^B(a) \in B$ is the best reply for follower. The pair of strategies (a_S, b_S) is a Stackelberg equilibrium. Notice that at this point the curve $b = R^B(a)$ is tangent to a contour line of Φ^A (after [18])

same time maximizes its expected return and minimizes its volatility for different levels of risk aversion, parametrized by λ.

In cases with asymmetry of information it is assumed there is a leader player A which announces her strategy in advance and a follower player B which makes his choice based on the leader's strategy. In this case, the game can be reduced to two consecutive optimization problems. The leader adopts strategy a. The follower maximizes his pay-off function $\Phi^B(a, b)$ by choosing the best reply $b^* \in R^B(a)$. As a response to that, the leader now has to maximize the composite function $\Phi^A(a, R^B(a))$. In case the follower has several optimal replies to $a \in A$ to choose from, he chooses the most favorable one to the leader. So, a pair of strategies $(a_S, b_S) \in A \times B$ is called a *Stackelberg equilibrium* if $b_S \in R^B(a_S)$ and $\Phi^A(a, b) \le \Phi^A(a_S, b_S)$ for every pair (a, b) where $b \in R^B(a)$. Under our assumptions of compactness of A and B as well as the continuity of Φ^A and Φ^B a Stackelberg equilibrium always exists. Figure 4.17 illustrates the calculation of a Stackelberg equilibrium.

In symmetric situations where the players cannot cooperate or share any information about their strategies, the pair of strategies (a^*, b^*) is called a *Nash equilibrium* of the game if, for every $a \in A$ and $b \in B$, $\Phi^A(a, b^*) \le \Phi^A(a^*, b^*)$ and $\Phi^B(a^*, b) \le \Phi^B(a^*, b^*)$. This means that players cannot increase their pay-offs by changing their strategies as long as the other player sticks to the equilibrium strategy. Note that a pair of strategies (a^*, b^*) is a Nash equilibrium if and only if it is a fixed point of the best reply map $a^* \in R^A(b^*), b^* \in R^B(a^*)$. One has to remember that a Nash equilibrium does not necessarily exist or be unique or be a Pareto optimum. Also, different Nash equilibria can yield different pay-offs to each player. Figure 4.18 illustrates the calculation of a Nash equilibrium.

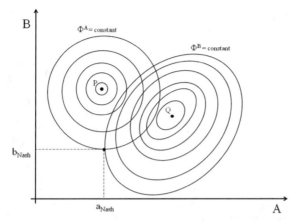

Fig. 4.18 There are no leaders or followers here. Player A chooses the horizontal coordinate, player B the vertical coordinate. The pay-off function Φ^A attains its global maximum at P, while Φ^B attains its global maximum at Q. The pair of strategies (a_{Nash}, b_{Nash}) is a Nash equilibrium. Notice that at this point the contour line of Φ^A has a horizontal tangent while the contour line of Φ^B has a vertical tangent (after [18])

Prisoner's Dilemma

Arguably the most famous example in game theory which goes back to the middle of the 20th century is the so-called prisoner's dilemma [25]. It helps understanding what governs the balance between cooperation and competition in business, politics and social settings. The problem is often presented as follows: Two persons are arrested in a crime scene and imprisoned. Each prisoner is in solitary confinement with no means of communicating with the other. The prosecutor lacks sufficient evidence to convict the prisoners on the principal charge, who hope to get both sentenced to a year in prison on a lesser charge. Each prisoner has the opportunity either to confess (C) and accuse the other, or not to confess (N) and remain silent. The pay-off bi-matrix is shown in Figure 4.19.

The negative pay-offs mean the number of years in jail faced by each prisoner depending on their actions. Looking from the perspective of prisoner A, one could argue as follows: If prisoner B confesses, my two options result in either 6 or 8 years in jail, hence confessing is the best choice. On the other hand, if prisoner B does not confess, then my two options result in either 0 or 1 years in jail. Again, confessing is the best choice. Prisoner B can argue exactly the same way. Therefore, the outcome of the game is that both prisoners confess, and get 6 years sentences. This is paradoxical, because an entirely rational argument results in the worst possible outcome: the total number of years in jail for the two prisoners is maximal (one could argue that this is the best outcome for the prosecutor, but let us not get into that). If they could have cooperated, they would both have achieved a better outcome, totaling only 2 years in jail. Note that the pair of strategies (C, C) is the unique Nash equilibrium, but it is not Pareto optimal. On the other hand, all three other pairs (C, N), (N, C), (N, N) are Pareto optimal.

Fig. 4.19 The pay-off
bi-matrix for the prisoner's
dilemma (after [18])

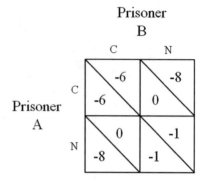

4.10.3 Zero-Sum Games

If $\Phi^A(a, b) + \Phi^B(a, b) = 0$ for every pair of strategies (a, b) the game is called a zero-sum game. Since $\Phi^A(a, b) = -\Phi^B(a, b)$ one can consider a single objective function $\Phi(a, b)$, which is further assumed to be continuous. One can, for instance, think of $\Phi(a, b)$ as the amount of money that changes hands in any kind of interaction between two players. Clearly, where the player A wants to maximize his income, player B wants to minimize his outlay. In a symmetric situation, each of the two players will have to make his choice without a priori knowledge of the action taken by his interlocutor. However, one may also consider cases where one player has this advantage of information.

Let us say the second player chooses a strategy $b \in B$, then the first player makes his choice, depending on this chosen b. This is a situation where player A has the advantage of knowing his opponent's strategy. The best reply of player A will be some $\alpha(b) \in A$ such that $\Phi(\alpha(b), b) = \max_{a \in A} \Phi(a, b)$. Therefore, the minimum payment that the second player can achieve is

$$V^+ := \min_{b \in B} \Phi(\alpha(b), b) = \min_{b \in B} \max_{a \in A} \Phi(a, b). \qquad (4.127)$$

Let us now say that the first player chooses a strategy $a \in A$, then the second player makes his choice depending on this chosen a. In this case player B has the advantage of knowing his opponent's strategy. The best reply of player B will be some $\beta(a) \in B$ such that $\Phi(a, \beta(a)) = \min_{b \in B} \Phi(a, b)$. Therefore, the maximum payment that the first player can achieve is

$$V^- := \max_{a \in A} \Phi(a, \beta(a)) = \max_{a \in A} \min_{b \in B} \Phi(a, b). \qquad (4.128)$$

It can be seen that $V^- \leq V^+$. Actually, the strict inequality $V^- < V^+$ may hold in general. In the special case where the equality holds $V := V^- = V^+$ is called the value of the game.

Furthermore, if there exists strategies $a^* \in A$ and $b^* \in B$ such that

$$\min_{b \in B} \Phi(a^*, b) = \Phi(a^*, b^*) = \max_{a \in A} \Phi(a, b^*), \qquad (4.129)$$

the pair (a^*, b^*) is a *saddle point* of the game. Naming the two sides of the equation (4.129) V it can be said that if A adopts the strategy a^*, he is guaranteed to receive no less than V and if B adopts the strategy b^*, he is guaranteed to pay no more than V. Therefore, for a zero-sum game, the concept of saddle point is the same as a Nash equilibrium. As mentioned before, a non-zero-sum game may have several Nash equilibria which provide different pay-offs to each player. However, for a zero-sum game, if a Nash equilibrium exists, then all Nash equilibria yield the same pay-off. This pay-off is the value of the game V.

4.10.4 The Co-Co Solution

Let us now go back to non-zero-sum games with two players. If the two players can cooperate, they can adopt a pair of strategies $(a^\sharp, b^\sharp)$ which maximizes their combined pay-offs:

$$V^\sharp := \Phi^A(a^\sharp, b^\sharp) + \Phi^B(a^\sharp, b^\sharp) = \max_{(a,b) \in A \times B} \Phi^A(a, b) + \Phi^B(a, b). \qquad (4.130)$$

This strategy can favor one player much more than the other (can you think of examples from real life?). In other words, one may have $\Phi^B(a^\sharp, b^\sharp) \ll \Phi^A(a^\sharp, b^\sharp)$. For emotional reasons, this might be unacceptable for player B. In this case player A needs to provide some incentive (a bribe?) to make player B adopt the strategy $b^\sharp$. In general, splitting the total pay-off $V^\sharp$ in two equal parts will not be acceptable, because it does not reflect the relative strength of the players and their personal contributions to the common achievement. A more realistic procedure to split the total pay-off among the two players goes as follows: given the two objective functions Φ^A, Φ^B, one can define

$$\Phi^\sharp(a, b) = \frac{\Phi^A(a, b) + \Phi^B(a, b)}{2}, \qquad \Phi^\flat(a, b) = \frac{\Phi^A(a, b) - \Phi^B(a, b)}{2}.$$
$$\qquad (4.131)$$

One can see that $\Phi^A = \Phi^\sharp + \Phi^\flat$ and $\Phi^B = \Phi^\sharp - \Phi^\flat$. Therefore, the original game can be split as the sum of a purely cooperative game, where both players have exactly the same pay-off $\Phi^\sharp$ and a purely competitive (zero-sum) game, where the players have opposite pay-offs: $\Phi^\flat$ and $-\Phi^\flat$. With $V^\flat$ being the value of the zero-sum game and $V^\sharp$ defined as in (4.130), the co-co value of the original game is defined as the pair of pay-offs $(\frac{V^\sharp}{2} + V^\flat, \frac{V^\sharp}{2} - V^\flat)$. A cooperative-competitive solution of the original game is defined as a pair of strategies $(a^\sharp, b^\sharp)$ *together with a side payment p from player B to player A*, such that $\Phi^A(a^\sharp, b^\sharp) + p = \frac{V^\sharp}{2} + V^\flat$ and $\Phi^B(a^\sharp, b^\sharp) - p = \frac{V^\sharp}{2} - V^\flat$. The co-co solution models a situation where the players join forces, implement a strategy $(a^\sharp, b^\sharp)$ which achieves their maximum combined pay-off. Then one of the two makes a side payment to the other, so that in the end acceptable pay-offs for both players are achieved.

4.10.5 Dynamic Games

This Subsection limits itself to the minimax solutions for the zero-sum differential games and the Nash solutions for the non-zero-sum games.

Let us consider the state equation[11]

$$\dot{x} = f(x, u_1, u_2, t), \qquad x(0) = x_0, \tag{4.132}$$

with x denoting the state, u_1 denoting the control variable for player 1 and u_2 denoting the control variable for player 2. Let the objective function be defined as

$$J(u_1, u_2) = \Psi(x(T)) + \int_0^T L(x, u_1, u_2, t)dt, \tag{4.133}$$

such that player 1 wants to maximize J, whereas player 2 wants to minimize it, meaning that any gain by player 1 corresponds to a loss by player 2 (hence, zero-sum games). The optimal control problem here is to find $u_1^*(t)$ and $u_2^*(t)$ such that

$$J(u_1^*, u_2) \geq J(u_1^*, u_2^*) \geq J(u_1, u_2^*) \qquad \text{where } u_1 \in U_1, \ u_2 \in U_2. \tag{4.134}$$

Extending the maximum principle from Section 4.7 gives us the necessary restrictions on u_1^* and u_2^* to satisfy these conditions, which are obtained by first forming the Hamiltonian as in (4.43)

$$H = L + \lambda f, \tag{4.135}$$

where the adjoint variable λ, just like in (4.44), is given by

$$\dot{\lambda} = -\frac{\partial H}{\partial x}, \qquad \lambda(T) = \Psi(x(T)). \tag{4.136}$$

To find the optimal control variables (4.45) is modified such that the minimax solution is given by

$$H(x^*, u_1^*, u_2^*, \lambda^*, t) = \min_{u_2 \in U_2} \max_{u_1 \in U_1} H(x^*, u_1, u_2, \lambda^*, t), \tag{4.137}$$

or equivalently

$$H(x^*, u_1^*, u_2, \lambda^*, t) \geq H(x^*, u_1^*, u_2^*, \lambda^*, t) \geq H(x^*, u_1, u_2^*, \lambda^*, t). \tag{4.138}$$

In case u_1 and u_2 are unconstrained (4.137) and (4.138) become

---

[11]All variables are assumed to be scalar for simplicity; the extension to multivariable systems is straight forward.

$$\frac{\partial H}{\partial u_1} = 0 \quad \text{and} \quad \frac{\partial H}{\partial u_2} = 0, \tag{4.139}$$

and

$$\frac{\partial^2 H}{\partial u_1^2} \leq 0 \quad \text{and} \quad \frac{\partial^2 H}{\partial u_2^2} \geq 0. \tag{4.140}$$

What is actually done here is minimizing the possible loss for a worst case scenario (minimizing the maximum loss), hence the term *minimax*. In zero-sum games, the minimax solution is the same as the so-called Nash equilibrium.

When one now looks at non-zero-sum differential games with not two but N players, one needs to make some adjustments. Let u_i, $i = 1, 2, \ldots, N$ be the control variable for the i-th player. Let the system be governed by the state equation

$$\dot{x} = f(x, u_1, u_2, \ldots, u_N, t), \qquad x(0) = x_0. \tag{4.141}$$

Further, let the objective function of the i-th player which he wants to maximize be defined as

$$J_i = \Psi_i(x(T)) + \int_0^T L_i(x, u_1, u_2, \ldots, u_N, t)dt. \tag{4.142}$$

In this case a Nash solution is defined by a set of N trajectories

$$\{u_1^*, u_2^*, \ldots, u_N^*\}, \tag{4.143}$$

which have the property

$$J_i(u_1^*, u_2^*, \ldots, u_N^*) = \max_{u_i} J_i(u_1^*, u_2^*, \ldots, u_{i-1}^*, u_i, u_{i+1}^*, \ldots, u_N^*), \qquad i = 1, 2, \ldots, N. \tag{4.144}$$

As already shown, to obtain the necessary conditions for a Nash solution for non-zero sum differential games, one has to distinguish between open-loop and closed-loop control.

Open-loop Nash solution is defined when (4.143) is given as time functions satisfying (4.144). To obtain maximum principle type conditions for such solutions to be a Nash solution, let us again define the Hamiltonians in (4.43) as

$$H_i = L_i + \lambda_i f, \qquad i = 1, 2, \ldots, N, \tag{4.145}$$

where adjoint variables λ_i satisfy the partial differential equation system

$$\dot{\lambda}_i = -\frac{\partial H_i}{\partial x}, \qquad \lambda_i(T) = \Psi_i(x(T)), \qquad i = 1, 2, \ldots, N. \tag{4.146}$$

For the control variables u_i^* obtained by maximizing the i-th Hamiltonian H_i with respect to u_i, u_i^* must satisfy

$$H_i(x^*, u_1^*, \ldots, u_N^*, \lambda^*, t) \geq H_i(x^*, u_1^*, \ldots, u_{i-1}^*, u_i, u_{i+1}^*, \ldots u_N^*, \lambda^*, t), \quad (4.147)$$

for all $u_i \in U_i$ with U_i being the set of all possible control variables the i-th player can choose from.

Closed-loop Nash solution is defined when optimal trajectories in (4.143) is defined in terms of the system state. $u_i^*(x, t) = \varphi_i(x, t)$, $i = 1, 2, \ldots, N$ nomenclature is used to avoid confusion. Remember that the other players' actions have an influence on the state x in this case. This needs to be taken into account in computing the Nash solution. Therefore the equation for the adjoint system in (4.146) needs to be modified as

$$\dot{\lambda}_i = -\frac{\partial H_i}{\partial x} - \sum_{j=1, j \neq i}^{N} \frac{\partial H_i}{\partial u_j} \frac{\partial \varphi_j}{\partial x}. \qquad (4.148)$$

This adjoint system becomes very difficult to solve by numerical integration due to the modification. However, it is possible to make use of dynamic programming as shown in the previous section for the solution of this problem.

Why is this disturbing summation term in (4.148) absent in the solution of the original optimal control problem's adjoint system (4.44)? Because $N = 1$ and therefore

$$\frac{\partial H}{\partial u} \frac{\partial u}{\partial x} = 0,$$

in the two player zero-sum game because $H_1 = -H_2$ so that

$$\frac{\partial H_1}{\partial u_2} \frac{\partial u_2}{\partial x} = -\frac{\partial H_2}{\partial u_2} \frac{\partial u_2}{\partial x} = 0,$$

and

$$\frac{\partial H_2}{\partial u_1} \frac{\partial u_1}{\partial x} = -\frac{\partial H_1}{\partial u_1} \frac{\partial u_1}{\partial x} = 0,$$

and in the open-loop non-zero-sum game because

$$\frac{\partial u_j}{\partial x} = 0 \,\forall j.$$

Therefore, it is only to be expected that open-loop and closed-loop Nash solutions are in general going to be different.

4.11 Exercises

A1: What is calculus of variations? What kind of problems can be solved by using calculus of variations?

A2: What is a functional?

A3: In which areas of automatic control do you find applications of calculus of variations?

A4: You are given a system by its state equations, initial states and an objective function which depends on the system states and its input. How do you find the optimal input signal which minimizes the objective function? How many differential equations do you have to solve? Specifically, how do you solve them numerically? Are there enough initial conditions? If not, how do you proceed?

A5: In practice, all controllers have limitations; the control signals are physically limited. What do you do, if the optimal input signal which you determined in the previous question turns out to be too large to be physically realizable? Are there formal solutions to the previous questions where you can consider the controller's limitations?

A6: What kind of a controller is used when controlling a second order system in a time optimal fashion?

A7: To solve some optimal control problems you make use of adjoint variables (Lagrange multiplier functions). What is the physical meaning of the adjoint variables? What are the initial values of the adjoint variables when solving the differential equations of the adjoint system?

A8: What is the mathematical difference between finding the optimal open-loop and optimal closed-loop control signals? How is the significantly larger effort for the latter justifiable?

A9: What are the assumptions for the linear quadratic optimization problem?

A10: How do you proceed to solve the Riccati differential equation numerically given that you do not have the initial states of the system?

A11: The solutions of the last two questions require that you know all of the system states at all times. In practice, however, you seldom know all the states. How do you deal with this situation?

A12: What is dynamic programming? What is the optimality principle?

A13: What are the differences between continuous, discrete and sampled system? What are the commonalities and differences in mathematical handling of such system?

A14: What are differential games and how do they relate to automatic control?

B1: The year is 2033... We begin by assuming that a spacecraft has been brought in a circular orbit around Mars. Further we make the following simplifications: the curvature of the Mars surface is negligible meaning that the gravitational attraction has constant direction during the landing maneuver; the height of the orbit is small compared to the radius of the planet meaning that the value of the gravitational attraction does not change either; the Martian atmosphere is negligible; the mass of the fuel is negligible compared to the total mass of the spacecraft; the amount of thrust is constant during the whole landing manoeuver, in other words, the only control variable is the direction of the thrust. We know, v_0, the initial speed of the space craft in orbit; h, the height of the orbit; m, the mass of the space craft; g, the gravitational constant of Mars; T, the amount of thrust. The equations of motion for the space craft are $m\ddot{x}_1 = -T\cos(\varphi)$ and $m\ddot{x}_2 = mg - T\sin(\varphi)$. x_1 is the horizontal distance. x_2 is the vertical distance. T is the thrust. φ is the angle between the direction of the thrust and the surface of the planet. Calculate $\varphi(t)$ which results in minimum fuel consumption for a smooth landing (vertical and horizontal speed zero) on Mars.

B2: The equation of movement of a space ship with respect to the horizon can be modeled as $\ddot{x}(t) = T(t)/J$ where $x(t)$ denotes the angle between the main axis of the ship and the horizon, J denotes the moment of inertia of the ship, and $T(t)$ the torque control where the limitation $|T(t)/J| \leq 1$ at all times. We can neglect any friction or other disturbances. The task is to calculate $T(t)$ which brings the ship from an angular position α and no angular velocity with respect to the horizon to zero angular position and zero angular velocity with respect to the horizon. (a) What is $T(t)$ if this manoeuver is to be completed in the shortest possible time? (b) What is $T(t)$ if this manoeuver is to be completed with minimal control energy? (c) What is $T(t)$ if this manoeuver is to be completed in a way which minimizes both time *and* fuel consumption (assume fuel is five times as costly as the time)? For each case sketch the phase portrait of the optimal trajectories as well as the time course of the adjoint variables.

B3: An armature controlled DC motor with constant excitation and negligible armature inductance can be modeled by its state equations $\dot{\alpha} = \omega$ and $\dot{\omega} = -\frac{1}{T_a}\omega + \frac{K_a}{T_a}u$ where α is the angular position, ω is the angular velocity, the physical parameters T_a, K_a can be taken as 1. Calculate the optimal closed-loop controller $u^*(t) = -\mathbf{k}^T \cdot [\alpha(t)\ \ \omega(t)]$ which minimizes the objective function $J = \int_0^\infty (\alpha^2 + u^2)dt$.

B4: A dynamic system is given by its difference equation as $x(k+1) = x(k) + u(k)$. Both the state variable x and the control variable u can have only integer values and are limited as $0 \leq x \leq 10$ and $-4 \leq u \leq 0$. The final state after $N = 5$ steps is not fixed. Use dynamic programming to find the control sequence $u(k)$ such that the objective function $Z_N = \sum_{k=1}^N [2x^2(k) + 3u^2(k)]$ is minimized.

B5: A linear dynamic process is given by its difference equation as $x(k+1) = x(k) + 2u(k)$, $k = 0, 1, \ldots, N-1$. You are looking for the optimal closed-loop controller $u(k) = u(x(k))$ which minimizes the objective function $Z_N = x^2(N) + 4\beta \sum_{i=1}^N u^2(i-1)$ where β is a positive constant. That goal can be achieved if you simply feed the system state back such that $u(k) = \alpha(N-k)x(k)$. Discuss this time variable feedback for the cases (a) $\beta = 0$, (b) $\beta \gg N$, (c) $N \to \infty$, d) $x(0) = 1$, $N = 10$, $\beta = 2$.

B6: You are given a sampled system by its state space description

$$\mathbf{x}_{k+1} = \begin{bmatrix} 1 & 3 \\ 0 & -1 \end{bmatrix} \mathbf{x}_k + \begin{bmatrix} 1 \\ 0 \end{bmatrix} u_k \qquad y_k = [1\ \ 2] \cdot \mathbf{x}_k .$$

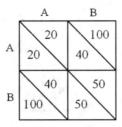

Fig. 4.20 Vendors net $1 per ice cream sold everywhere

The system is initially at the state $\mathbf{x}_0 = [1\ 1]^T$ and you want to bring it to its final state $\mathbf{x}_F = [0\ 0]^T$. How do you proceed?

B7: Let u_1, u_2 be the strategies implemented by the two players A and B, and let the corresponding pay-offs be

$$\Phi^A(u_1, u_2) = a_1 u_1 + a_2 u_2 - \left[a_{11} \frac{u_1^2}{2} + a_{12} u_1 u_2 + a_{22} \frac{u_2^2}{2} \right],$$

$$\Phi^B(u_1, u_2) = b_1 u_1 + b_2 u_2 - \left[b_{11} \frac{u_1^2}{2} + b_{12} u_1 u_2 + b_{22} \frac{u_2^2}{2} \right],$$

with $a_{11} > 0$, $b_{11} > 0$. (a) Compute a Nash equilibrium solution. (b) Compute a Stackelberg equilibrium solution. (c) For which values of the coefficients is the Stackelberg equilibrium better than the Nash equilibrium, for Player A?

B8: Consider two ice cream vendors P1 and P2, located in a town with two selling venues: the airport, A, and the beach, B. The demand at A is for 40 ice creams, while the demand at B is for 100 ice creams. If they choose different locations, they each sell the quantity demanded at their respective locations; and if they choose the same location, they split the local demand equally. Each vendor has to choose a location without any knowledge of the opponent's choice. Suppose they each net $1 per ice cream sold. The game is given by the bi-matrix in Figure 4.20.

(a) What is the Nash equilibrium of this game? (b) What is the co-co value of this game? (c) Can you find several agreements between the vendors that will attain this co-co value? (d) Given that everything is more expensive at the airport, the game changes as shown in Figure 4.21 if we assume that the ice cream vendor at the airport nets $2 per ice cream sold. How do things change now? (After [65]).

B9: Let x_i be the market share of the company i, and u_i be its advertising budget. For $i = 1, 2$ the state equations are given as

$$\dot{x}_1 = b_1 u_1 (1 - x_1 - x_2) + e_1 (u_1 - u_2)(x_1 + x_2) - a_1 x_1, \qquad x_1(0) = x_{10},$$
$$\dot{x}_2 = b_2 u_2 (1 - x_1 - x_2) + e_2 (u_2 - u_1)(x_1 + x_2) - a_2 x_2, \qquad x_2(0) = x_{20},$$

where a_i, b_i, e_i are given positive constants. The companies want to maximize their objective functions

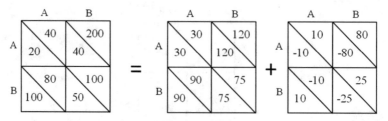

Fig. 4.21 The vendor at the airport nets \$2 per ice cream sold. The game is given on the left, and the decomposition on the right

$$J_i = w_i e^{-\rho T} x_i(T) + \int_0^T (c_i x_i - u_i^2) e^{-\rho T} dt \,,$$

where w_i, c_i, ρ are given positive constants. Derive the necessary conditions for the open-loop Nash solution and formulate the resulting boundary value problem.

B10: Reconsider Exercise B3 at the end of Chapter 2. Formulate and solve it as a static game.

C1: We have seen bang-bang control as a solution to the minimum time control with constrained control signals. A conjugate time theory has already been developed for such problems [115, 127]. Develop solutions that involve both bang-bang and singular arcs.

C2: Usually, optimal control problems are solved numerically using techniques based on the gradient vector or the Hessian matrix [5]. Study other numerical techniques like adaptive finite element approximation using adaptive multi meshes [76] and tabulate their efficiencies according to problems and parameters like system order, etc.

C3: In section "Differential Games" we have assumed that the system at hand is deterministic. Develop solutions for Nash equilibria if the system is a stochastic system, governed by the stochastic differential equation

$$dx = f(x, u_1, u_2, t)dt + \sigma(t, x)dW$$

where W denotes a Brownian motion. It might help to begin by studying Chapter 5 on Stochastic Analysis.

Chapter 5
Stochastic Analysis

> *I turn aside with a shudder of horror from this lamentable*
> *plague of functions which have no derivatives.*
> — Charles Hermite
>
> *Information: the negative reciprocal value of probability.*
> — Claude Shannon

5.1 Introduction

This chapter begins by studying white noise closely and by introducing the concept of stochastic differential equations (SDE). Different ways of solving such equations are then discussed. Stochastic integration and Itô integrals are shown together with Itô's lemma for scalar and vector processes. Various stochastic models used in financial applications are illustrated. The connection between deterministic partial differential equations and SDE's is indicated. The chapter aims to give just a taste of stochastic calculus/stochastic control and whet the appetite for this broad field. The reader is referred to standard books of reference for an in depth study of the subject matter.

5.2 White Noise

Let us recall from Chapter 3 that Brownian motions are pathological functions in the sense that they are continuous but nowhere differentiable.[1] Nevertheless, engineers often write

---

[1] This chapter follows the structure of [46].

© Springer International Publishing AG 2018
S. S. Hacısalihzade, *Control Engineering and Finance*, Lecture Notes in Control and Information Sciences 467, https://doi.org/10.1007/978-3-319-64492-9_5

Fig. 5.1 Continuous-time measurement $y(t)$ of, for instance, the third state variable $x_3(t)$ of system S is corrupted by an additive white noise $v(t)$ such that $y(t) = x_3(t) + v(t)$

$$v(t) = \frac{dW(t)}{dt},$$

which defines a random process $v(t)$ called *stationary white Gaussian noise* as the derivative of a Brownian motion $W(t)$ with the drift μ and the variance σ^2. The white noise $v(t)$ is assumed to be defined on the infinite time interval $(-\infty, \infty)$, thus making it stationary and Gaussian for all t. This stationary white noise is characterized uniquely by its expected value $E\{v(t)\} \equiv \mu$ and its autocovariance function $\Sigma(\tau) = E\{[v(t+\tau) - \mu][v(t) - \mu]\} \equiv \sigma^2 \delta(\tau)$ where $\delta(\cdot)$ denotes the Dirac impulse function.

White noise can now be used as the model of a completely unpredictable random process which, as shown in Figure 5.1, corrupts a state measurement.

Remembering that $W(t)$ is actually not differentiable, one can use a Brownian motion to describe the additive noise correctly and say that the integral of the continuous-time measurement $y(t)$ of the third state variable $x_3(t)$ is corrupted by an additive Brownian motion W:

$$\int_0^t y(\tau)\,d\tau = \int_0^t x_3(\tau)\,d\tau + W(t) .$$

One can also say that $\bar{y}(t)$ is the third state variable $x_3(t)$ averaged over ΔT and corrupted by an additive increment of a Brownian motion $W(t)$:

$$\bar{y}(t) = \frac{1}{\Delta T}\int_{t-\Delta T}^t y(\tau)\,d\tau = \frac{1}{\Delta T}\int_{t-\Delta T}^t x_3(\tau)\,d\tau + \frac{W(t) - W(t-\Delta T)}{\Delta T} .$$

As ΔT approaches zero, in the limit case the last term becomes the differential form of the Brownian motion $dW(t)$. This means that the Brownian motion $W(\cdot)$ on the time interval $[0, \infty)$ can be retrieved from the stationary white noise $v(\cdot)$ by integration:

$$W(t) = \int_0^t v(\tau)\,d\tau = \int_0^t \frac{dW(\tau)}{d\tau}\,d\tau = \int_0^t dW(\tau) .$$

Accordingly, a Brownian motion Y with the drift μ, variance σ^2, and the initial time $t = 0$ satisfies the stochastic differential equation (SDE)

$$\frac{dY(t)}{dt} = \mu + \sigma v(t) \quad \text{or}$$
$$dY(t) = \mu dt + \sigma dW(t), \quad Y(0) = 0,$$

where W is a standard Brownian motion.

This stochastic differential equation can now be generalized in several ways:

A Brownian motion is called *non-stationary Brownian motion* if the drift and/orvariance parameters are time dependent as in

$$dY(t) = \mu(t)dt + \sigma(t)dW(t) .$$

One calls it a *locally Brownian motion* if the drift and variance parameters depend on the random process itself in addition to being time dependent as in

$$dY(t) = \mu(Y(t), t)dt + \sigma(Y(t), t)dW(t) .$$

In a special case of a locally Brownian motion where both the drift $\mu(Y(t))$ and the volatility $\sigma(Y(t))$ are proportional to the value $Y(t)$ of the random process as in

$$dY(t) = \mu Y(t)dt + \sigma Y(t)dW(t),$$

it is called a *geometric Brownian motion*. This model is very useful in the area of finance and is employed later in this chapter and in Chapter 8.

Another special case is the so-called *Ornstein-Uhlenbeck process*:

$$dY(t) = \kappa (\mu - Y(t)) dt + \sigma dW(t) \quad \text{with } \kappa > 0 .$$

This mean reverting process is used to model interest rates, currency exchange rates and commodity prices stochastically. μ represents the mean value supported by market fundamentals, σ the degree of volatility around this mean value, and κ the rate by which market shocks dissipate and the variable reverts towards its mean.

5.3 Stochastic Differential Equations

An ordinary differential equation (ODE) can be written in the symbolic differential form as

$$dx(t) = \varphi(t, x)dt ,$$

or, more accurately, as an integral equation

$$x(t) = x_0 + \int_{t_0}^t \varphi(\tau, x(\tau)) d\tau , \tag{5.1}$$

and, without loss of generality, taking $t_0 = 0$ and where $x(t) = x(t, x_0, t_0)$ is the particular solution which satisfies the given initial condition $x(t_0) = x_0$. An ODE can be generalized to an SDE by simply adding noise to the system under consideration. Consider the system

$$\frac{dx(t)}{dt} = a(t)x(t), \quad x(0) = x_0, \tag{5.2}$$

where $a(t)$ is subjected to some random effects such that, for instance, $a(t) = \varphi(t) + \xi(t)v(t)$ with $v(t)$ representing a white noise process. The differential equation (5.2) can now be rewritten as

$$\frac{dX(t)}{dt} = \varphi(t)X(t) + \xi(t)X(t)v(t). \tag{5.3}$$

Note that this chapter distinguishes between random variables (stochastic processes) and deterministic variables by using capital, respectively, lower case letters.

(5.3) can be rewritten in the differential form as

$$dX(t) = \varphi(t)X(t)dt + \xi(t)X(t)dW(t),$$

where $dW(t)$ is the differential form of a standard Brownian motion $W(t)$ as defined earlier in Chapter 3, *i.e.*, $dW(t) = v(t)dt$. More generally, an SDE can be written as

$$dX(t, \omega) = f(t, X(t, \omega))dt + g(t, X(t, \omega))dW(t, \omega), \tag{5.4}$$

where $X = X(t, \omega)$ is a stochastic process with the deterministic initial condition $X(0, \omega) = X_0$. For the time being it is assumed that $f(\cdot), g(\cdot)$ are scalar functions and $W(\cdot)$ is a one dimensional process. Commonly, $f(t, X(t))$ in (5.4) is called *drift* and $g(t, X(t))$ is called *diffusion*. As in (5.1), (5.4) can be written as

$$X(t, \omega) = X_0 + \underbrace{\int_0^t f(\tau, X(\tau, \omega)) d\tau}_{I_1} + \underbrace{\int_0^t g(\tau, X(\tau, \omega)) dW(\tau, \omega)}_{I_2}. \tag{5.5}$$

I_1 is the ordinary Riemann integral but it is not clear at this stage how to interpret the stochastic integral I_2. Therefore, in the next Section, the stochastic integral concept is defined with an approach similar to the Riemann integral.

5.4   Stochastic Integration

Before beginning to calculate the stochastic integral I_2 in (5.5) let us review the Riemann integral briefly as discussed in [93].

Fig. 5.2 Illustration of a Riemann sum shows that the sum of the rectangular areas approximates the area between the graph of $f(t)$ and the t-axis which, in exact form is equal to $\int_0^{t_5} f(t)dt$ in this example

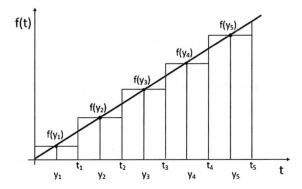

Suppose that $f(t)$ is a real valued function defined on $[0, 1]$. Let us partition this interval as

$$\tau_N \quad : \quad 0 = t_0 < t_1 < t_2 < \ldots < t_{N-1} < t_N = 1,$$

and define the lengths

$$\Delta_i = t_i - t_{i-1}, \quad i = 1, \ldots, N.$$

An intermediate partition ρ_N of τ_N is given by any values of y_i such that $t_{i-1} \leq y_i \leq t_i$ for $i = 1, \ldots, N$. For given partitions ρ_N and τ_N the Riemann sum S_N can be defined as

$$S_N = S_N(\tau_N, \rho_N) = \sum_{i=1}^{N} f(y_i)(t_i - t_{i-1}) = \sum_{i=1}^{N} f(y_i)\Delta_i. \tag{5.6}$$

One can see that a Riemann sum is an average of function values $f(y_i)$ weighted by the corresponding lengths of the intervals Δ_i. Figure 5.2 illustrates that S_N is also an approximation of the area between the graph of the function $f(t)$ and the t-axis (assuming $f(t)$ is non-negative in the interval of interest).

The bases of the rectangles in Figure 5.2 can be made smaller and smaller such that the norm of the partition tends to zero

$$\|\tau_N\| \to 0,$$

where

$$\|\tau_N\| := \max_{i=1,\ldots,N} \Delta_i = \max_{i=1,\ldots,N} (t_i - t_{i-1}).$$

If the limit

$$S = \lim_{N \to \infty} S_N = \lim_{N \to \infty} \sum_{i=1}^{N} f(y_i)\Delta_i,$$

exists as $\|\tau_N\| \to 0$ and S is independent of the choice of the partitions τ_N and their intermediate partitions ρ_N, then S is called the Riemann integral of $f(t)$ on $[0, 1]$. The Riemann integral exists if $f(t)$ is sufficiently smooth.

Example: Calculate, as an illustrative example, the integral $\int_0^1 t \, dt$ as the limit of Riemann sums as shown in Figure 5.3. $\int_0^1 t \, dt = 0.5$ and it represents the triangular area between the graph of $f(t) = t$ and the t-axis between 0 and 1. Although not strictly necessary, choosing an equidistant partition of $[0, 1]$ is convenient:

$$t_i = \frac{i}{N}, \qquad i = 0, \ldots, N.$$

First, taking the left end points of the intervals $[(i-1)/N, \ i/N]$ for the intermediate partition:

$$S_N^{left} = \sum_{i=1}^{N} \frac{i-1}{N} \Delta_i = \frac{1}{N} \sum_{i=1}^{N} \frac{i-1}{N}.$$

Remembering that $\sum_{k=1}^{n} k = \frac{n(n+1)}{2}$,

$$S_N^{left} = \frac{1}{N^2} \frac{(N-1)N}{2}.$$

Therefore, as $N \to \infty$,

$$S_N^{left} = \frac{1}{2}.$$

Analogously, taking the right end points of the intervals $[(i-1)/N, \ i/N]$ for the intermediate partition:

$$S_N^{right} = \sum_{i=1}^{N} \frac{i}{N} \Delta_i = \frac{1}{N} \sum_{i=1}^{N} \frac{i}{N}$$

$$= \frac{1}{N^2} \frac{N(N+1)}{2}.$$

Also here as $N \to \infty$

$$S_N^{right} = \frac{1}{2}.$$

Choosing as y_i the middle points of the intervals $[(i-1)/N, \ i/N]$ and denoting the corresponding Riemann sums by S_N^{middle}, since $S_N^{left} \leq S_N^{middle} \leq S_N^{right}$, it can be concluded that $\lim_{N \to \infty} S_N^{middle} = 0.5$.   □

With this refreshed understanding of the Riemann integral one can now proceed to calculate the stochastic integral I_2 in (5.5). Beginning by assuming that $g(t, \omega)$ changes only at the times t_i ($i = 1, \ldots, N-1$) as shown in Figure 5.4, with

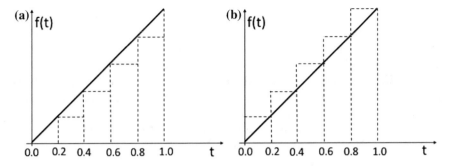

Fig. 5.3 Two Riemann sums for the integral $\int_0^1 t\,dt$. The partition is $t_i = i/5$, $i = 1, 2, 3, 4, 5$. In **a** the left end points and in **b** the right end points of the intervals $[(i-1)/5, \, i/5]$ are taken as points y_i of the intermediate partition

Fig. 5.4 To calculate the stochastic integral the function $g(t, \omega)$ is assumed to be piece-wise constant

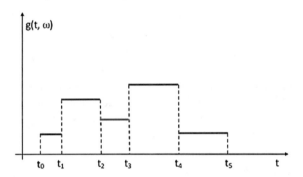

$$0 = t_0 < t_1 < t_2 < \ldots < t_{N-1} < t_N = T.$$

The stochastic integral I_2 in (5.5) $S = \int_0^T g(t, \omega)dW(t, \omega)$ can now be defined as the limit of Riemann sums for $N \to \infty$ of

$$S_N(\omega) = \sum_{i=1}^{N} g(t_{i-1}, \omega)\Big(W(t_i, \omega) - W(t_{i-1}, \omega)\Big). \tag{5.7}$$

However, this limit cannot be treated as one would treat a deterministic limit. In the case of stochastic calculus, one must apply the concept of mean-square convergence. As discussed above for Riemann integrals, regardless of the point in the interval $[t_{i-1}, t_i]$ where the value of $g(t, \omega)$ is approximated, the limit converges to the same value. However, this is not the case for stochastic integrals; the limit depends on the choice of the point at which the approximation is considered. ForItô calculus,[2] $g(t, \omega)$ is approximated at t_{i-1}. This choice has some concrete consequences for the stochastic properties of the Itô integral which are discussed later on. Because t_{i-1} is chosen—in other words the beginning of the interval—for approximating $g(t, \omega)$,

---

[2] Kiyoshi Itô, Japanese mathematician (1915–2008).

this approximation is called *non-anticipative*. An alternative to the Itô integral is the Stratonovich integral[3] which obeys the usual chain rule when performing change of variables. It can therefore be easier to use for some calculations. The Itô integral, however, is a martingale,[4] which lends it some convenient theoretical properties and many available martingale theorems for taking advantage. This comes at the cost of the chain rule in classical calculus having to be modified for the stochastic case as will become apparent in the next Section [101].

A random variable S is called the Itô integral of a stochastic process $g(t, \omega)$ with respect to the Brownian motion $W(t, \omega)$ on the interval $[0, T]$ if

$$\lim_{N \to \infty} \mathrm{E}\bigg[\underbrace{\bigg(S - \sum_{i=1}^{N} g(t_{i-1}, \omega)\Big(W(t_i, \omega) - (W(t_{i-1}, \omega)\Big)\bigg)}_{S_N}\bigg] = 0, \qquad (5.8)$$

for each sequence of partitions $(t_0, t_1, \ldots, t_N)$ of the interval $[0, T]$ such that $\|\tau_N\| \to 0$. In this case, $S = \int_0^T g(t, \omega) dW(t, \omega)$ is defined and can be computed.

The limit of S_N in (5.8) converges to the stochastic integral in the mean-square sense with a certain probability. Therefore, the stochastic integral is a random variable. Samples of this random variable depend on the individual realizations of the paths $W(., \omega)$. At time $t = 0$, the stochastic integral is deterministic with value 0; however, at time $t = T$, all realizations of W have been observed and have contributed to the random value of the integral.

Example: As an illustrative case, let us look at a degenerate stochastic process where the argument is a deterministic variable which does not vary (!) such that $g(t, \omega) = g(t) = c$. With the definition in (5.8), and with $t_0 = 0$ one gets

$$\int_0^T c \, dW(t, \omega) = c \lim_{N \to \infty} \sum_{i=1}^{N} \Big(W(t_i, \omega) - W(t_{i-1}, \omega)\Big)$$

$$= c \lim_{N \to \infty} [(W(t_1, \omega) - W(t_0, \omega)) + (W(t_2, \omega) - W(t_1, \omega)) + \ldots$$

$$+ (W(t_N, \omega) - W(t_{N-1}, \omega))$$

$$= c \, (W(T, \omega) - W(0, \omega)),$$

where $W(T, \omega)$ and $W(0, \omega)$ are random variables with standard Gaussian distributions. This is not surprising and matches what one would expect from standard calculus, *i.e.*, $c \, (W(T, \omega) - W(0, \omega))$. Assuming that the Brownian motion starts from zero, *i.e.*, $W(0, \omega) = 0$ for simplicity's sake, the last result becomes, as one would expect,

---

[3]Ruslan Stratonovich, Russian physicist (1930–1997).

[4]A martingale is a stochastic process whose conditional expectation of the next value, given the current and preceding values, is the current value. Formally, $\mathrm{E}[X_{n+1}|X_1, X_2, \ldots, X_N] = X_n \; \forall n$. For instance, a fair coin game is a martingale but a blackjack game is not.

$$\int_0^T c\, dW(t, \omega) = c\, W(T, \omega).$$

$\square$

Brownian Motion Revisited

If $W(t)$ is a Brownian motion, then $W(t) - W(0)$ is a normally distributed random variable with mean μt and variance $\sigma^2 t$. Therefore, the density function of the Brownian motion is

$$f_{W(t)}(x) = \frac{1}{\sqrt{2\pi\sigma^2 t}} e^{-\frac{(x-\mu t)^2}{2\sigma^2 t}}.$$

The sample paths of the Brownian motion are continuous but not differentiable. Given the fractal nature of Brownian motion, this is intuitively clear and can also be verified mathematically:

$$E\left[\left(\frac{W(t + \Delta t) - W(t)}{\Delta t}\right)^2\right] = \frac{E[(W(t + \Delta t) - W(t))^2]}{\Delta t^2} = \frac{\sigma^2}{\Delta t}.$$

Since this expected value diverges for $\Delta t \to 0$, $W(t)$ is not differentiable in the mean-square sense. Further,

- $W(0) = 0$,
- $E[W(t)] = \mu$,
- $E[W^2(t)] = \sigma^2 t$,
- $E[dW(t)] = 0$,
- $E[(W(t) - \mu t)(W(s) - \mu s)] = \sigma^2 \min(t, s)$,
- The total variation of a Brownian motion over a finite interval $[0, T]$ is infinite,
- The "sum of squares" of a Brownian motion is deterministic:

$$\lim_{N \to \infty} \sum_{i=1}^{N} \left(W(k\frac{T}{N}) - W((k-1)\frac{T}{N})\right)^2 = \sigma^2 T,$$

$$\Rightarrow \quad dW^2 = \sigma^2 dt.$$

As will become apparent soon, this last property is very useful and applied frequently in this Book.

Let $g(t, \omega)$ now be a 'truly random' function which can be approximated by a random step function

$$g(t, \omega) \approx g(t_{i-1}, \omega), \quad t \in [t_{i-1}, t_i].$$

Example: Let $g(t, \omega) = W(t, \omega)$, with $W(\cdot)$ denoting the standard Brownian motion ($\mu = 0, \sigma = 1$), so that the integrand itself is a random variable. With standard calculus one would expect $\int_0^T W(t, \omega) \, dW(t, \omega)$ to be equal to $\frac{1}{2} W^2(T, \omega)$. But, is it? Let us do the calculation step by step to see the result.

Here the Riemann sum in (5.7) is used and the definition in (5.8) is applied. With the help of the nifty algebraic identity $y(x - y) = yx - y^2 + \frac{1}{2}x^2 - \frac{1}{2}x^2 = \frac{1}{2}x^2 - \frac{1}{2}y^2 - \frac{1}{2}(x - y)^2$ one has,

$$
\int_0^T W(t, \omega) \, dW(t, \omega) = \lim_{N \to \infty} \sum_{i=1}^N W(t_{i-1}, \omega) \Big(W(t_i, \omega) - W(t_{i-1}, \omega) \Big)
$$

$$
= \underbrace{\lim_{N \to \infty} \left[\frac{1}{2} \sum_{i=1}^N \Big(W^2(t_i, \omega) - W^2(t_{i-1}, \omega) \Big) \right]}_{A}
$$

$$
\underbrace{- \frac{1}{2} \lim_{N \to \infty} \left[\sum_{i=1}^N \Big(W(t_i, \omega) - W(t_{i-1}, \omega) \Big)^2 \right]}_{B}
$$

$$
= \underbrace{- \frac{1}{2} \lim_{N \to \infty} \sum_{i=1}^N \Big(W(t_i, \omega) - W(t_{i-1}, \omega) \Big)^2}_{C}
$$

$$
\underbrace{+ \frac{1}{2} W^2(T, \omega)}_{D} \, .
$$

because, $A = D$ as seen in the example with $g(t) = c$ above and $B = C$.

Furthermore, due to the sum-of-squares property of the normalized Brownian motion W, C is deterministic and equal to $-\frac{1}{2}T$, because, for normalized or standard Brownian motion $\sigma = 1$. Thus, finally

$$
\int_0^T W(t, \omega) \, dW(t, \omega) = \frac{1}{2} W^2(T, \omega) - \frac{1}{2} T \, . \tag{5.9}
$$

The Itô integral differs by the term $-\frac{1}{2}T$ from what one would expect to get with standard calculus! This example shows that the rules of integration need to be re-formulated for stochastic calculus. Although this might seem annoying at first glance, the very martingale property of the Itô integral, as opposed to Stratonovich integral, makes it more suitable for dealing with most stochastic process problems [101].   □

5.5 Properties of Itô Integrals

Let us begin by studying the statistical properties of Itô integrals, namely the mean and the variance of the stochastic integral.

$$
\mathrm{E}\left[\int_0^T g(t,\omega)dW(t,\omega)\right] = \mathrm{E}\left[\lim_{N\to\infty}\sum_{i=1}^N g(t_{i-1},\omega)\Big(W(t_i,\omega) - W(t_{i-1},\omega)\Big)\right]
$$

$$
= \lim_{N\to\infty}\sum_{i=1}^N \mathrm{E}\big[g(t_{i-1},\omega)\big]\,\mathrm{E}\left[\Big(W(t_i,\omega) - W(t_{i-1},\omega)\Big)\right]
$$

$$
= 0\,,
$$

where $W(\cdot)$ denotes a Brownian motion which is statistically independent of $g(\cdot)$. The *expected value* of stochastic integrals is, as one would expect, zero.

$$
\mathrm{Var}\left[\int_0^T g(t,\omega)\,dW(t,\omega)\right] = \mathrm{E}\left[(\int_0^T g(t,\omega)\,dW(t,\omega))^2\right]
$$

$$
= \mathrm{E}\left[\Big(\lim_{N\to\infty}\sum_{i=1}^N g(t_{i-1},\omega)\Big(W(t_i,\omega) - W(t_{i-1},\omega)\Big)\Big)^2\right]
$$

$$
= \lim_{N\to\infty}\sum_{i=1}^N\sum_{j=1}^N \mathrm{E}[g(t_{i-1},\omega)g(t_{j-1},\omega)\cdot
$$

$$
(W(t_i,\omega) - W(t_{i-1},\omega))(W(t_j,\omega) - W(t_{j-1},\omega))]
$$

$$
= \lim_{N\to\infty}\sum_{i=1}^N \mathrm{E}[g^2(t_{i-1},\omega)]\,\mathrm{E}[\Big(W(t_i,\omega) - W(t_{i-1},\omega)\Big)^2]
$$

$$
= \lim_{N\to\infty}\sum_{i=1}^N \mathrm{E}[g^2(t_{i-1},\omega)]\,(t_i - t_{i-1})
$$

$$
= \int_0^T \mathrm{E}[g^2(t,\omega)]\,dt\,.
$$

In other words, the *variance* of stochastic integrals is the definite integral of the expected value of the integrand's square, also as one would expect.

The question whether stochastic integrals are always well defined and solvable has not yet been addressed. Therefore, one has to find the conditions for stochastic integrals to be well defined and computable.[5] A natural requirement for $g(t,\omega)$ in connection with causality is that it does not depend on future values of the Brownian motion $W(t,\omega)$. This means that at any time t, the stochastic variable $g(t,\omega)$ depends only on the earlier values of the stochastic variable $\{W(t-h,\omega) \mid h \geq 0\}$ or other variables which are independent of the Brownian motion. This is the reason why the integrand of the Itô integral is approximated in a non-anticipating manner by

---

[5] For an exact discussion of the conditions for the stochastic integral to be well defined see [33].

$g(t_{i-1}, \omega)$. Another requirement is that, thevariance of the stochastic integral should be bounded. In other words, $g(t, \omega)$ should only attain "large values" with "low probability". Boundedness means that there exists a c such that

$$P(|g(t, \omega)| > c) < \epsilon \qquad \forall \epsilon, \forall t .$$

Consequently

$$\int_0^T E[g^2(t, \omega)]dt < \infty \qquad \forall T.$$

A further property of the stochastic integral is its *linearity, i.e.,*

$$\int_0^T [a_1\, g_1(t, \omega) + a_2\, g_2(t, \omega)]dW(t, \omega)$$

$$= a_1 \int_0^T g_1(t, \omega)dW(t, \omega) + a_2 \int_0^T g_2(t, \omega)dW(t, \omega) ,$$

must hold for all real numbers a_1, a_2 and all stochastically bounded functions $g_1(t, \omega)$, $g_2(t, \omega)$.

5.6   Itô Calculus

Due to the sum-of-squares property of the Brownian motion, the rules of differentiation in the stochastic case differ from those in the deterministic case. The problem can be stated as follows: Given a stochastic differential equation for the process $X(t, \omega)$[6]

$$dX(t) = f(t, X(t))dt + g(t, X(t))dW(t) , \qquad X(t_0) = X_0, \qquad (5.10)$$

and another process $Y(t)$ which is a function of $X(t)$,

$$Y(t) = \phi(t, X(t)) ,$$

where the function $\phi(t, X(t))$ is continuously differentiable in t and twice continuously differentiable in X, find the stochastic differential equation for the process $Y(t)$:

$$dY(t) = \tilde{f}(t, X(t))dt + \tilde{g}(t, X(t))dW(t) .$$

---

[6]$X(t, \omega)$ is a random process, but for brevity $X(t)$ or just X is used.

So what is $\tilde{f}(\cdot)$ and $\tilde{g}(\cdot)$? In the case where $g(t, X(t)) = 0$, the answer is the chain rule for standard calculus given by[7]

$$dY(t) = \left(\phi_t(t, X) + \phi_x(t, X)f(t, X)\right)dt \,.$$

To calculate the answer for the general case, first calculate the Taylor expansion of $\phi(t, X(t))$ up to second order terms:

$$dY(t) = \phi_t(t, X)dt + \phi_x(t, X)dX(t) + \frac{1}{2}\phi_{tt}(t, X)dt^2$$
$$+ \frac{1}{2}\phi_{xx}(t, X)(dX(t))^2 + \phi_{xt}(t, X)dX(t)dt + \text{higher order terms}\,.$$

Substituting (5.10) for $dX(t)$ in (5.11) results in

$$dY(t) = \phi_t(t, X)dt + \phi_x(t, X)[f(t, X(t))dt + g(t, X(t))dW(t)]$$
$$+ \frac{1}{2}\phi_{tt}(t, X)dt^2 + \frac{1}{2}\phi_{xx}(t, X)[f(t, X(t))dt + g(t, X(t))dW(t)]^2$$
$$+ \phi_{xt}(t, X)[f(t, X(t))dt + g(t, X(t))dW(t)]dt + \text{higher order terms}\,.$$

The second order differentials are much smaller than the first order differentials dt and dW. Therefore dt^2 and $dt\,dW(t)$ can be neglected just like the higher order terms. But, due to the sum-of-squares property of the Brownian motion W, $dW^2(t) = dt$ and cannot be neglected. Thus, neglecting higher order terms, results in

$$dY(t) = \phi_t(t, X)dt + \phi_x(t, X)\left[f(t, X(t))dt + g(t, X(t))dW(t)\right] + \frac{1}{2}\phi_{xx}(t, X)g^2(t, X(t))dt$$
$$= \left[\phi_t(t, X) + \phi_x(t, X)f(t, X(t)) + \frac{1}{2}\phi_{xx}(t, X)g^2(t, X(t))\right]dt + \phi_x(t, X)g(t, X(t))dW(t)\,.$$

To recapitulate: let $\phi(t, X(t))$ be a suitably differentiable function. For the stochastic process defined by the stochastic differential equation

$$dX(t) = f(t, X(t))dt + g(t, X(t))dW(t)\,,$$

the transformed stochastic process $Y(t, X(t)) = \phi(t, X(t))$ satisfies the stochastic differential equation

$$dY(t) = \tilde{f}(t, X(t))dt + \tilde{g}(t, X(t))dW(t)\,, \qquad (5.11)$$

where

---

[7] ϕ_x is used for $\frac{\partial \phi}{\partial X}$, ϕ_{xx} for $\frac{\partial^2 \phi}{\partial X^2}$ and ϕ_{xt} for $\frac{\partial^2 \phi}{\partial X \partial t}$.

$$\tilde{f}(t, X(t)) = \phi_t(t, X) + \phi_x(t, X)f(t, X(t)) + \frac{1}{2}\phi_{xx}(t, X)g^2(t, X(t))$$

$$\tilde{g}(t, X(t)) = \phi_x(t, X)g(t, X(t)) .$$
(5.12)

This important result is known as *Itô's lemma* and it is central to financial applications as shown in Chapter 9.

The term $\frac{1}{2}\phi_{xx}(t, X)g^2(t, X(t))$ which appears in the expression for $\tilde{f}(t, X(t))$ is sometimes called the Itô correction term since it is absent in the deterministic case. This term vanishes if ϕ is linear in $X(\cdot)$.

Note that the SDE of the process $Y(t)$ can also be written as

$$dY(t) = \phi_t(t, X)dt + \phi_x(t, X)dX + \frac{1}{2}\phi_{xx}(t, X)dX^2 ,$$
(5.13)

where dX^2 is computed according to the rules $dt^2 = 0$, $dt \cdot dW = 0$, and $dW^2 = dt$.

Example: Given the SDE $dX(t) = dW(t)$ for the process $X(\cdot)$ and another process $Y(t) = \phi(t, X(t)) = X^2(t)$. From the SDE it can be concluded that $X(t) = W(t)$. The required partial derivatives are: $\phi_x(t, X) = 2X$, $\phi_{xx}(t, X) = 2$, and $\phi_t(t, X) = 0$. Applying Itô's lemma

$$\underbrace{\phi_t dt}_{0} + \underbrace{\phi_x dX}_{2X dX} + \underbrace{\frac{1}{2}\phi_{xx} dX^2}_{\frac{1}{2} \cdot 2 \cdot dX^2} .$$

Substituting W for X makes the last term dW^2 and because $dW^2 = dt$ yields

$$d(W^2(t)) = 2W(t)dW(t) + dt.$$

Rewriting the equation in integral form with $W(0) = 0$ yields

$$W^2(t) = 2\int_0^t W(\tau)dW(\tau) + t ,$$

$$\text{or } \int_0^t W(\tau)dW(\tau) = \frac{1}{2}W^2(t) - \frac{1}{2}t .$$

□

This is the result already obtained earlier by solving the stochastic integral in (5.9) which made us conclude that the intuition based on standard calculus does not work. This conclusion is now confirmed by the Itô correction term.

In the more general case where the stochastic process $\mathbf{X}(t)$ and the standard Brownian motion $\mathbf{W}(t)$ are n and m dimensional vectors respectively[8] and therefore, $\mathbf{f}(\cdot)$ has the dimension n, $\mathbf{G}(\cdot)$ has the dimensions $n \times m$ and the function $\phi(\cdot)$ is

---

[8]The components of the vector $\mathbf{W}(\cdot)$, $W_i(\cdot)$ are independent scalar Brownian motions.

scalar-valued, the generalized form of Itô's lemma becomes

$$dY(t) = \tilde{f}(t, X(t))dt + \tilde{G}(t, X(t))dW(t),$$

with

$$\tilde{f}(t, X(t)) = \phi_t(t, X(t)) + \phi_x(t, X(t))f(t, X(t)) + \frac{1}{2}\text{tr}\left(\phi_{xx}(t, X(t))G(t, X(t))G^T(t, X(t))\right)$$

$$= \phi_t(t, X(t)) + \phi_x(t, X(t))f(t, X(t)) + \frac{1}{2}\text{tr}\left(G^T(t, X(t))\phi_{xx}(t, X(t))G(t, X(t))\right),$$

and

$$\tilde{G}(t, X(t)) = \phi_x(t, X(t))G(t, X(t)),$$

where "tr" denotes the trace operator. $\tilde{f}(\cdot)$ is an n dimensional vector and $\tilde{G}(\cdot)$ is an $n \times m$ dimensional matrix. Let us see how to use Itô's formula with the help of several classical examples.

Example: Consider the SDE of a geometric Brownian motion:

$$dS(t) = \mu S(t)dt + \sigma S(t)dW(t).$$

What is the SDE for the process Y which is the logarithm of S?

$$Y(t) = \phi(t, S(t)) = \log(S(t)). \tag{5.14}$$

So, $f = \mu S$, $g = \sigma S$. The relevant partial derivatives are $\phi_S(t, S) = \frac{1}{S}$, $\phi_{SS}(t, S) = -\frac{1}{S^2}$, and $\phi_t(t, S) = 0$. Therefore,

$$\tilde{f} = 0 + \frac{1}{S} \cdot \mu \cdot S - \frac{1}{2} \cdot \frac{1}{S^2} \cdot \sigma^2 S^2, \quad \tilde{g} = \frac{1}{S}\sigma S$$

$$\Rightarrow \quad dY(t) = (\mu - \frac{1}{2}\sigma^2)dt + \sigma dW(t). \tag{5.15}$$

$Y(\cdot)$ is a non-standard Brownian motion with

$$E[Y(t)] = (\mu - \frac{1}{2}\sigma^2)t \quad \text{and}$$
$$\text{Var}[Y(t)] = \sigma^2 t.$$

So, for the sample path of $Y(\cdot)$, one has the following closed-form solution:

$$Y(t) = Y_0 + (\mu - \frac{1}{2}\sigma^2)t + \sigma W(t).$$

Hence, for the sample path of the geometric Brownian motion $S(\cdot) = e^{Y(\cdot)}$, one gets the closed-form solution

$$S(t) = e^{Y_0} e^{(\mu - \frac{1}{2}\sigma^2)t + \sigma W(t)} . \tag{5.16}$$

The geometric Brownian motion is a popular stock price model and (5.16) gives the exact solution for this price process. This result will form the foundation for Modern Portfolio Theory in Chapter 8.   □

Example: Suppose that two processes $X_1(t)$ and $X_2(t)$ are given by the coupled[9] SDE's

$$\begin{bmatrix} dX_1(t) \\ dX_2(t) \end{bmatrix} = \begin{bmatrix} f_1(t, \mathbf{X}(t)) \\ f_2(t, \mathbf{X}(t)) \end{bmatrix} dt + \begin{bmatrix} g_1(t, \mathbf{X}(t)) & 0 \\ 0 & g_2(t, \mathbf{X}(t)) \end{bmatrix} \begin{bmatrix} dW_1(t) \\ dW_2(t) \end{bmatrix},$$

where $W_1(t)$ and $W_2(t)$ are two independent Brownian motions. Let us now compute a stochastic differential equation for the product $Y = X_1 X_2$. Thus, $\phi(t, X_1, X_2) = X_1 X_2$. An interpretation of ϕ could be that X_1 describes the evolution of an American stock in US dollars, whereas X_2 describes the evolution of the exchange rate CHF/USD and thus ϕ describes the evolution of the price of the American stock measured in Swiss Francs. Let us now apply the multivariate form of the Itô calculus. The partial derivatives are

$$\phi_t(t, X_1, X_2) = 0$$
$$\phi_{x_1}(t, X_1, X_2) = X_2$$
$$\phi_{x_2}(t, X_1, X_2) = X_1$$
$$\phi_{x_1 x_1}(t, X_1, X_2) = \phi_{x_2 x_2}(t, X_1, X_2) = 0$$
$$\phi_{x_1 x_2}(t, X_1, X_2) = \phi_{x_2 x_1}(t, X_1, X_2) = 1 .$$

Having computed the elements of the Jacobian and the Hessian, one can calculate the SDE for $Y(t)$:

$$\tilde{f}(t, \mathbf{X}(t)) = \phi_t(t, \mathbf{X}) + \phi_{\mathbf{x}}(t, \mathbf{X}) f(t, \mathbf{X}(t)) + \frac{1}{2}\mathrm{tr}\left(\mathbf{G}^T(t, \mathbf{X}(t))\phi_{\mathbf{xx}}(t, \mathbf{X})\mathbf{G}(t, \mathbf{X}(t))\right) .$$

Leaving out the arguments of the functions for simplicity's sake

---

[9]These equations are coupled, because $\mathbf{X} = [X_1 \ X_2]^T$, hence $X_1 = f_1(t, X_1, X_2), X_2 = f_2(t, X_1, X_2)$.

$$\tilde{f} = 0 + [\phi_{x_1} \quad \phi_{x_2}] \begin{bmatrix} f_1 \\ f_2 \end{bmatrix} + \frac{1}{2} \text{tr} \left(\begin{bmatrix} g_1 & 0 \\ 0 & g_2 \end{bmatrix} \begin{bmatrix} \phi_{x_1 x_1} & \phi_{x_1 x_2} \\ \phi_{x_2 x_1} & \phi_{x_2 x_2} \end{bmatrix} \begin{bmatrix} g_1 & 0 \\ 0 & g_2 \end{bmatrix} \right)$$

$$= 0 + \phi_{x_1} f_1 + \phi_{x_2} f_2 + \frac{1}{2} \text{tr} \left(\begin{bmatrix} g_1 & 0 \\ 0 & g_2 \end{bmatrix} \begin{bmatrix} 0 & 1 \\ 1 & 0 \end{bmatrix} \begin{bmatrix} g_1 & 0 \\ 0 & g_2 \end{bmatrix} \right)$$

$$= 0 + X_2 f_1 + X_1 f_2 + \frac{1}{2} \text{tr} \left(\begin{bmatrix} g_1 & 0 \\ 0 & g_2 \end{bmatrix} \begin{bmatrix} 0 & g_2 \\ g_1 & 0 \end{bmatrix} \right)$$

$$= X_2 f_1 + X_1 f_2 + \frac{1}{2} \text{tr} \underbrace{\begin{bmatrix} 0 & g_1 g_2 \\ g_2 g_1 & 0 \end{bmatrix}}_{0}$$

$$= X_2(t) f_1(t, \mathbf{X}(t)) + X_1(t) f_2(t, \mathbf{X}(t)) .$$

$$\tilde{\mathbf{G}} = \phi_{\mathbf{x}} \mathbf{G}$$

$$= [\phi_{x_1} \quad \phi_{x_2}] \begin{bmatrix} g_1 & 0 \\ 0 & g_2 \end{bmatrix}$$

$$= [\phi_{x_1} g_1 \quad \phi_{x_2} g_2]$$

$$= [X_2 g_1 \quad X_1 g_2] .$$

$$dY(t) = X_2 f_1 dt + X_1 f_2 dt + X_2 g_1 dW_1 + X_1 g_2 dW_2$$
$$= X_2 \underbrace{(f_1 dt + g_1 dW_1)}_{dX_1} + X_1 \underbrace{(f_2 dt + g_2 dW_2)}_{dX_2}$$
$$= X_1(t) dX_2(t) + X_2(t) dX_1(t) .$$

This SDE for $Y(t)$ shows that the investor faces two sources of uncertainties: the first from the uncertainty of the American stock and the second from the uncertainty of the USD/CHF exchange rate. The risk is represented by the two Brownian motions, which drive the uncertainty of the SDE for $Y(t)$. □

Example: This time two processes $X_1(t)$ and $X_2(t)$ which are given by the *uncoupled* SDE's are considered

$$\begin{bmatrix} dX_1(t) \\ dX_2(t) \end{bmatrix} = \begin{bmatrix} f_1(t, X_1(t)) \\ f_2(t, X_2(t)) \end{bmatrix} dt + \begin{bmatrix} g_1(t, X_1(t)) \\ g_2(t, X_2(t)) \end{bmatrix} dW(t) ,$$

Or in vector form,

$$d\mathbf{X} = \mathbf{f} dt + \mathbf{g} dW ,$$

with

$$\mathbf{f} = \begin{bmatrix} f_1 \\ f_2 \end{bmatrix}, \quad \mathbf{g} = \begin{bmatrix} g_1 \\ g_2 \end{bmatrix} .$$

To compute a stochastic differential for the product $Y = X_1 X_2$ ($\phi(t, X_1, X_2) = X_1 X_2$) one needs to apply the multivariate form of the Itô calculus. The partial derivatives are

$$\phi_t(t, X_1, X_2) = 0$$
$$\phi_{x_1}(t, X_1, X_2) = X_2$$
$$\phi_{x_2}(t, X_1, X_2) = X_1$$
$$\phi_{x_1 x_1}(t, X_1, X_2) = \phi_{x_2 x_2}(t, X_1, X_2) = 0$$
$$\phi_{x_1 x_2}(t, X_1, X_2) = \phi_{x_2 x_1}(t, X_1, X_2) = 1 .$$

Again leaving out the arguments of the functions for simplicity's sake

$$\tilde{f} = \phi_t + \phi_{\mathbf{x}}\mathbf{f} + \frac{1}{2}\text{tr}\left(\mathbf{g}^T \phi_{\mathbf{xx}}\mathbf{g}\right)$$

$$= 0 + \phi_{x_1} f_1 + \phi_{x_2} f_2 + +\frac{1}{2}\text{tr}\left([g_1 \quad g_2]\begin{bmatrix}0 & 1\\1 & 0\end{bmatrix}\begin{bmatrix}g_1\\g_2\end{bmatrix}\right)$$

$$= X_2 f_1 + X_1 f_2 + \frac{1}{2}\text{tr}\left([g_1 \quad g_2]\begin{bmatrix}g_2\\g_1\end{bmatrix}\right)$$

$$= X_2 f_1 + X_1 f_2 + \frac{1}{2}\text{tr}[g_1 g_2 + g_1 g_2]$$

$$= X_2 f_1 + X_1 f_2 + g_1 g_2$$

$$\tilde{g} = \phi_{\mathbf{x}}\mathbf{g}$$

$$= [\phi_{x_1} g_1 + \phi_{x_2} g_2]$$

$$= [X_2 g_1 + X_1 g_2]$$

$$dY = X_2 f_1 dt + X_1 f_2 dt + g_1 g_2 dt + X_2 g_1 dW + X_1 g_2 dW$$

$$= X_2 \underbrace{(f_1 + g_1 dW)}_{dX_1} + X_1 \underbrace{(f_2 + g_1 dW)}_{dX_2} + g_1 g_2 dt$$

or,

$$dY(t) = X_1(t)dX_2(t) + X_2(t)dX_1(t) + g_1(t, X_1(t))g_2(t, X_2(t))dt .$$

$\square$

5.7   Solving Stochastic Differential Equations

One distinguishes also here between non-linear SDE's and linear SDE's and further between scalar linear and vector-valued linear SDE's. Examples from finance and engineering for these different classes of SDE's are discussed in this Section.

5.7.1 Linear Scalar SDE's

In the case of linear SDE's, one can derive exact solutions and compute moments [101]. A stochastic differential equation

$$dX(t) = f(t, X(t))dt + \mathbf{g}^T(t, X(t))d\mathbf{W}(t), \qquad X(0) = X_0$$

for a one-dimensional stochastic process $X(t)$ is called a linear (scalar) SDE if and only if the functions $f(t, X(t))$ and $g(t, X(t))$ are affine functions[10] of $X(t)$. Therefore,

$$f(t, X(t)) = A(t)X(t) + a(t),$$
$$g(t, X(t)) = [B_1(t)X(t) + b_1(t), \cdots, B_m(t)X(t) + b_m(t)],$$

where $A(t), a(t), B_i(t), b_i(t)$ are scalar valued functions and $\mathbf{W}(t)$ is an m-dimensional Brownian motion. Hence, $f(t, X(t))$ is a scalar and $\mathbf{g}^T(t, X(t))$ is an m-dimensional vector.

A scalar linear SDE can also be written in the form

$$dX(t) = (A(t)X(t) + a(t))dt + \sum_{i=1}^{m}(B_i(t)X(t) + b_i(t))dW_i(t). \qquad (5.17)$$

This linear SDE has the solution

$$X(t) = \Phi(t)\left(X_0 + \int_0^t \Phi^{-1}(s)\left[a(s) - \sum_{i=1}^{m} B_i(s)b_i(s)\right]ds \right.$$
$$\left. + \sum_{i=1}^{m}\int_0^t \Phi^{-1}(s)b_i(s)\,dW_i(s)\right), \qquad (5.18)$$

where the fundamental matrix

$$\Phi(t) = \exp\left(\int_0^t \left[A(\tau) - \sum_{i=1}^{m}\frac{B_i^2(\tau)}{2}\right]d\tau + \sum_{i=1}^{m}\int_0^t B_i(\tau)dW_i(\tau)\right), \qquad (5.19)$$

is the solution of the SDE:

---

[10]An affine mapping is the composition of translation and a stretching. Vector algebra uses matrix multiplication to represent linear mappings (stretching) and vector addition to represent translations. Formally, in the finite-dimensional case, if the linear mapping is represented as a multiplication by a matrix $\mathbf{A}$ and the translation as the addition of a vector $\mathbf{b}$ an affine mapping $\mathbf{f}$ acting on a vector $\mathbf{x}$ can be represented as $\mathbf{f}(\mathbf{x}) = \mathbf{Ax} + \mathbf{b}$.

$$d\Phi(t) = A(t)\Phi(t)dt + \sum_{i=1}^{m} B_i(t)\Phi(t)dW_i(t), \qquad (5.20)$$

with the initial condition $\Phi(0) = 1$. For a proof the reader is referred to Chapter 8 of [6].

Example: Let us assume that $W(t)$ is scalar, $a(t) = 0$, $b(t) = 0$, $A(t) = A$, $B(t) = B$. Under these simplifying assumptions the SDE

$$dX(t) = AX(t)dt + BX(t)dW(t)$$
$$X(0) = X_0 .$$

can be solved using (5.18) and (5.19) as

$$X(t) = X_0 e^{(A - \frac{1}{2}B^2)t + BW(t)},$$

as already calculated in (5.16).   □

For a random process $X(t)$ given by a scalar linear SDE's as in (5.17), the expected value $m(t) = E[X(t)]$ and $P(t) = E[X^2(t)]$ can be calculated by solving the following system of ODE's:

$$\dot{m}(t) = A(t)m(t) + a(t)$$
$$m(0) = X_0$$
$$\dot{P}(t) = \left(2A(t) + \sum_{i=1}^{m} B_i^2(t)\right)P(t) + 2m(t)\left(a(t) + \sum_{i=1}^{m} B_i(t)b_i(t)\right) + \sum_{i=1}^{m} b_i^2(t)$$
$$P(0) = X_0^2 .$$

The ODE for the expectation is derived by applying the expectation operator on both sides of (5.17). Using the rules for the expectation operator yields

$$E[dX(t)] = E[(A(t)X(t) + a(t))dt + \sum_{i=1}^{m}(B_i(t)X(t) + b_i(t))dW_i(t)]$$

$$\underbrace{E[dX(t)]}_{dm(t)} = (A(t)\underbrace{E[X(t)]}_{m(t)} + a(t))dt + \sum_{i=1}^{m} E[(B_i(t)X(t) + b_i(t))]\underbrace{E[dW_i(t)]}_{0}$$
$$dm(t) = (A(t)m(t) + a(t))dt .$$

To compute the second moment, first the SDE for $Y(t) = X^2(t)$ is derived with Itô's lemma, where $X(t)$ is governed by (5.17):

$$dY(t) = \left[2X(t)(A(t)X(t) + a(t)) + \sum_{i=1}^{m}\left(B_i(t)X(t) + b_i(t)\right)^2\right]dt$$

$$+2X(t)\sum_{i=1}^{m}\left(B_i(t)X(t) + b_i(t)\right)dW_i(t)$$

$$= \left[2A(t)X^2(t) + 2X(t)a(t) + \sum_{i=1}^{m}\left(B_i^2(t)X^2(t) + 2B_i(t)b_i(t)X(t)\right.\right.$$

$$\left.\left.+b_i^2(t)\right)\right]dt + 2X(t)\sum_{i=1}^{m}\left(B_i(t)X(t) + b_i(t)\right)dW_i(t) . \qquad (5.21)$$

Then, the expected value of (5.21) is calculated and the substitutions $P(t) = E[X^2(t)] = E[Y(t)]$ and $m(t) = E[X(t)]$ employed. This yields

$$\underbrace{E[dY(t)]}_{dP(t)} = \left[2A(t)\underbrace{E[X^2(t)]}_{P(t)} + 2a(t)\underbrace{E[X(t)]}_{m(t)} + \sum_{i=1}^{m}\left(B_i^2(t)\underbrace{E[X^2(t)]}_{P(t)}\right.\right.$$

$$\left.\left.+2B_i(t)b_i(t)\underbrace{E[X(t)]}_{m(t)} + b_i^2(t)\right)\right]dt$$

$$+E\left[2X(t)\sum_{i=1}^{m}\left(B_i(t)X(t) + b_i(t)\right)\right]\underbrace{E\left[dW_i(t)\right]}_{0}$$

$$dP(t) = \left[2A(t)P(t) + 2a(t)m(t) + \sum_{i=1}^{m}\left(B_i^2(t)P(t) + 2B_i(t)b_i(t)m(t) + b_i^2(t)\right)\right]dt .$$

In the special case of (5.17) where $B_i(t) = 0$, $i = 1, \ldots, m$, one can explicitly give the exact probability density function of $P(t)$: The solution of this scalar linear SDE

$$dX(t) = (A(t)X(t) + a(t))dt + \sum_{i=1}^{m}b_i(t)dW_i(t) , \qquad (5.22)$$

with the initial condition $X(0) = X_0$ is normally distributed:

$$P(X(t)) \sim \mathcal{N}(m(t), V(t)) . \qquad (5.23)$$

The mean $m(t)$ and the variance $V(t) = P(t) - m^2(t)$ are the solutions of the following ODE's:

$$\dot{m}(t) = A(t)m(t) + a(t)$$
$$m(0) = X_0$$
$$\dot{V}(t) = 2A(t)V(t) + \sum_{i=1}^{m} b_i^2(t)$$
$$V(0) = 0.$$

Let us now look at some specific scalar linear SDE's which are commonly used in practice and see what types of asset prices they could represent. Further examples of price processes can be found in [99, 101].

- Brownian motion:
  The simplest case of a stochastic differential equation is where the drift and the diffusion coefficients are independent of the information received over time

$$dS(t) = \mu dt + \sigma dW(t, \omega), \quad S(0) = S_0.$$

This model is sometimes used to simulate commodity prices. The mean is $E[S(t)] = \mu t + S_0$ and the variance $\mathrm{Var}[S(t)] = \sigma^2 t$. $S(t)$ fluctuates around the straight line $S_0 + \mu t$. For each instant, the process is normally distributed with the given mean and variance. The parameter σ is also called volatility. The volatility is used in many asset models as an indicator of the risk an investor is taking by buying a certain asset. A typical realization is shown in Figure 5.5. One can see that the variance grows with time. A serious limitation for using this process for modeling any prices is that the process might attain negative values.
- Geometric Brownian motion:
  The geometric Brownian motion

$$dS(t) = \mu S(t)dt + \sigma S(t)dW(t, \omega), \quad S(0) = S_0,$$

is the standard model used for (logarithmic) stock prices as discussed in detail in Chapter 8, Modern Portfolio Theory. Its mean is $E[S(t)] = S_0 e^{\mu t}$ and its variance is $\mathrm{Var}[S(t)] = S_0^2 e^{2\mu t}(e^{\sigma^2 t} - 1)$. A typical realization is shown in Figure 5.6. This model is also the starting point for the Black-Scholes formula for option pricing as shown in Chapter 9 (Derivative Instruments). The geometric Brownian motion is very popular for stock price models, because all returns are in scale with the current price and $S(t) > 0$ for all $t \in [0, T]$. This process has a log-normal probability density function. For $\mu > 0$, the moments of the geometric Brownian motion become infinite.
- Mean reverting process:
  Another very popular class of SDE's are mean reverting linear SDE's which is given by

$$dS(t) = \kappa[\mu - S(t)]dt + \sigma dW(t), \quad S(0) = S_0, \tag{5.24}$$

Fig. 5.5 A possible realization of a Brownian motion described by the SDE $dS(t) = \mu dt + \sigma dW(t, \omega)$ with $S_0 = 300, \mu = 1, \sigma = 1.07$

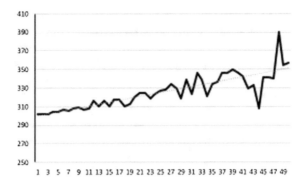

and used to model, for instance, interest rates (see Section 8.7 on Bond Portfolios). Processes modeled by this type of SDE naturally fall back to their equilibrium levels indicated by μ. The parameter $\kappa > 0$ determines how fast the process moves back to μ. When the price $S(t)$ is above μ, then $\kappa(\mu - S(t)) < 0$ and the probability that $S(t)$ decreases is high and when $S(t)$ is below μ, then $\kappa(\mu - S(t)) > 0$ and the probability is high that $S(t)$ increases. The expected price is $E[S(t)] = \mu - (\mu - S_0)e^{-\kappa t}$ and the variance is $\text{Var}[S(t)] = \frac{\sigma^2}{2\kappa}\left(1 - e^{-2\kappa t}\right)$. In the long run, following approximations are valid $\lim_{t \to \infty} E[S(t)] = \mu$ and $\lim_{t \to \infty} \text{Var}[S(t)] = \frac{\sigma^2}{2\kappa}$. This means that the process fluctuates around μ and has a variance of $\frac{\sigma^2}{2\kappa}$. This process is in the limit case a stationary process which is normally distributed. There are many variations of the mean reverting process. A popular extension is where the diffusion term is in scale with the current value, *i.e.*, the geometric mean reverting process

$$dS(t) = \kappa[\mu - S(t)]dt + \sigma S(t)dW(t, \omega), \quad S(0) = S_0.$$

- Engineering model:
  The most important scalar case for control engineers is

$$dX(t) = (a(t)X(t) + c(t)u(t)) dt + \sum_{i=1}^{m} b_i(t) dW_i, \qquad (5.25)$$

where $u(t)$ is the control variable. $X(t)$ is normally distributed, because the Brownian motion is just multiplied by time-dependent factors. The certainty equivalence principle which is a special case of the separation principle mentioned in Chapter 4, states that the optimal control law for a system governed by this SDE is identical with the optimal control law for the deterministic case where the Brownian motion is absent [62]. Therefore, the stochastic nature of the systems to be controlled is often ignored in control engineering as derived in the previous chapter [3].

Fig. 5.6 A possible realization of a geometric Brownian motion described by the SDE $dS(t) = \mu S(t)dt + \sigma S(t)dW(t, \omega)$ with $S_0 = 0.5$, $\mu = 0.04$, $\sigma = 0.05$

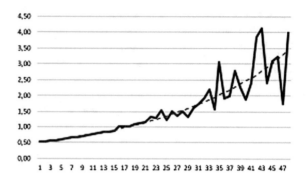

5.7.2   Vector-Valued Linear SDE's

In this subsection the results of the previous subsection are simply extended to the case where the random process $\mathbf{X}(t)$ is an n-dimensional vector.

A stochastic vector differential equation

$$dX(t) = \mathbf{f}(t, \mathbf{X}(t))dt + \mathbf{G}(t, \mathbf{X}(t))d\mathbf{W}(t), \qquad \mathbf{X}(0) = \mathbf{X}_0,$$

for an n-dimensional stochastic process $\mathbf{X}(t)$ is called a linear SDE if the vector $\mathbf{f}(t, \mathbf{X}(t))$ and the matrix $\mathbf{G}(t, \mathbf{X}(t))$ are affine functions of $\mathbf{X}(t)$ as

$$\mathbf{f}(t, \mathbf{X}(t)) = \mathbf{A}(t)\mathbf{X}(t) + \mathbf{a}(t),$$
$$\mathbf{G}(t, \mathbf{X}(t)) = [\mathbf{B}_1(t)\mathbf{X}(t) + \mathbf{b}_1(t), \cdots, \mathbf{B}_m(t)\mathbf{X}(t) + \mathbf{b}_m(t)],$$

where $\mathbf{A}(t)$ is an $n \times n$ and $\mathbf{B}_i(t)$ are $n \times m$ matrices, $\mathbf{a}(t)$ is an n-dimensional vector and $\mathbf{b}_i(t)$ are m-dimensional vectors. $\mathbf{W}(t)$ is an m-dimensional Brownian motion.

Therefore, the vector-valued linear SDE can be written as[11]

$$dX(t) = (\mathbf{A}(t)\mathbf{X}(t) + \mathbf{a}(t))dt + \sum_{i=1}^{m} (\mathbf{B}_i(t)\mathbf{X}(t) + \mathbf{b}_i(t))dW_i(t). \qquad (5.26)$$

The solution of (5.26) is given by:

---

[11] If the SDE in (5.26) is controlled externally by the control vector $\mathbf{u}(t)$ it becomes

$$dX(t) = (\mathbf{A}(t)\mathbf{X}(t) + \mathbf{C}(t)\mathbf{u}(t) + \mathbf{a}(t))dt + \sum_{i=1}^{m} (\mathbf{B}_i(t)\mathbf{X}(t)\mathbf{D}_i(t)\mathbf{u}(t) + \mathbf{b}_i(t))dW_i$$

where $\mathbf{u}(t)$ is the k-dimensional control vector, $\mathbf{C}(t)$ and $\mathbf{D}_i(t)$ are $n \times k$ matrices.

$$\mathbf{X}(t) = \boldsymbol{\Phi}(t)\left(\mathbf{X}_0 + \int_0^t \boldsymbol{\Phi}^{-1}(\tau)\left[\mathbf{a}(\tau) - \sum_{i=1}^m \mathbf{B}_i(\tau)\mathbf{b}_i(\tau)\right]d\tau\right.$$

$$\left.+ \sum_{i=1}^m \int_0^t \boldsymbol{\Phi}^{-1}(\tau)\mathbf{b}_i(\tau)dW_i(\tau)\right), \tag{5.27}$$

where the $n \times n$ dimensional fundamental matrix $\boldsymbol{\Phi}(t)$ is the solution of the homogeneous stochastic differential equation

$$d\boldsymbol{\Phi}(t) = \mathbf{A}(t)\boldsymbol{\Phi}(t)dt + \sum_{i=1}^m \mathbf{B}_i(t)\boldsymbol{\Phi}(t)dW_i(t), \tag{5.28}$$

with the initial condition $\boldsymbol{\Phi}(0) = \mathbf{I}$. For a proof see [101]. Note that these matrix equations degenerate to equations (5.17), (5.18) and (5.20) in the scalar case.

One can calculate the moments of vector valued stochastic processes by solving ODE's. The expected value and the second moment can be computed as:

$$\dot{\mathbf{m}}(t) = \mathbf{A}(t)\mathbf{m}(t) + \mathbf{a}(t)$$
$$\mathbf{m}(0) = \mathbf{X}_0$$
$$\dot{\mathbf{P}}(t) = \mathbf{A}(t)\mathbf{P}(t) + \mathbf{P}(t)\mathbf{A}^T(t) + \mathbf{a}(t)\mathbf{m}^T(t) + \mathbf{m}(t)\mathbf{a}^T(t)$$
$$+ \sum_{i=1}^m \left[\mathbf{B}_i(t)\mathbf{P}(t)\mathbf{B}_i^T(t) + \mathbf{B}_i(t)\mathbf{m}(t)\mathbf{b}_i^T(t)\right.$$
$$\left. + \mathbf{b}_i(t)\mathbf{m}^T(t)\mathbf{B}_i^T(t) + \mathbf{b}_i(t)\mathbf{b}_i(t)^T\right]$$
$$\mathbf{P}(0) = \mathbf{X}_0\mathbf{X}_0^T,$$

where $\mathbf{m}(t)$ is an n-dimensional vector and $\mathbf{P}(t)$ is an $n \times n$ dimensional matrix. The covariance matrix for $\mathbf{X}(t)$ is

$$\boldsymbol{\Sigma}(t) = \mathbf{P}(t) - \mathbf{m}(t)\mathbf{m}^T(t).$$

In the special case where $\mathbf{B}_i(t) = 0$, $i = 1, \ldots, m$, an explicit solution for the first two moments as probability density functions can be found. The solution of the linear vector SDE

$$d\mathbf{X}(t) = (\mathbf{A}(t)\mathbf{X}(t) + \mathbf{a}(t))dt + \sum_{i=1}^m \mathbf{b}_i(t)dW_i(t), \qquad \mathbf{X}(0) = \mathbf{X}_0,$$

is normally distributed, i.e., $P(\mathbf{X}(t)) \sim \mathcal{N}(\mathbf{m}(t), \boldsymbol{\Sigma}(t))$. The expected value $\mathbf{m}(t)$ and the covariance matrix $\boldsymbol{\Sigma}(t)$ are the solutions of the ODE's:

$$\dot{\mathbf{m}}(t) = \mathbf{A}(t)\mathbf{m}(t) + \mathbf{a}(t)$$

$$\mathbf{m}(0) = \mathbf{X}_0$$

$$\dot{\boldsymbol{\Sigma}}(t) = \mathbf{A}(t)\boldsymbol{\Sigma}(t) + \boldsymbol{\Sigma}(t)\mathbf{A}^T(t) + \sum_{i=1}^{m} \mathbf{b}_i \mathbf{b}_i^T(t)$$

$$\boldsymbol{\Sigma}(0) = \mathbf{0}.$$

Let us now look at a couple of vector-valued linear price models commonly used in finance.

- Multi-dimensional geometric Brownian motion:
  The most popular stock price model is the geometric Brownian motion. In order to model the price of n stocks which are correlated and each have volatilities which can be modeled as geometric Brownian motions, the following system of equations suggests itself:

$$dS_1(t) = \mu_1 S_1(t)dt + S_1(t)\Big(\sigma_{11}dW_1(t) + \sigma_{12}dW_2(t) + \ldots + \sigma_{1n}dW_n(t)\Big)$$

$$dS_2(t) = \mu_2 S_2(t)dt + S_2(t)\Big(\sigma_{21}dW_1(t) + \sigma_{22}dW_2(t) + \ldots + \sigma_{1n}dW_n(t)\Big)$$

$$\vdots$$

$$dS_n(t) = \mu_n S_n(t)dt + S_n(t)\Big(\sigma_{n1}dW_1(t) + \sigma_{n2}dW_2(t) + \ldots + \sigma_{nn}dW_n(t)\Big).$$

The price processes are correlated if $\sigma_{ij} = \sigma_{ji} \neq 0$, $i \neq j$. This linear system of SDE's is the underlying model for Modern Portfolio Theory (see Chapter 8).

- Linear SDE with stochastic volatility:
  The observed volatility for stock price processes in real life is *not* constant as most models assume. Rather, the volatility itself is a stochastic process. Numerous studies have shown that during times of crises, the volatilities tend to be much higher than at other times [117]. Also, the distribution of the logarithmic prices are not normal: the tails of the distributions are "fat" [119]. In order to capture these empirical facts, the stock prices are better modeled by a set of two SDE's. The first SDE describes the logarithm of the price $P(t)$ and the second SDE describes the change of the volatility $\sigma(t)$ over time as:

$$dP(t) = \mu dt + \sigma(t)dW_1(t)$$

$$P(0) = P_0$$

$$d\sigma(t) = \kappa(\theta - \sigma(t))dt + \sigma_1 dW_2(t)$$

$$\sigma(0) = \sigma_0,$$

where θ is the average volatility, σ_1 the "volatility of the volatility", and κ the mean reversion rate of the volatility process $\sigma(t)$. Stock prices $S(t)$ are obtained from these equations with the transformation $P(t) = \log(S(t))$. Note that, in general, the two Brownian motions $dW_1(t)$ and $dW_2(t)$ are correlated.

5.7.3 Non-linear SDE's and Pricing Models

There is no general solution theory for non-linear SDE's. Also, there are no explicit formulas for calculating the moments of stochastic processes governed by non-linear SDE's. In this subsection, an example of non-linear SDE's in the case of square root processes is discussed.

In general, a scalar square root process can be formulated as

$$dX(t) = f(t, X(t))dt + g(t, X(t))dW(t),$$

with

$$f(t, X(t)) = A(t)X(t) + a(t)$$
$$g(t, X(t)) = B(t)\sqrt{X(t)},$$

where $A(t)$, $a(t)$, and $B(t)$ are real scalars. A special case of square root processes is a mean reverting square root process given as

$$dS(t) = \kappa[\mu - S(t)]dt + \sigma\sqrt{S(t)}\,dW(t)$$
$$S(0) = S_0.$$

Such a process shows a less volatile behavior than a linear geometric process and has a non-central chi-square distribution. Such representations are useful, for instance, in modeling short-term interest rates or volatilities as described in the previous subsection.

Another square root process similar to the geometric Brownian motion, but with a square root diffusion term is given as

$$dS(t) = \mu S(t)dt + \sigma\sqrt{S(t)}dW(t), \quad S(0) = S_0.$$

In this special case, the mean and the variance can be given explicitly as $E[S(t)] = S_0 e^{\mu t}$ and $\mathrm{Var}[S(t)] = \frac{\sigma^2 S_0}{\mu}(e^{2\mu t} - e^{\mu t})$. The variance of square root processes grow much slower than geometric processes with time.

5.8   Partial Differential Equations and SDE's

Stochastic differential equations (SDE's) and deterministic partial differential equations (PDE's) are related mathematically, what might come as a surprise. In this Section, the Feynman-Kac theorem which establishes this connection is introduced and applied to a financial problem. A proof of this profound theorem can be found in [70] or [125].

Given the deterministic PDE

$$\frac{\partial F(t, x)}{\partial t} + \mu(t, x)\frac{\partial F(t, x)}{\partial x} + \frac{1}{2}\sigma^2(t, x)\frac{\partial^2 F(t, x)}{\partial x^2} + L(t, x) = 0, \quad (5.29)$$

subject to the terminal condition $F(T, x) = \psi(x)$ where μ, σ, ψ and L are known real valued scalar functions. What real function F defined in $[0, T]$ is the solution of (5.29)? The Feynman-Kac theorem states that the solution can be calculated as the conditional expected value[12]

$$F(t, x) = E_{t,x}\left[\int_t^T L(s, X(s))ds \,\Big|\, X(t) = x\right], \quad (5.30)$$

where X is driven by the SDE

$$dX = \mu(t, X)\,dt + \sigma(t, X)\,dW, \quad X(t) = x. \quad (5.31)$$

Example: Find the average volatility of a stock in a given time period $[0, T]$.[13] In other words, find the expected average volatility with respect to the given initial time $t = 0$ and the fixed initial state x as described by

$$E_{0,x}\left[\int_0^T \frac{1}{T}x(\tau)\,d\tau\right],$$

where

$$dx(t) = \kappa(\theta - x(t))dt + \sigma\sqrt{x(t)}\,dW(t), \quad x(0) = x_0.$$

Now, recognizing that the expected value here can be the function $F(\cdot)$ in the Feynman-Kac theorem leads to the PDE

---

[12]The mathematical concept of conditional expected value is rather subtle. An introduction to its intricacies can be found in [66] or [93].

[13]It is possible to buy a derivative product (see Chapter 9) called a volatility swap which pays off a profit proportional to the average volatility of the stock in a given period.

$$\frac{\partial F(t,x)}{\partial t} + \kappa(\theta - x)\frac{\partial F(t,x)}{\partial x} + \frac{1}{2}\sigma^2\frac{\partial^2 F(t,x)}{\partial x^2} + \frac{x}{T} = 0 \tag{5.32}$$

$$F(T,x) = 0.$$

An *Ansatz* for $F(t,x)$ is needed to solve this PDE. Let us say

$$F(t,x) = a_1(t) + a_2(t)x(t),$$

and compute the corresponding partial derivatives as

$$F_t(t,x) = \dot{a}_1(t) + \dot{a}_2(t)x(t)$$
$$F_x(t,x) = a_2(t)$$
$$F_{xx}(t,x) = 0,$$

with the terminal conditions $a_1(T) = 0$ and $a_2(T) = 0$. Substituting these partial derivatives in (5.32) yields

$$\dot{a}_1(t) + \dot{a}_2(t)x(t) + \kappa(\theta - x)a_2(t) + \frac{x(t)}{T} = 0$$

$$\Rightarrow \underbrace{\left[\dot{a}_1(t) + \kappa\theta a_2(t)\right]}_{\alpha(t)} + \underbrace{\left[(\dot{a}_2(t) - \kappa a_2(t) + \frac{1}{T})\right]}_{\beta(t)}x(t) = 0.$$

For $\alpha(t) + \beta(t)x(t) = 0$ to hold always, both $\alpha(t)$ and $\beta(t)$ must always be 0. This leads to the two ordinary differential equations in $a_1(t)$ and $a_2(t)$:

$$\dot{a}_1(t) + \kappa\theta a_2(t) = 0$$
$$\dot{a}_2(t) - \kappa a_2(t) + \frac{1}{T} = 0.$$

The solution can be calculated analytically as

$$a_1(t) = \theta + \frac{\theta T}{\kappa}\left(1 - e^{-\kappa T}\right),$$

$$a_2(t) = \frac{1}{\kappa T}\left(1 - e^{-\kappa(t-T)}\right).$$

Hence, the average volatility $E_{t,x}$ can be computed as

$$F(t,x) = \theta + \frac{\theta T}{\kappa}\left(1 - e^{-\kappa T}\right) + \frac{1}{\kappa T}\left(1 - e^{-\kappa(t-T)}\right)x(t).$$

□

5.9 Solutions of Stochastic Differential Equations

In this Section, three different methods of computing the solution of SDE's are shown. The first method is based on the Itô integral (this method was already discussed for the solution of linear SDE's). The second method is the numerical solution based on computing path-wise solutions of SDE's. The third method is based on finding the probability density function of the solution of the SDE by solving a deterministic partial differential equation.

5.9.1 Analytical Solutions of SDE's

The integral form of the SDE

$$dX(t) = f(t, X(t))dt + g(t, X(t))dW(t), \qquad X(0) = X_0,$$

is given by

$$X(t, \omega) = X_0 + \int_0^t f(\tau, X(\tau)) d\tau + \int_0^t g(\tau, X(\tau)) dW(\tau),$$

as already seen in (5.5) where the first integral is an ordinary Riemann integral and the second integral is an Itô integral. The existence of this solution is guaranteed if $f(t, X(t))$ and $g(t, X(t))$ are sufficiently smooth.

One way of finding analytical solutions is by guessing them (the *Ansatz* method) and by using Itô calculus to verify the results as the following example shows.

Example: Use the *Ansatz*

$$X(t) = (W(t) + \sqrt{X_0})^2, \tag{5.33}$$

as the solution of the non-linear SDE

$$dX(t) = dt + 2\sqrt{X(t)} dW(t). \tag{5.34}$$

To verify that (5.33) is indeed a solution of (5.34) begin by defining $Z(t) = W(t) \Rightarrow dZ(t) = dW(t)$. This is an SDE with $f = 0$ and $g = 1$ in the nomenclature of this chapter. To use Itô's lemma as already shown in Section 5.6 on Itô Calculus above, let us write $X(t) = \phi(Z)$ where $\phi(Z) = (W(t) + \sqrt{X_0})^2$. The derivatives are $\phi_z(Z) = 2(Z(t) + \sqrt{X_0})$, $\phi_{zz}(Z) = 2$ and $\phi_t(Z) = 0$. Using Itô's lemma (5.12) yields

$$dX(t) = \widetilde{f}(t, X)dt + \widetilde{g}(t, X)dW(t)$$
$$\widetilde{f}(t, X) = \phi_z(W)0 + \frac{1}{2}\phi_{zz}(W)1 = 1$$
$$\widetilde{g}(t, X) = \phi_z(W)1 = 2(W(t) + \sqrt{X_0}).$$

Since $X(t) = (W(t) + \sqrt{X_0})^2$, $(W(t) + \sqrt{X_0}) = \sqrt{X(t)}$. Thus the calculation with Itô's lemma generated the original SDE.   □

As with integration by substitution in classical calculus, another way of finding analytical solutions is to use Itô calculus to transform the SDE in such a way that the resulting SDE can readily be integrated.

Example: Let us apply this method on the SDE

$$dX(t) = \left(-\frac{1}{2}X(t) - \frac{1}{8X^3(t)}\right)dt + \frac{1}{2X(t)}dW(t).$$

By using the transformation $Y(t) = X^2(t)$, so that $\phi_x(X(t)) = 2X(t)$, $\phi_{xx}(X(t)) = 2$ and $\phi_t(X(t)) = 0$. Using Itô's lemma yields

$$dY(t) = \widetilde{f}(t, X(t))dt + \widetilde{g}(t, X(t))dW(t)$$

with

$$\widetilde{f}(t, X(t)) = \phi_x(X(t))\left(-\frac{1}{2}X(t) - \frac{1}{8X^3(t)}\right) + \frac{1}{2}\frac{1}{4X^2(t)}\phi_{xx}(X(t))$$
$$= -X^2(t)$$
$$\widetilde{g}(t, X(t)) = \phi_x(X(t))\frac{1}{2X(t)} = 1$$
$$dY(t) = -X^2(t)dt + dW(t)$$
$$= -Y(t)dt + dW(t).$$

The last equation is a well-known SDE, the solution of which $Y(t)$, is an Ornstein-Uhlenbeck process. Thus the original SDE is solved with the transformation $X(t) = \sqrt{Y(t)}$.   □

At first glance, it might look as if one needs a hefty portion of luck to find a solution to SDE's by both of these analytical methods (*Ansatz* and substitution). However, as with factoring polynomials in high school algebra or integration by substitution in first year calculus, one gets quite apt in finding the right *Ansatz* or substitution with some exercise.

5.9.2   *Numerical Solution of SDE's*

Many a frustrated student knows the difficulty, even the impossibility of solving most ODE's or PDE's. It is not any different with SDE's. Also here, if no analytical solution can be found for an SDE, one is left with numerical approximations. Literature is rich with many reference books on this subject, *e.g.,* [20, 56, 86, 94, 114]. In this subsection two commonly used numerical methods are introduced following the path of [114].

The Euler–Maruyama Method is based on the time axis being discretized. Begin by defining the approximate solution path at discrete points on the time axis

$$a = t_0 < t_1 < t_2 < \ldots < t_n = b,$$

and assigning approximate y-values

$$w_0 < w_1 < w_2 < \ldots < w_n$$

at the respective t points. Given the SDE

$$dX(t) = f(t, X)dt + g(t, X)dW(t), \qquad X(a) = X_a, \tag{5.35}$$

the solution is calculated approximately as

$$w_{i+1} = w_i + f(t_i, w_i)\Delta t_i + g(t_i, w_i)\Delta W_i, \tag{5.36}$$

beginning with $w_0 = X_a$ and where

$$\begin{aligned} \Delta t_i &= t_{i+1} - t_i \\ \Delta W_i &= W_{t_{i+1}} - W_{t_i}. \end{aligned} \tag{5.37}$$

The question now is how to model the Brownian motion W_i. Let us define $N(0, 1)$ as the standard random variable that is normally distributed with mean 0 and standard deviation 1. Each random number W_i is then computed as

$$\Delta W_i = z_i \sqrt{\Delta t_i}, \tag{5.38}$$

where $z_i \sim N(0, 1)$

Each set of $\{w_0, \ldots, w_n\}$ produced this way is an *approximate* realization of the stochastic process $X(t)$, which depends on the random numbers z_i. Since $W(t)$ is a stochastic process, each realization will be different, and so will be the approximations.

Example: Let us now apply the Euler–Maruyama Method to the geometric Brownian motion

$$dX(t) = \mu X(t)dt + \sigma X(t)dW(t). \tag{5.39}$$

Applying (5.36) on this SDE yields

$$w_0 = X_0$$
$$w_{i+1} = w_i + \mu w_i \Delta t_i + \sigma w_i \Delta W_i .$$

(5.40)

One can now compute this approximation and compare it with the analytically correct realization

$$X(t) = X_0 e^{(\mu - \frac{1}{2}\sigma^2)t + \sigma W(t)} .$$

(5.41)

Lamperti Transform

Let $X(t)$ be a stochastic process defined by

$$dX(t) = f(X(t), t)dt + \sigma(X(t), t)dW(t) .$$

(5.42)

Lamperti transform for bijective functions (one to one) is

$$Z(t) = \Psi(X(t), t) = \int \frac{1}{\sigma(X(t), t)} dx \Big|_{x=X(t)} .$$

(5.43)

If $\Psi(\cdot)$ is not bijective it is defined as

$$Z(t) = \Psi(X(t), t) = \int_\xi^x \frac{1}{\sigma(u, t)} du ,$$

(5.44)

where ξ is some point inside the state space of $X(t)$.
$Z(t)$ has unit diffusion and is governed by the SDE

$$dZ(t) = \left(\Psi\left(\Psi^{-1}(Z(t), t) + \frac{f(\Psi^{-1}(Z(t), t))}{\sigma(\Psi^{-1}(Z(t), t))} - \frac{1}{2} \frac{\partial \sigma}{\partial x} (\Psi^{-1}(Z(t), t)) \right) \right) dt + dW(t) .$$

(5.45)

This transform is a very useful approach for removal of level dependent noise and is commonly used in finance. It is particularly useful with separable volatility functions both in analytic and numerical contexts. Even though the Lamperti transform is limited by the user's ability of finding the inverse, it is still possible to use transformations that remove level dependent noise for quite general classes of diffusion processes [97].

To get a better approximation, one can use the so-called Milstein scheme which includes information about the derivative of the drift part of the SDE:

$$w_{i+1} = w_i + f(t_i, w_i)\Delta t_i + g(t_i, w_i)\Delta W_i + \frac{1}{2}g(t_i, w_i)\frac{\partial g}{\partial X}(t_i, w_i)(\Delta W_i^2 - \Delta t_i).$$

$$(5.46)$$

Applying the Milstein method with constant step size Δt on the geometric Brownian motion in (5.39) yields

$$w_{i+1} = w_i + \mu w_i \Delta t + \sigma w_i \Delta W_i + \frac{1}{2}\sigma^2 w_i(\Delta W_i^2 - \Delta t). \qquad (5.47)$$

One can again compute this approximation and compare it with the known solution (5.41).   □

One says that a numerical solution of an SDE has the order m if the expected value of the approximation error is of m-th order in the step size; *i.e.*, if for any time T, $\mathrm{E}[|X(T) - w(T)|] = O(\Delta t^m)$ as the step size $\Delta t \to 0$. In that sense, the orders of the numerical approximation methods introduced above are 1/2 for Euler–Maruyama and 1 for Milstein as can be seen in Table 5.1. Higher-order methods can be developed for SDE's, but are much more complicated as the order grows.

A word of caution is certainly due here: convergence is a significant issue while looking for numerical solutions of SDE. This is even more problematic for coupled multidimensional SDE. For an in depth study of this and related issues see [69].

5.9.3   Solutions of SDE's as Diffusion Processes

In the last subsection, numerical ways of solving SDE's were discussed. However, one has to keep in mind that, given a realization of a Brownian motion, these methods

Table 5.1 Comparison of the approximation error ϵ as a function of the step size Δt for Euler–Maruyama and Milstein methods averaged over 100 realizations for the solution of (5.39) shows that reducing the step size by a factor of 4 is required to reduce the error by a factor of 2 with the Euler–Maruyama Method. For the Milstein Method, cutting the step size by a factor of 2 achieves the same result. This demonstrates the respective orders of 1/2 and 1 for the two methods

t	Euler–Maruyama	Milstein
2^{-1}	0.169369	0.063864
2^{-2}	0.136665	0.035890
2^{-3}	0.086185	0.017960
2^{-4}	0.060615	0.008360
2^{-5}	0.048823	0.004158
2^{-6}	0.035690	0.002058
2^{-7}	0.024277	0.000981
2^{-8}	0.016399	0.000471
2^{-9}	0.011897	0.000242
2^{-10}	0.007913	0.000122

can only deliver *one sample path* of the given process. To understand the process described by the SDE better, one needs to know the statistical distribution of the possible outcomes of the stochastic process at all times.

To be able to do that one needs a theorem as the starting point which says that under certain weak assumptions, for any fixed initial value $X_0 = X(t_0)$ the solution $X(t)$ of the SDE (5.4) is a diffusion process on $[t_0, T]$ with the drift coefficient $f(t, x)$ and the diffusion coefficient $g(t, x)$ (for a complete proof of this theorem, the reader is referred to [103]).

For many stock and other financial instrument price models, solutions of the SDE's are Markov processes. This is significant, because it allows the application of powerful tools to solutions of these types of SDE's. Generally speaking, a diffusion process is a kind of Markov process with continuous realizations which have the transition probability $P(s, x, t, B)$. So, the interesting question here is what is the connection between the transition probability and the drift and diffusion coefficients? Can $P(\cdot)$ be described as a function of $f(\cdot)$, $g(\cdot)$?

Firstly one needs to understand what the transition probability really means. Let $X(t)$ be a Markov process. The function $P(s, x, t, B)$ is the conditional distribution of the probability $P(X(t) \in B \mid X(s) = x)$ and is called the transition probability. It can be seen as the probability that the process is found inside the set B at time t, when at time $s < t$ it was found to be in state $X(s) = x$. The associated transition probability density $p(\cdot)$ is given by

$$P(s, x, t, B) = \int_B p(s, x, t, y)\, dy.$$

A Markov process is called shift-invariant if its transition probability $P(s, x, t, B)$ is stationary, *i.e.*,

$$P(s + u, x, t + u, B) \equiv P(s, x, t, B).$$

In this case, the transition probability is only a function of x, $t - s$, and B and can be written as

$$P(s, x, t, B) = P(t - s, x, B),$$

A d-dimensional Markov process $\mathbf{X}(t)$ is called a diffusion process, if its transition probability $P(s, \mathbf{x}, t, \mathbf{B})$ satisfies the following conditions for $s \in [t_o, T)$ and $\varepsilon > 0$

- It is continuous:

$$\lim_{t \downarrow s} \frac{1}{t - s} \int_{|\mathbf{y} - \mathbf{x}| > \varepsilon} P(s, \mathbf{x}, t, \mathbf{dy}) = 0.$$

- There exists a d-dimensional function $\mathbf{f}(t, \mathbf{x})$ with

$$\lim_{t \downarrow s} \frac{1}{t - s} \int_{|\mathbf{y} - \mathbf{x}| \leq \varepsilon} (\mathbf{y} - \mathbf{x}) P(s, \mathbf{x}, t, \mathbf{dy}) = \mathbf{f}(s, \mathbf{x}).$$

- There exists a symmetric $d \times d$ matrix $\boldsymbol{\Sigma}(t, \mathbf{x})$ with

$$\lim_{t \downarrow s} \frac{1}{t-s} \int_{|\mathbf{y}-\mathbf{x}| \leq \varepsilon} (\mathbf{y}-\mathbf{x})(\mathbf{y}-\mathbf{x})^T P(s, \mathbf{x}, t, \mathbf{dy}) = \boldsymbol{\Sigma}(s, \mathbf{x}).$$

So, what is the transition probability $P(s, \mathbf{x}, t, \mathbf{B})$ of the process?

The functions $\mathbf{f}(t, \mathbf{x})$ and $\boldsymbol{\Sigma}(t, \mathbf{x}) = \mathbf{g}(t, \mathbf{x})\mathbf{g}^T(t, \mathbf{x})$ are called the coefficients of the diffusion process. Namely, $\mathbf{f}(\cdot)$ is the drift vector and $\boldsymbol{\Sigma}(\cdot)$ the diffusion matrix.

Assuming that the first and second moments of $\mathbf{X}$ exist and with the boundary condition $\mathbf{X}(s) = \mathbf{x}$, the first condition (continuity) makes it unlikely for the process $\mathbf{X}$ to have large changes in values in a short time interval. It thereby rules out any Poisson type effects in the diffusion process. With $o(t - s)$ denoting a term in the order of $t - s$:

$$P(| \mathbf{X}(t) - \mathbf{X}(s) | \leq \varepsilon \mid \mathbf{X}(s) = \mathbf{x}) = 1 - o(t - s).$$

For the second condition one can write

$$\lim_{t \downarrow s} E_{s,\mathbf{x}}(\mathbf{X}(t) - \mathbf{X}(s)) = \mathbf{f}(s, \mathbf{x})(t - s) + o(t - s),$$

and for the third condition

$$\lim_{t \downarrow s} E_{s,\mathbf{x}}(\mathbf{X}(t) - \mathbf{X}(s))(\mathbf{X}(t) - \mathbf{X}(s))^T = \boldsymbol{\Sigma}(s, \mathbf{x})(t - s) + o(t - s),$$

or

$$\lim_{t \downarrow s} \mathrm{Cov}_{s,\mathbf{x}}(\mathbf{X}(t) - \mathbf{X}(s)) = \boldsymbol{\Sigma}(s, \mathbf{x})(t - s) + o(t - s).$$

As a consequence, one can say that $\mathbf{f}$ is the average speed of the stochastic movement and $\boldsymbol{\Sigma}$ describes the magnitude of the fluctuation $\mathbf{X}(t) - \mathbf{X}(s)$ around the mean value of $\mathbf{X}$ with respect to the transition probability $P(s, \mathbf{x}, t, \mathbf{B})$. One can now choose an arbitrary matrix $\mathbf{g}(t, \mathbf{x})$ such that $\boldsymbol{\Sigma} = \mathbf{g}\mathbf{g}^T$ and can write the following equation:

$$\mathbf{X}(t) - \mathbf{X}(s) = \mathbf{f}(s, \mathbf{X}(s))(t - s) + \mathbf{g}(s, \mathbf{X}(s))\boldsymbol{\xi},$$

where $E_{s,\mathbf{x}}[\boldsymbol{\xi}] = 0$ and $\mathrm{Cov}_{s,\mathbf{x}}[\boldsymbol{\xi}] = (t - s)\mathbf{I}$. One can replace this by a Brownian motion, because $\mathbf{W}(t) - \mathbf{W}(s)$ has the distribution $\mathcal{N}(0, (t - s)\mathbf{I})$, i.e., exactly the distribution of $\boldsymbol{\xi}$. So,

$$\mathbf{X}(t) - \mathbf{X}(s) = \mathbf{f}(s, \mathbf{X}(s))(t - s) + \mathbf{g}(s, \mathbf{X}(s))(\mathbf{W}(t) - \mathbf{W}(s)),$$

or, in differential form,

$$d\mathbf{X}(t) = \mathbf{f}(t, \mathbf{X}(t))dt + \mathbf{g}(t, \mathbf{X}(t))d\mathbf{W}(t). \tag{5.48}$$

This, of course, is the SDE familiar to the reader! Note that this form of the SDE (5.48) has been derived directly from the transition probabilities and the characteristics of diffusion processes without knowing the general solution of (5.4) in advance. This shows how the same process can be described either through transition probabilities or equivalently through a stochastic differential equation.

So far the link between the transition probability and the stochastic process defined by the SDE was established. But how does one get the transition probability P as a function of the drift $\mathbf{f}$ and the set of processes $\mathbf{B}$ in which the stochastic process $\mathbf{X}$ can be found at time t if it's value was $\mathbf{x}$ at time s?

To answer this interesting question let us again begin with $\mathbf{X}$ as the solution of a d-dimensional diffusion process

$$dX(t) = \mathbf{f}(t, \mathbf{X}(t))dt + \mathbf{g}(t, \mathbf{X}(t))d\mathbf{W}(t).$$

Assume $\mathbf{f}$ and $\mathbf{g}$ to be sufficiently smooth. Then, the transition probabilities $P(s, \mathbf{x}, t, \mathbf{B}) = P(\mathbf{X}(t) \in \mathbf{B} \mid \mathbf{X}(s) = \mathbf{x})$ are given as the solution of the equation

$$\frac{\partial P(s, \mathbf{x}, t, \mathbf{B})}{\partial s} + \mathbf{f}^T(s, \mathbf{x})\frac{\partial P(s, \mathbf{x}, t, \mathbf{B})}{\partial \mathbf{x}} + \frac{1}{2}\text{tr}\{\mathbf{g}(s, \mathbf{x})\mathbf{g}^T(s, \mathbf{x})\frac{\partial^2 P(s, \mathbf{x}, t, \mathbf{B})}{\partial \mathbf{x}^2}\} = 0,$$

where $P(s = t, \mathbf{x}, t, \mathbf{B}) = I_{\mathbf{B}}$ the indicator function of the set $\mathbf{B}$. It is also known that

$$P(s, \mathbf{x}, t, \mathbf{B}) = \mathrm{E}_{s,\mathbf{x}}\big[I_{\mathbf{B}}(\mathbf{X}(t))\big] = P\big(\mathbf{X}(t)] \in \mathbf{B} \mid \mathbf{X}(s) = \mathbf{x}\big).$$

The transition probability $P(s, \mathbf{x}, t, \mathbf{B})$ has the density $p(s, \mathbf{x}, t, \mathbf{y})$ and the so-called Kolmogorov backward equation for the transition probability density function $p(s, \mathbf{x}, t, \mathbf{y})$ can be written as

$$\frac{\partial p(s, \mathbf{x}, t, \mathbf{y})}{\partial s} + \mathbf{f}^T(s, \mathbf{x})\frac{\partial p(s, \mathbf{x}, t, \mathbf{y})}{\partial \mathbf{x}} + \frac{1}{2}\text{tr}\{\mathbf{g}(s, \mathbf{x})\mathbf{g}^T(s, \mathbf{x})\frac{\partial^2 p(s, \mathbf{x}, t, \mathbf{y})}{\partial \mathbf{x}^2}\} = 0,$$

$$(5.49)$$

and in the limit case

$$\lim_{s\uparrow t} p(s, \mathbf{x}, t, \mathbf{y}) = \delta(\mathbf{y} - \mathbf{x}).$$

The reason for calling this equation a backward equation is that the differential operators are considered with regard to the backward variables $(s, \mathbf{x})$; i.e., the differential equation gives the backward evolution with respect to the initial state, given the final state $(t, \mathbf{y})$.

The Fokker-Planck equation, on the other hand, gives the forward evolution with respect to the final state $(t, \mathbf{y})$ starting at the initial state $(s, \mathbf{x})$. Therefore, it is also called the Kolmogorov forward equation.

$$\frac{\partial p(s, \mathbf{x}, t, \mathbf{y})}{\partial t} + \mathbf{f}^T(t, \mathbf{y})\frac{\partial p(s, \mathbf{x}, t, \mathbf{y})}{\partial \mathbf{y}} - \frac{1}{2}\mathrm{tr}\{\mathbf{g}(t, \mathbf{y})\mathbf{g}^T(t, \mathbf{y})\frac{\partial^2 p(s, \mathbf{x}, t, \mathbf{y})}{\partial \mathbf{y}^2}\} = 0,$$

(5.50)

and in the limit case

$$\lim_{t \downarrow s} p(s, \mathbf{x}, t, \mathbf{y}) = \delta(\mathbf{y} - \mathbf{x}).$$

For a formal derivation of the Kolmogorov backward and forward equations see [72].

Let us now apply this on the Wiener process. Some relevant properties of the Wiener process $W(t)$ are $E[W(t)] = 0$, $W(t)$ is stationary and homogeneous, i.e., $p(W(t + h) \mid W(t)) = p(W(h) \mid W(0))$, it is a solution of the SDE $dX = 0 \cdot dt + 1 \cdot dW$, hence $f = 0$, $g = 1$. Therefore, Kolmogorov forward and backward equations are given by

$$\frac{\partial p}{\partial t} - \frac{1}{2}\frac{\partial^2 p}{\partial y^2} = 0, \qquad \frac{\partial p}{\partial s} - \frac{1}{2}\frac{\partial^2 p}{\partial x^2} = 0.$$

As a more advanced example, let us look at the scalar linear SDE

$$dX(t) = \big(AX(t) + a\big)dt + bdW(t), \quad X(0) = c, \quad t \ge 0,$$

which, according to (5.18) and (5.19), has the following solution

$$X(t) = ce^{At} + \frac{a}{A}(e^{At} - 1) + b\int_0^t e^{A(t-s)}dW(s).$$

Notable special cases are the Ornstein-Uhlenbeck process ($a = 0$), deterministic differential equations ($b = 0$), and the standard Brownian motion ($A = a = 0$, $b = 1$, $c = 0$). With $b \ne 0$ one can find a solution for the transition probability density $p(s, x, t, y)$ using the Kolmogorov forward equation

$$\frac{\partial}{\partial t}p(s, x, t, y) + (Ay + a)\frac{\partial}{\partial y}p(s, x, t, y) - \frac{1}{2}b^2\frac{\partial^2}{\partial y^2}p(s, x, t, y) = 0.$$

As boundary conditions one can assume that p vanishes (including all its partial derivatives) for $|x| \to \infty$ and $|y| \to \infty$.

5.10   Exercises

A1: What is a stochastic differential equation?

A2: What are the differences between Riemann and Itô integrals?

A3: What does Itô's lemmasay?

A4: Why are stochastic differential equation relevant for finance?

A5: Describe some scalar linear stochastic differential equations and explain where they are commonly used. **A6**: Explain what multi-dimensional geometric Brownian motion is and how such vector-valued linear price models are used in Modern Portfolio Theory. What happens if the volatility is not constant but itself a stochastic process?

A7: What are square root processes? Why does it sometimes make more sense to use them instead of geometric processes?

A8: How are stochastic differential equations related to deterministic partial differential equations?

A9: Most stochastic differential equations cannot be solved analytically and have to be approached by numerical tools. Explain how this is done.

A10: Explain how to analyze stochastic processes by transforming them to diffusion processes.

B1: Given the scalar SDE

$$dX(t) = -\frac{1}{2}e^{-2X(t)}\,dt + e^{-X(t)}\,dW(t).$$

The stochastic process $Y(t)$ is a function of $X(t)$ with the relationship $Y(t) = \phi(t, X(t)) = e^{X(t)}$. Calculate the SDE for $Y(t)$. What is its distribution? What are its first two moments?

B2: Given the following SDE

$$dX(t) = \mu X(t)\,dt + \sigma X(t)\,dW(t).$$

Calculate the stochastic process $Y(t) = \phi(t, X(t)) = \sqrt{X(t)}$. What are the different statistical properties of $X(t)$ and $Y(t)$?

B3: Use Itô's lemma to prove for $k \geq 2$ that

$$E[W(t)^k] = \frac{1}{2}k(k-1)\int_0^t E[W(s)^{k-2}]\,ds.$$

B4: Show by using Itô's lemma that

$$E[e^{\theta W(t)}] = \frac{1}{2}\theta^2 \int_0^t E[e^{\theta W(\tau)}]\,d\tau.$$

B5: Use Itô's lemma to show that for a deterministic function $h(t)$ the Itô Integral can be calculated as

$$\int_0^t h(\tau)dW(\tau) = h(t)W(t) - \int_0^t h'(\tau)W(\tau)\, d\tau,$$

just as in the deterministic case.

B6: The Cox-Ross-Ingersoll model is widely used for predicting the evolution of interest rates. It is a mean-reverting, stochastic, square root process which only has positive values. Clearly, its applicability is reduced in the negative interest environments like Europe in 2016. The corresponding SDE is

$$dR(t) = a(b - cR(t))\, dt + \sigma\sqrt{R(t)}\, dW(t) \qquad a, b, c, \sigma, R(0) > 0.$$

$W(t)$ is a Brownian motion. Calculate the expected value and the variance or $R(t)$ by using Itô's lemma.

B7: Simulate the Cox-Ingersoll-Ross process from the previous exercise. Take the parameter values $a, b, c, \sigma = 1$ and use $\Delta T = 0.01$ as step size. First simulate the process for $R(0) = 2$ and $t \in [0, 10]$ and generate a plot for $R(t)$ which should show the expected value as well as the band $E[R(t)] \pm \sqrt{\mathrm{Var}(R(t))}$ then repeat the same exercise for $R(0) = 1$ and $t \in [0, 100]$.

B8: Given the linear, stochastic differential equation:

$$d\mathbf{X}(t) = (\mathbf{A}\mathbf{X} + \mathbf{a})\, dt + \mathbf{B}\mathbf{X}\, d\mathbf{W}(t),$$

$$\mathbf{X} = \begin{bmatrix} X_1 \\ X_2 \end{bmatrix}, \quad \mathbf{A} = \begin{bmatrix} -1 & 4 \\ 0 & -1.5 \end{bmatrix}, \quad \mathbf{a} = \begin{bmatrix} -3 \\ 1.5 \end{bmatrix}, \quad \mathbf{B} = \begin{bmatrix} 1 & 0 \\ 0 & 1 \end{bmatrix},$$

where $\mathbf{X}(0) = [2, 2]^T$.

- Calculate the expected value and covariance of $\mathbf{X}$ using a numerical ODE solver for $t = [0, 10]$.
- Simulate the SDE using Euler discretization ($\Delta t = 0.01$) and calculate 500 sample paths.
- Compare the solution for mean and covariance obtained from the numerical solutions of the ODE with the empirical value obtained from the simulation.
- Analyze the distribution (histogram) at time $t = 1$ and $t = 10$ for X_1 and X_2. Plot the difference to normal distribution. Is the normal distribution an acceptable approximation?

B9: Given the SDE for X_1,

$$dX_1(t) = \mu_1 X_1(t)\, dt + \sigma_1 X_1(t)\, dW(t),$$

and the SDE for X_2

$$dX_2(t) = \mu_2 X_2(t)\, dt + \sigma_2 X_2(t)\, dW(t).$$

Calculate the SDE for the stochastic process $Z(t) = \frac{X_1(t)}{X_2(t)}$. What is your interpretation of X_1, X_2 and Z?

B10: Given the SDE for X_1,

$$dX_1(t) = \mu_1 X_1(t)\, dt + \sigma_1 X_1(t)\, dW_1(t),$$

and the SDE for X_2

$$dX_2(t) = \mu_2 X_2(t)\, dt + \sigma_2 X_2(t)\, dW_2(t).$$

Calculate the SDE for the stochastic process $Z(t) = \log\left(X_1(t)X_2(t)\right)$. What is your interpretation of X_1, X_2 and Z?

B11: Calculate the fourth central moment (kurtosis) of the Brownian motion.

B12: Given the SDE

$$dX(t) = -\beta^2 X(t)(1 - X^2(t))\, dt + \beta(1 - X^2(t))\, dW(t), \qquad X(0) = X_0$$

Show that
$$X(t) = \frac{(1 + X_0)e^{2\beta W(t)} + X_0 - 1}{(1 + X_0)e^{\beta W(t)} + 1 - X_0}$$

is a solution of the SDE.

B13: Given the SDE with real constants a, b

$$dX(t) = \frac{b - X(t)}{1 - t} + dW(t), \qquad X(0) = a, \qquad 0 \le t < 1.$$

Show that

$$X(t) = a(1 - t) + bt + (1 - t)\int_0^t \frac{1}{1 - \tau}\, dW(\tau), \qquad 0 \le t < 1.$$

Also show that $\lim_{t \to 1}(X(t)) = b$ is (almost certainly) true. Discuss your results.

B14: Given the following system of non-linear SDE's

$$dX_1 = (a_1 - b_1 X_1 + b_{12} X_2)\, dt + \sqrt{X_2}\, dW_1, \quad X_1(0) = X_{10}$$
$$dX_2 = (a_2 - b_2 X_2)dt + \sigma\rho\sqrt{X_2}\, dW_1 + \sigma\sqrt{1 - \rho^2}\sqrt{X_2}\, dW_2, \quad X_2(0) = X_{20}.$$

Calculate the conditional expected value $E[X_1(t)|\{X_{10}, X_{20}\}]$ and the conditional covariance
$E[X_1(t)X_2(t)|X_{10}, X_{20}]$.

B15: Consider the following non-linear SDE:

$$dX(t) = \alpha X^3(t)\, dt + \beta X(t)\, dW(t), \quad X_0 > 0,$$

where $\alpha \neq 0$ and $\beta \neq 0$ are real constants. Calculate the closed-form solution of $X(t)$.

B16: Simulate the following linear SDE:

$$dX(t) = (a - bX(t))\, dt + \sigma X(t)\, dW(t), \quad X_0 = \frac{a}{b},$$

with $\Delta t = 0.05$, $t = [0, 25]$ and different combinations of a, b, σ. Discuss your results.

C1: Use Itô calculus to derive the non-linear extended Kalman filter. What are the insights you gained by taking this route?

C2: Study the Lotka-Volterra type non-linearities in competition-diffusion systems, connect them to stochastic systems and apply Itô calculus to analyze such systems. What financial applications can you envisage?

C3: Most mathematical models discussed in this chapter assume that stochastic processes applied to finance problems are of Brownian nature. However, as it has become clear in the past decade, and painfully so for some, the real financial world *cannot* correctly be modeled this way. Extend some important results shown in this chapter dropping the Brownian assumption.

Chapter 6
Financial Markets and Instruments

The mathematical expectancy of the speculator is zero.

— Louis Bachelier

Successful investing is anticipating the anticipations of others.

— John Maynard Keynes

6.1 Introduction

The Chapter begins by introducing the concept of time value of money, fixed income instruments like savings accounts, certificates of deposit and bonds. It then moves on to cover a number of topics, from common stocks to various types of funds like common stock funds, balanced funds and exchange traded funds. Commodities, real estate and some other instruments for investing are mentioned. Fundamental concepts relating to risk, return and utility are also reviewed in this Chapter. Finally, the role and importance of the banks in economies are presented.

6.2 Time Value of Money

You will be earning money for the rest of your life. With this money, you will buy things you need (and things you do not need, too), and you will use it to help others, but you will still be left with a surplus which you will save. You will then face the

© Springer International Publishing AG 2018
S. S. Hacısalihzade, *Control Engineering and Finance*, Lecture Notes in Control
and Information Sciences 467, https://doi.org/10.1007/978-3-319-64492-9_6

inevitable question: what do I do with my savings? Hopefully, you will make your decisions rationally and maximize the utility of your income.[1]

You can, of course, put your savings under your mattress until such time in the future when your general consumption desires exceed your income at that time. Similarly, you might think that it is better to save now and accumulate your surplus income for a particular large ticket item like a car or a home. This exchange of present consumption with a higher level of future consumption is the basis of saving and investment. Conversely, if your current income is less than your current consumption desires, you exchange part of your future income for a higher level of current consumption by getting a credit from a bank, a credit institute, your friends or your family. This negative savings is referred to as borrowing. Note that the money you borrow can be used for current consumption or theoretically be invested at a rate of return above the cost of borrowing.

Clearly, if you forgo a part of your current income, say $1'000, you will expect to get more than $1'000 in the future, say in one year. Similarly, if you consume or invest more than your current income, again, say, $1'000, you should be willing to pay more than $1'000 in one year. The rate of exchange between certain future consumption and certain current consumption is called the *pure time value of money* (*e.g.,* 2%). This rate is determined in the capital market based on the supply of excess income available to be invested at that time and the demand for borrowing. In other words, if you do not spend your $1'000 today, you are certain to earn $20 with it in one year (at this point you might want to reconsider the wisdom of putting the money under your mattress). However, if you are not certain that you will get your money back in one year, you will require a higher return on your investment than the pure time value of money. This difference is called the *risk premium* (*e.g.,* 5%). You would than be willing to lend $1'000 if you think you would get $1'070 in one year.

One can thus define an investment as the current commitment of funds for a period of time in order to derive a future flow of funds that will compensate the investor for the time the funds are committed, for the uncertainty involved in the future flow of funds [109]. The discussion so far assumed no inflation. If inflation is present, it has to be accounted for in the calculations. A common property of all investments, be they financial investments or investments in production means like machinery, is that the investor trades a known amount of money today for some expected but uncertain future amount.

---

[1] Several researchers, most notably Amos Nathan Tversky (1937–1996) and Daniel Kahneman (1934–), who "shared" the Nobel Prize in Economics in 2002, have studied the correlation of income and happiness. They were able to confirm empirically that people do not make their economic decisions rationally [131]. Kahneman who is not an economist but a psychologist, also found that once the yearly income exceeds $50'000, the effect of additional income on increasing happiness drops very fast [63].

According to Robbins,[2] economics is the science which studies human behavior as a relationship between ends and scarce means which have alternative uses [110]. Hence, money certainly qualifies as a subject of economics. On the other hand, finance deals with two specific activities: the study of how money is managed and the actual process of acquiring needed funds. Individuals, businesses and governments all need funding to operate. The field is often separated into three subcategories: personal finance, corporate finance and public finance. However, this and the following Chapters, deal solely with the management of funds.

Haves and Have-Nots

Probably it is not an overstatement to say that one of the biggest and most pressing issues the world faces today, together with global warming, is the gross inequality among its inhabitants in terms of income and wealth. As can be seen in Figure 6.1, the top 0.01% of the world population has the same wealth as the rest of the world! As if this were not bad enough, it is getting even worse: in 2010 388 individuals had as much combined wealth as the poorest 3.5 billion people; in 2015 only 62 individuals had as much combined wealth as the poorest 3.5 billion people. One does not need to be an oracle to predict the social and political problems this might cause in the coming years. Figure 6.2 illustrates this point from the viewpoint of an artist.

Fig. 6.1 Global distribution of wealth in 2015. *Source* Oxfam Report, January 2016

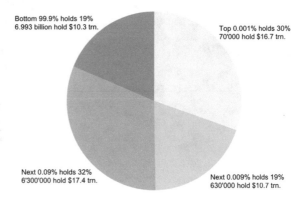

Bottom 99.9% holds 19%
6.993 billion hold $10.3 trn.

Top 0.001% holds 30%
70'000 hold $16.7 trn.

Next 0.09% holds 32%
6'300'000 hold $17.4 trn.

Next 0.009% holds 19%
630'000 hold $10.7 trn.

---

[2]Lionel Robbins, British economist (1898–1984); mostly famous for his instrumental efforts in shifting Anglo-Saxon economics from its earlier neoclassical direction as shaped by Alfred Marshall.

Fig. 6.2 "Survival of the Fattest" is a sculpture of a small starved African boy, carrying a fat western woman, made by Jens Galschiøt in 2002, as a symbol of the imbalanced distribution of the world's resources. The inscription on the sculpture reads "I'm sitting on the back of a man. He is sinking under the burden. I would do anything to help him. Except stepping down from his back"

6.3   Financial Investment Instruments

There are numerous investment instruments available and investors should at least be aware of various alternatives. The reason lies in the advantage of diversification of assets from which one can derive substantial benefits. *Diversification*, in this context, means investing in different classes of instruments with different return patterns over time. Specifically, the aim of diversification is to invest in instruments with negatively correlated returns with the objective of attaining relatively stable earnings for the totality of the investments, often referred to as the investment portfolio or simply *portfolio*. In other words, diversification should help reduce the portfolio risk (see next Section). This Section reviews major instrument classes which an investor should consider for portfolio diversification.

Bonds and Derivative/ Structured Products as investment instrument classes each merit their own Chapters and are dealt with in much greater detail further on in the Book.

6.3.1   Fixed Income Investments

This category of investments comprise instruments that have a fixed payment schedule. Fixed income instruments are often seen as instruments which have a fixed yield. Although true in most cases, in general the incomes resulting from such instruments are by no means constant. The owner of such securities is promised specific payments at predetermined times.

6.3.1.1   Savings Accounts

When you deposit money in a *savings account*, you are actually lending money to the bank to derive a fixed return. Such accounts are considered to be very low risk investments, because they are insured (up to a maximum amount), and liquid. Therefore, the rate of return on savings accounts are generally lower than other investment types. Also, usually, there is no lower limit on the size of such accounts.

6.3.1.2   Certificates of Deposit

Investors with larger amounts of money (typically above $1'000) who are willing to give up the use of their funds for a specified amount of time, typically ranging from one month to one year, can make use of *certificates of deposit* or time deposit accounts. Such instruments are also insured up to a maximum amount. Normally, the returns on such deposits are higher than onsavings accounts and under normal circumstances they increase with the length of the time for which the funds are deposited. If you want to cash in your certificate of deposit before its agreed expiration date you cannot get the agreed interest rate and have to pay a hefty penalty.

6.3.1.3 Bonds

A bond is a long term public debt of an issuer. Bonds are considered fixed income instruments, because the debt service obligations of the issuer are fixed. The issuer agrees to pay a certain amount of periodic interest to the holder and repay a fixed amount of principal at the date of maturity. Usually, interest on US Dollar denominated bonds is paid out every six months. The principal, the par value of the issue, is due at maturity. Short term debt with maturities below a year is known as money market investments. Long term debt with maturities above seven years is simply called bonds. Intermediate debt with maturities between short and long term are often referred to as notes (see Chapter 7 for a detailed treatment of Bonds).

6.3.2 Common Stocks

Many investors think of equity or *common stock* or shares when they think about investing their money. Owning such stock is paramount to owning a part of the company. This means the investor who owns shares of a company shares the successes and problems of the company. If the company does well, the value of its shares increases, some times even phenomenally so (*e.g.,* the price of Microsoft shares have increased 100 fold from 1990 to 2000). On the other hand if the company goes bankrupt, the value of its shares goes practically to zero (*e.g.,* Eastman Kodak lost 99.8% of its share value from 2000 to 2013). Also, companies which are making a profit, often pay out a part of this profit as *dividends* to their shareholders.

There are different dimensions along which one can classify stocks. One such classification is by business sector like industrials, banks, transportation, etc. Another important classification is according to their internal operating performance like growth companies (companies which can invest capital at return rates that far exceed their cost of capital; such companies typically pay no or little dividend and use their earnings for rapid growth; although such companies can attain significant growth rates, they can also be very risky, especially if their growth rates decrease; examples are high-tech companies), cyclical companies (financial success of such companies are closely tied to economic cycles and typically outperform the aggregate economy; examples are steel producers or automotive) or defensive companies (the earnings of such companies are expected to move countercyclically to the economy, especially during downturns; even in economical growth periods, the performance of such companies is not expected to be spectacular; retail, food, health or telecoms operators are examples of defensive companies).

6.3.3 Funds

The instruments like money market instruments, shares or bonds, are all individual securities. However, rather than buying equity or debt issued by governments or corporations directly, one might want to buy shares of an investment company which itself owns various stocks or bonds. Such investment companies that fulfill certain regulatory conditions are often called *funds* and identified by the types of instruments they hold. The rationale behind such funds is diversification of instruments to reduce risks. An illustrative (but definitely not realistic) example is a fund owning shares of a company which sells ice cream as well as shares of a company which sells skiing gear. The idea behind this fund's diversification is when the weather is warm, the company selling ice cream will make a lot of money and when the weather is cold, the company selling skis and anoraks will make a lot of money.

Such funds, because they are diversified over several instruments, reduce the risk of loss for the investors. This advantage comes with a price, though: the management fee. Fund managers actively manage their funds by buying and selling instruments on a daily basis based on their analysis of economic, corporate and market data. This is hard and difficult work (some call it an art) and good fund managers belong to the best paid class of people in the world. The management fee required for a fund is typically around 2–3% of the invested amount. Some funds have more complex fee structures based on their level of success.

A *mutual fund* is basically a managed portfolio of stocks and/or bonds. They can be categorized as *open-end funds* and *closed-end funds*. Open-end funds, do not have a limit as to how many shares they can issue. So, when an investor purchases shares in an open-end fund, more shares are created (conversely, when an investor sells his shares, these shares are taken out of circulation). Such funds are priced usually on a daily basis, based on the amount of shares bought and sold and their price is based on the total value of the fund or the so-called net asset value (NAV). Open-end funds must maintain cash reserves to cover redemptions. Otherwise, the fund may have to sell some of its investments in order to pay the redeeming investor. On the other hand, closed-end funds are launched through an initial public offering (IPO) in order to raise money and then trade in the open market just like a stock (some even pay dividends). They only issue a set amount of shares and, although their value is also based on their NAV, the actual price of the fund is affected, like shares, by supply and demand of the fund itself. An important caveat here is that closed-end funds often use leverage to produce higher gains and are thus subject to higher risk than open-end funds [60].

6.3.3.1 Money Market Funds

Money market funds invest in high quality, low risk money market instruments like treasury bills with varying maturity dates, certificates of deposits from large banks or repurchase agreements (repo). The yields of money market funds should always be higher than individual short term certificates of deposit, because of the higher volumes involved and longer maturity dates (in practice this is not always the case because of the fund management fees). The major advantage of investing in a money market over a certificate of deposit is its liquidity irrespective of timing.

6.3.3.2 Bond Funds

Bond funds generally invest in medium to long term bonds issued by governments and/or corporations. Typically, a bond fund is diversified over issuers, maturities and credit ratings. Such instruments are examined in detail in the final Section of Chapter 8.

6.3.3.3 Common Stock Funds

Common stock funds or share funds typically invest in a variety of common stocks depending on the declared investment objective of the fund. For instance, a fund which observes the aging of the population might invest in pharma companies which produce drugs for the elderly, companies that produce hearing aids or insulin pumps. Another example is a high-tech fund which invests in companies that produce components using cloud technology or big data applications and might include risky start-ups with great potential growth. It might be too risky for an individual investor to invest in any single such company but by investing in a fund that invests in such companies, her risk is significantly reduced through diversification.

6.3.3.4 Balanced Funds

Balanced funds invest both in bonds and in shares, perhaps with a small portion in money market instruments. More aggressive funds have higher percentages of their capital invested in shares and more conservative funds have higher percentages of their capital invested in bonds and/or money market instruments.

6.3.3.5 Exchange Traded Funds (ETF)

An exchange traded fund (ETF) is an instrument that tracks an index (*e.g.,* Dow Jones Industrial Index) or a commodity (*e.g.,* copper with different delivery dates), but trades like a stock on a stock exchange. ETF's experience price changes throughout the day as they are bought and sold. ETF's have become very popular during the last couple of decades. Since all an ETF has to do is track an index, its management consists of daily adjustments of its components in a straight forward mechanistic manner. Therefore, their management fees, typically below 0.5%, are much lower than those of managed funds.

Table 6.1 Classes of commodities and some representative examples. Just for oil, there are about 100 different grades traded on commodities exchanges around the world

Energy	Metals	Livestock	Agriculture
Crude oil (WTI)	Gold	Cattle LV	Wheat
Crude oil (Brent)	Silver	Cattle FD	Corn
Kerosene (TOCOM)	Palladium	Lean hogs	Sugar
NYMEX Natural gas	Copper		Soy beans
PJM Electricity	Platinum		Orange juice
Coal	Aluminum		Coffee
Uranium			Cocoa
			Cotton
			Lumber

6.3.4 Commodities

Marx[3] defines a commodity as "a marketable item produced to satisfy wants or needs" [85]. This admittedly abstract definition is too general to be of much use here. In the sense of financial instruments, commodities are liquidly tradeable raw materials that come out of the earth and are strictly specified in terms of origin, quality and so on. Some of the most traded commodities are listed in Table 6.1.

When you want to buy a commodity for current delivery you go to a *spot market* and acquire the available supply. There is a spot market for all commodities and prices in these markets vary according to current supply and demand. However, most of the trading in commodity exchanges is in *future contracts* (see Chapter 9) which are contracts to deliver a commodity at some time in the future. The current price reflects what the market players think the price will be in the future. This way, *e.g.*, if in March, you think that the price of oil will rise, say in six months, you can buy a contract for September and sell the contract on that date. This is a bit different than buying stocks. Maybe the most important difference is in the leverage one can employ by actually putting up only a small proportion of the contract, typically 10%. Another important difference is that the maturity of a future contract is less than a year, whereas you can hold on to as stock forever.

6.3.5 Forex

It should not come as a surprise to the reader that foreign currencies can also be seen and treated as commodities. As an example, think of an investor whose reference

---

[3]Karl Marx, German philosopher and economist (1818–1883); famous for his theories about society, economics and politics, known as Marxism, an ideology which affected the 20th century to a great extent.

Table 6.2 Some commonly traded currencies and their international acronyms

Country	Currency	Acronym	Country	Currency	Acronym
USA	Dollar	USD	European Union	Euro	EUR
United Kingdom	Pound	GBP	Canada	Dollar	CAD
Japan	Yen	JPY	Switzerland	Franc	CHF
Australia	Dollar	AUD	Norway	Kroner	NOK
Turkey	Lira	TRY	China	Yuan	CNY
Hungary	Forint	HUF	India	Rupee	INR
Mexico	Peso	MXN	South Africa	Rand	ZAR

currency[4] is Euro. She might predict that British Pounds will gain in value during the coming months due to a certain conjectural juxtaposition. Therefore, she might buy a future contract on GBP/EUR to make a profit on her foresight without actually selling her Euros to buy Pounds. Such operations are called foreign exchange or Forex operations. Clearly, unlike commodities, specification are not necessary here. Also, mostly because of higher liquidity and volume, one can work with a much higher *leverage*: it is often sufficient to put up only 1% of the traded amount meaning that a single percent change in value of the traded currency can either double the initial investment or totally annihilate it.

In Forex trading all trade-able currencies have three letter acronyms. Some of the commonly traded currencies and their acronyms are listed in Table 6.2. Accordingly, *e.g.,* if you want to buy American Dollars against Turkish Lira you are trading USD/TRY. Sometimes the 'slash' is left out so that if you want to sell Swiss Francs against South African Rands you trade CHFZAR.

6.3.6   Derivative/Structured Products

A derivative financial instrument is a security where the value depends explicitly on other variables called underlyings. Most common derivative products are forward contracts, futures and options. Structured products combine classic investments like bonds with derivatives. Chapter 9 treats such products in detail.

---

[4]Many portfolios comprise instruments in different currencies. In such cases it is important to define a reference currency and follow the performance of the portfolio in that currency. It makes sense to choose as the reference currency the currency in which the investor expects to have most of his spending in the future. In most cases the reference currency will, therefore, be the currency of the country where the investor resides.

6.3.7 Real Estate

One of the more traditional investment areas is real estate, especially in countries or regions with a continuous and rapid growth of population coupled with migration from rural to urban centers. This results in the demand for real estate in urban centers to keep rising. This in turn affects the prices in or around urban areas. Therefore, long term investments in real estate have usually been profitable.

The most common direct real estate investment is the purchase of a home. The financial commitment consists of a down payment of around 20% and specific payments over a long period of time which can extend to 30 years. Since you have to live somewhere, buying a home can indeed be more profitable than renting over the years. One can, of course, also buy raw land as a real estate investment and sell it in the future with a profit. However, an important drawback of investing in real estate is the lack of liquidity of such assets compared to the liquidity of financial instruments.

Another difficulty with direct investments in real estate is that it is in the domain of investors with large capital bases. There are ways around this large capital requirement like equity trusts, construction trusts, development trusts, mortgage trusts which, together, are called real estate investment trusts, REIT. An REIT is a fund that invests in various real estate properties. This is very similar to a mutual stock fund and the shares of an REIT are traded as common stock in the market. Naturally, the regulatory constraints on REIT are quite different from those of bond or share funds.

6.3.8 Other Instruments

For the sake of completeness, let us mention some other investment types, albeit with lower liquidity than financial instruments. These are artworks [87] like paintings and sculpture (the current record price was paid for a painting by Paul Gauguin named 'Nafea Faa Ipoipo', which was sold for $300 million in 2011, antiques [112], precious stones like diamonds (perhaps highest value density in terms of dollars per cubic centimeters can be found in high quality and high carat gems) and collector's items like coins [15] or stamps [34] (the Swedish stamp called Tre Skilling Banco Yellow was sold for $2.3 million in 1996).

6.4 Return and Risk

It follows from the previous Section that the purpose of investing is to defer current consumption for an increased future consumption by increasing wealth in the meanwhile. The rate of return, or simply return, R, is defined as

Table 6.3 Various rates of return and their probabilities of occurring

Probability	0.05	0.1	0.1	0.1	0.1	0.1	0.1	0.1	0.1	0.1	0.05
Rate of return	−50%	−30%	−25%	−10%	0%	10%	20%	25%	30%	40%	60%

$$R = \frac{W_e - W_i + c}{W_i}, \tag{6.1}$$

where W_i is the wealth at the beginning of the investment period and W_e is the wealth at the end of the investment period. c denotes the cash flow generated by the investment during that period.

Risk is defined as the uncertainty regarding the expected rate of return of an investment. One has to remember that *any* investment is associated with some risk. Imagine that an investor is investing in shares of a company for an investment horizon of one year and she expects the rate of return to be 10% under current conditions. Further, she expects the rate of return to be 20% if the economy picks up speed and −20% if the economy slows down. The economists do not expect a major change in the economy during the coming year but there is a 15% chance of a recession and 10% chance of an economic boom. What is the expected rate of return of her investment?

To answer this question, let us make use of the expected value concept and compute the expected rate of return as the sum of all possible returns R_i weighted by their probabilities of occurring P_i

$$E[R] = \sum_i P_i R_i . \tag{6.2}$$

With the given numerical values, the answer is

$$E[R] = 0.15 \cdot (-0.20) + 0.10 \cdot (0.20) + (1 - 0.15 - 0.10) \cdot (0.10) = 6.5\% .$$

Let us now assume that the investor is advised by an oracle who can tell her the probabilities of various economic growth scenarios and how the return on her investment will be in each case as shown in Table 6.3.

In this case the expected rate of return calculated using (6.2) also yields 6.5%. In both cases the expected rates of return are the same. However, in the second case, the investor is much less certain that she will actually get that yield, because there are many more possible outcomes. Therefore, such an investment is seen as a higher risk investment.

How can one generalize the last statement? Can one define a measure of risk? There are several ways of measuring the risk of an investment ranging from the simplest (the range of the distribution) to the very sophisticated (*e.g.*, conditional value at risk, CVaR[5]). It is, however, common to use the standard deviation (square root of the variance) as the measure of risk where, with $\bar{R}$ denoting the expected value of the rate of return,

$$\text{Var}[R] = \sigma^2 = \sum_i P_i (R_i - \bar{R})^2 . \tag{6.3}$$

When one compares the risk of the two investments one can see that in the first case $\sigma = 0.040$ and in the second case $\sigma = 0.063$. Empirical studies have shown the investors to be, in general, risk averse. Consequently, if investors are given a choice between two investments with the same expected return but different standard deviations, they prefer the investment with the smaller standard deviation.

6.5 Utility

Economics studies human behavior as a relationship between ends and scarce means with alternative uses. Because means are scarce, economies need to allocate their resources efficiently. The concept of *utility* is implicit in the laws of demand and supply [124], and it represents the fulfillment a person receives from consuming a good or service. In other words, utility explains how individuals and economies optimize their satisfaction under the constraints dictated by the scarcity of resources. The admittedly abstract concept of utility does not represent an observable quantity but rather an arbitrary relative value.

Total utility is the sum of satisfaction that an individual gains from consuming a given amount of goods or services. A person's total utility, therefore, corresponds to that person's level of consumption. *Marginal utility* is the additional satisfaction gained from each extra unit of consumption. Total utility usually increases as more of a good is consumed, as one knows from daily life, but marginal utility usually decreases with each additional increase in the consumption of a good. This decrease is a manifestation of *the law of diminishing marginal utility*. In other words, total utility increases at a slower pace as the consumer increases his consumption.

Example: After eating a chocolate bar a person's craving for chocolate will be satisfied to a large extent. The marginal utility (and total utility) after eating one chocolate bar will be very high. But if the person keeps eating more chocolate bars, the pleasure she gets from each additional bar will be less than the pleasure she received from eating the one before, because her taste buds will start getting saturated

---

[5]Conditional value at risk is a risk assessment technique often used to reduce the probability that a portfolio will incur large losses. This is performed by assessing the likelihood (at a specific confidence level) that a specific loss will exceed the value at risk. Mathematically speaking, CVaR is derived by taking a weighted average between the value at risk and losses exceeding the value at risk.

Table 6.4 Utility concept explained using chocolate bars. Note that the first chocolate bar gives a total utility of 70 arbitrary units but the next three chocolate bars together increase total utility by only 18 additional units (adapted from Investopedia)

Chocolate bars eaten	Marginal chocolate utility	Total chocolate utility
0	0	0
1	70	70
2	10	80
3	5	85
4	3	88

and she will start feeling full. Table 6.4 shows that total utility increases at a slower rate as marginal utility diminishes with each additional bar.

The law of diminishing marginal utility is related with the negative slope of the demand curve: the less of something a consumer has, the more satisfaction he gains from each additional unit he consumes. The marginal utility he gains by consuming that product is therefore higher and it gives him a higher willingness to pay more for it. Prices are lower at a higher quantity, because the consumer's additional satisfaction diminishes as he demands and consumes more.

In order to determine what a consumer's utility and total utility are, economists employ demand theory which studies the relationship between consumer behavior and satisfaction. Classical economists still assume that consumers are rational and therefore aim at maximizing their total utilities by purchasing a combination of different products rather than more of one particular product. In this example, instead of spending all of her money on three chocolate bars having a total utility of 85, she should instead purchase one chocolate bar, which has a utility of 70, and perhaps a glass of milk, which has a utility of 50. This combination will give her a maximized total utility of 120 at the same cost as the three chocolate bars.   □

Behavioral Economics

A relatively new approach to economics, called Behavioral Economics, studies the effects of emotional and social factors on the decision making of individuals and institutions. Not surprisingly, departing from the assumption of *homo economicus* has profound consequences on resource allocation. Behavioral Economics studies the bounds of rational thinking of participants in the economy, broadly called *agents*. It also deals with the question of how market decisions are made and the mechanisms that drive public choice. Behavioral finance posits that (a) people often make decisions based on rules of thumb and not strict logic (heuristics), (b) there is a collection of anecdotes that constitute the emotional framework which agents rely on to understand and respond to issues and events (framing), and (c) there are market inefficiencies based on irrational decision making and resulting in mispricings.

Historically, neoclassical economists like Vilfredo Pareto (1848–1923) and Irving Fisher (1867–1947) began using psychological explanations to deal with the predictions of classical economics. This was followed by economic psychologists like George Katona (1901–1981) and Laszlo Garai (1935–) producing testable hypotheses about decision making under uncertainty. Finally, Behavioral Economics was born by challenging these hypotheses, most notably by Maurice Allais (1911–2010) and it further gained acceptance through the works of cognitive scientists like Amos Nathan Tversky (1937–1996) and Daniel Kahneman (1934–).

6.6 Utility Functions

As stated in the Section on Return and Risk, the outcomes of investments are uncertain, in other words random. Therefore, the allocation of funds in an investment portfolio with various instruments, each with their own expected returns and risks, is not a trivial problem. Utility functions may be used to provide a ranking to judge uncertain situations.

For a risk averse investor, a utility function U for wealth must be monotonously increasing ($U' > 0$) and concave ($U'' < 0$). The first property makes sure that an investor always prefers more wealth to less wealth. The second property is the manifestation of risk aversion. Some commonly used utility functions include [46]

1. the exponential function ($a > 0$)

$$U(x) = -e^{-ax},$$

2. the logarithmic function

$$U(x) = \log(x),$$

3. the power functions ($b < 1$ and $b \neq 0$)

$$U(x) = \frac{1}{b}x^b.$$

All of these functions have the same qualitative course: they are concave and increase monotonously.

Example: To illustrate this point, let us consider two investment opportunities. The first investment will either yield, with equal probability, x_1 or x_2 ($x_2 > x_1$). The second investment will yield $\frac{1}{2}(x_1 + x_2)$ with certainty. To rank both options, the expected utility is computed. The first option yields $\frac{1}{2}[U(x_1) + U(x_2)]$ and the second $U(\frac{1}{2}(x_1 + x_2))$. Since U is concave, the line between $U(x_1)$ and $U(x_2)$ is below the function of $U(x)$. Therefore, as shown in Figure 6.3, the second option has a higher utility. Both options have the same expected value, but the first option involves more

Fig. 6.3 A risk averse utility
function (adapted from [46])

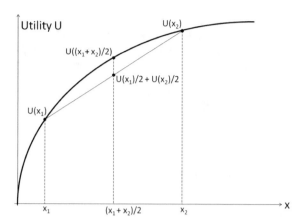

risk, since one might end up with $U(x_1)$. The second option is risk-free, since the
payoff is deterministic.   □

Since the shapes of utility functions are more interesting than their absolute values,
Pratt and Arrow have developed measures for risk aversion [7, 105]. These are

• the Arrow-Pratt measure of absolute risk aversion:

$$A(x) = -\frac{U''(x)}{U'(x)},$$

• the Arrow-Pratt measure of relative risk aversion:

$$R(x) = -x\,\frac{U''(x)}{U'(x)}.$$

Clearly, the higher the curvature of $U(x)$, the higher the risk aversion. The most
straight forward implications of increasing or decreasing risk aversion occur in the
context of forming a portfolio. As discussed in Chapter 8 on Modern Portfolio Theory,
risk aversion is measured as the additional marginal reward an investor requires to
accept additional risk.

Actually the concept of utility arose as an answer to a question posed by Nicolaus
Bernoulli[6] in 1713 and answered by his younger brother Daniel Bernoulli[7] in 1738.
Imagine a casino offers a game of chance for a single player in which a fair coin
is tossed at each stage. The pot starts at 1 dollar and is doubled every time a head
appears. The first time a tail appears, the game ends and the player wins whatever is
in the pot. Thus the player wins 1 dollar if a tail appears on the first toss, 2 dollars if
a head appears on the first toss and a tail on the second, 4 dollars if a head appears
on the first two tosses and a tail on the third, 8 dollars if a head appears on the first

---

[6]Nicolaus Bernoulli, Swiss mathematician (1687–1759).
[7]Daniel Bernoulli, Swiss mathematician and physicist (1700–1782).

three tosses and a tail on the fourth, and so on. In short, the player wins 2^{k-1} dollars if the coin is tossed k times until the first tail appears. The question is what would be a fair price to pay the casino for entering the game? The expected gain of this game can be calculated as

$$
\begin{aligned}
E &= \frac{1}{2} \cdot 1 + \frac{1}{4} \cdot 2 + \frac{1}{8} \cdot 4 + \frac{1}{16} \cdot 8 + \cdots \\
&= \frac{1}{2} + \frac{1}{2} + \frac{1}{2} + \frac{1}{2} + \frac{1}{2} + \cdots.
\end{aligned}
$$

In other words, the expected value grows without bound. Therefore, the expected win for the player, at least in this idealized form, is an infinite amount of money. Consequently, a rational person should play the game at *any* price if offered the opportunity. However, empirically, almost anybody who is asked this question, is not willing to pay more than a couple of dollars. This discrepancy is known as the St. Petersburg paradox.

The solution of this paradox lies in the irrational nature of people while making economic decisions. So Daniel Bernoulli says

> The determination of the value of an item must not be based on the price, but rather on the utility it yields. There is no doubt that a gain of one thousand ducats is more significant to the pauper than to a rich man though both gain the same amount.

6.7 Role of the Banks in Capitalist Economies

Banks are financial institutions whose operations consist of accepting deposits from the public and issuing loans. The receiving of deposits distinguishes banks from other financial institutions and is heavily regulated. Banks act as intermediaries when they mobilize the savings of the depositors (surplus units) to finance productive activities of borrowers (shortage units) [54].

Banks thus contribute to the functioning of an economy in a most significant way. They also reduce friction costs by achieving economies of scale in transaction costs. A small saver who would have to make do without a bank's services has to search and contact a borrower, negotiate with the borrower on terms of the loan, has to diversify among several borrowers to reduce his risk, monitor the performance of his loans and enforce his rights in case of tardy borrowers. A bank does all that for the saver and given the large number of savings and deposits by banks, related transaction costs are smaller. On top of that, the banks enjoy informational advantages over individual investors because of their access to privileged data on borrowers with accounts at the bank. Banks reduce transaction costs also by provision of convenient places of business, standardized products and less costly expertise through the use of tested procedures.

The existence of supervisory and regulatory bodies which ensure conformity with acceptable codes of behavior frees the saver from the burden of collecting information and monitoring the banks. Lewellyn's broadly cited work on this subject [77] lists eight significant elements in this context:

- Information advantages,
- Imperfect information,
- Delegated monitoring,
- Control,
- Insurance role of banks,
- Commitment theories,
- Regulatory subsidies, and
- Special role of the banks in the payment systems.

Another important role of the banks is to provide liquidity. Borrowers and lenders have different liquidity preferences. Banks pool funds together and rely on the law of large numbers to be able to offer liquidity to their customers. Actually, just by looking at the balance sheet of a bank you can see that the bank accepts deposits and in turn provide transaction services (liabilities side) and that it issues loans, thereby providing liquidity (assets side).

A large company can fund itself through the capital markets either by selling equity (common stock) or selling debt (bonds). However, smaller companies cannot do that for several reasons. Therefore, for them, bank lending (commercial credits) is vital. Banks build relationships with their customers that give them valuable information about the details of their operations and in difficult times (like cash flow problems or even recessions), companies with close relationships with the bank are better able to obtain financing to endure the crisis (at least in principle).

As the Bank for International Settlements (BIS) in Basel likes to point out, payment and settlement systems are to economic activity what roads are to traffic: necessary but typically taken for granted, unless they cause an accident or bottlenecks develop. Indeed, banks administer payment systems that are central to an economy. By means of these systems, they execute customers' payment instructions by transferring funds between their accounts. Also that way, customers can receive payments and pay for the goods and services by checks, orders, credit cards or debit cards. Ultimately, the funds flow between individuals, retail and wholesale business quickly and safely thanks to the payment systems.

6.8   Exercises

A1: Explain the concept of pure time value of money in your own words.

A2: What is risk premium?

A3: Define economics in your own words.

A4: How does finance relate to economics?

A5: Name some fixed income investments. Why are they called "fixed" income instruments?

A6: What are common stocks?

A7: What are exchange traded funds (ETF)?

A8: Why are the management fees for ETF's lower than the fees for managed funds?

A9: What commodity classes do you know?

A10: Make a competition with your colleagues to see who can name more commodities.

A11: What are the problems related to investing in real estate? How can those problems be mitigated?

A12: How do return and risk relate to each other conceptually and formally?

A13: Define utility in your own words with an example from your daily life.

A14: Why are banks so important for the economy and well being of a country?

A15: You have made an invention. You want to start and run a company to make a product of that invention, produce it, market it and sell it. List all the difficulties you would run into if there were no banks.

B1: What is the rate of return of the financial instrument B1 which pays over its life time of five years a quarterly interest of 1% and the total invested amount at the end of its life time?

B2: What is the rate of return of another instrument B2 which pays no interest but pays a bonus of 20% together with the total invested amount at the end of its life time?

B3: What is the rate of return of the financial instrument B3 which pays over its life time of five years a yearly interest of 2% above the inflation and the total invested amount at the end of its life time? Economists expect a low inflation of 2% per year over the next five years.

B4: Would you rather invest in B1, B2 or B3? Explain your reason carefully!

B5: Imagine that a risk averse, rational investor is given two different investment opportunities A and B to choose from. In A the investor invests in a fund which pays a 4% return in one year with certainty. In B the investor invests in a fund which pays a 3% return in one year with 70% probability and 10% return in one year with 29% probability and has adefault probability of 1%. Which investment will he choose and why?

B6: You have the choice of investing $100 in a financial instrument B4 which pays an interest of 2% at the end of the year together with your total invested amount.

You also have the option of buying a lottery ticket with $100 which, in one year, either pays nothing or $1 million with a probability of 0.01%. Which choice would be more rational?

B7: Forget about being rational for a moment. Would you have chosen the investment B4 in the previous exercise or the lottery ticket? Would your choice have been the same ten years ago? Explain your answers using the concept of utility.

C1: There are a number of folk wisdoms in the financial markets. Most famous of these are "Markets are never wrong—opinions often are", or "Never average losses", or "Accepting losses is the most important single investment device to insure safety of capital". Study some of these wisdoms and see how well they are backed by empirical data.

C2: Stock markets move cyclically. The underlying cycles are business cycles of expansion and recession which take around six years to complete. Superimposed on that, there are believed to be shorter cycles which take only months to complete. One saying in the markets is "Sell in May and go away". Study the wisdom of this saying empirically in different markets and find conditions and qualifying statements to improve on it like "Sell in May and go away but you may stray".

C3: Insider trading is the trading of a public company's stock by individuals with access to non-public information about the company. In most countries this is illegal. Other than ethical considerations, there are studies which claim that insider trading raises the cost of capital thus decreasing the economic growth. However, some economists have argued the opposite, claiming that insider trading is actually good for the economy. Study both arguments and defend your opinion in a paper.

Chapter 7
Bonds

> *The United States can pay any debt it has because we can always print money to do that. So, there is zero probability of default.*
>
> — Alan Greenspan

> *An investment in knowledge pays the best interest.*
>
> — Benjamin Franklin

7.1 Introduction

This Chapter defines various parameters of bonds like coupon, maturity, par value, ownership type and indentures. Types of bonds differentiated by the type of collateral behind the issue are introduced. Bond returns and valuations are derived heuristically based on the return concept defined in the previous Chapter. Fundamental determinants of interest rates and bond yields, together with the Macaulay duration are discussed with examples. The yield curve is presented together with its implications for the economy.

A bond is a long term public debt of an issuer which is marketed in an affordable denomination. An understanding of bonds is essential, because they make up a significant part of investment opportunities and are necessary for diversification of investment portfolios. Bonds are considered fixed income instruments, because the debt service obligations of the issuer are fixed.

The issuer agrees to pay a certain amount of periodic interest to the holder and repay a fixed amount of principal at the date of maturity. Traditionally, interest on US Dollar denominated bonds is paid out every six months. Euro denominated bonds pay interest in yearly cycles. The principal, the par value of the issue, is due at maturity. It can be as high as a million dollars but is seldom below $1'000.

© Springer International Publishing AG 2018
S. S. Hacısalihzade, *Control Engineering and Finance*, Lecture Notes in Control and Information Sciences 467, https://doi.org/10.1007/978-3-319-64492-9_7

Another important characteristic of bonds is their term to maturity. Short term debt with maturities below a year is known as money market instruments. Long term debt with maturities above seven years is simply called bonds. Intermediate debt with maturities between short and long term are often referred to as notes.

Usually, the largest borrower in any country is the state. The state needs large amounts of funds to cover its running costs (*e.g.,* salaries of state employees, educational or military expenses, financial expenses, etc.) as well as investments into the future (*e.g.,* building roads, dams, universities, etc.). The main sources of income for the state are taxes (*e.g.,* income tax, corporate tax, sales tax, VAT, etc.). However, most of the time, the incomes are not sufficient to cover the expenses. This budget deficit has to be financed somehow. The state can, of course print money to cover this deficit. However, due its inflationary nature, this is often not the preferred course of action. Another option for the state is to sell its assets. However, "family silver" can be sold only once. Therefore, the only remaining solution is to borrow the missing amount. This is done through the issuing of short term *treasury bills* or longer term *government bonds*. A popular version of government bonds are so-called *zero coupon bonds* also known as *zero bonds*. As their name suggests, such bonds do not pay out any interest but are issued at a lower price than their nominal face value. Their value increases with time and reach the face value at maturity.

Just like the state, private corporations which need financing but do not want to (or cannot) get a commercial loan from a bank, can also issue bonds. Such instruments, known as *corporate bonds*, are very similar in their nature to government bonds. However, because of their higher risk, they have to yield more interest than government bonds.

7.2 Bond Parameters

One can characterize a bond in different ways. Each bond has intrinsic features that relate to the issue itself. There are different types of bonds with various indenture provisions that affect the yield of a bond.

The essential parameters of a bond are its coupon, maturity, par value and its ownership type. The *coupon* indicates the periodic income that the owner will receive over the holding period of the bond. It also goes under the names of coupon income, interest income or nominal yield (see Figure 7.1). The *maturity* of an issue specifies the date when it will be redeemed. The *par value* represents the original principal value of the bond. The principal value is not necessarily the market value of the bond. The market value varies with time according to the prevailing capital market conditions. Another defining characteristic of a bond is its *ownership type* as bearer versus registered. The bulk of bonds are *bearer bonds* and the issuer does not keep track who owns the bond as they are traded. The current owner is entitled to the coupon payment and gets it through financial institutions like a commercial bank in a routine manner. The issuers of *registered bonds*, on the other hand, keep track of owners of record and at the time of interest or principal payment pay the owner of record by check.

Fig. 7.1 In earlier times, bonds were issued on colorful paper with hard to counterfeit designs on them. Clearly written on those bond issues were the characteristics of the bond including whether it was a bearer or registered issue, its par value, the issue and maturity dates as well as the interest rate and coupon payment dates. At the bottom of the bonds were coupons which could be cut and exchanged for cash on coupon payment dates. Bonds no longer physically entail coupons but the term stuck. Here is a picture of a $1'000 bond of Pacific Rail Road Board in San Francisco, issued on May 1, 1892 with a maturity date of May 1895, an interest rate of 7% and coupon payment dates of November 1 and May 1. One can also see the six coupons at the bottom

7.3 Types of Bonds

Unlike common stocks, bonds come in a multitude of flavors. Perhaps the most significant differentiator is the type of collateral behind the issue. Senior bonds are secured by legal claim on specific physical properties (*e.g.,* real estate, factories, ships, etc.) of the issuer. Unsecured or junior bonds, on the other hand, lack this kind of security. Perhaps the most common type of junior debt is a debenture which is secured by the general credit of the issuer and not backed by any specific assets. Subordinated debentures represent a claim on income of the issuer that is subordinated to the claim of another debenture (this is the most junior type of bond, because, legally, interest on such bonds need to be paid only if sufficient income is earned).

Although most bonds have to be serviced by the issuer until its maturity date, some bonds have a call feature, which either allows the issuer to retire the bond (pay it back) anytime during the life of the bond with perhaps just a month's notice or retire it after a certain date specified at issue. The investor must always be aware of this feature, because it might influence the expected yield significantly. Callable bonds are popular by the issuers during times of high interest rates. For instance, an issuer might not be able to issue a 10 year bond below a coupon of 15%. However, depending on where the economy is on the current business cycle, he expects the interests to drop significantly after five years. Therefore, the issuer stipulates that the bond *may* be called after five years. If, indeed, interest rates are lower by that time, the issuer might issue a new bond which has a coupon of, say, 7%, call the first bond (fully or partially) and use the proceeds of the new issue to pay pack the principal of the first bond.

Another type of bonds have variable coupons. In environments of unpredictable interest rates, bonds which tie the coupon rate to a price index are called inflation adjusted bonds (*e.g.,* Turkish 10 year government bonds which pay coupons 3% above the consumer price index, thus protecting the principal against inflation). There are also bonds that tie the coupon rate to LIBOR[1] (*e.g.,* European credit agencies' bonds which pay 1.5% on top of the current LIBOR rate in Euro) or to the yield of reference bonds (*e.g.,* bonds of Indian banks which pay a coupon 1% on top of a reference Indian government bond) that are popular with the investors. A not so uncommon rarity among bonds is the class of perpetual bonds. A perpetual bond is a bond with no maturity date. Therefore, it may actually be treated as common stock (equity)) and not as debt. Issuers pay coupons on perpetual bonds forever (which, admittedly, is a long time), and they do not have to redeem the principal. Most perpetual bonds are deeply subordinated bonds issued by banks. They are considered as Tier 1 capital,[2] and help the banks fulfill their capital requirements. Most perpetual bonds are callable with a first call date of more than five years from the date of issue.

7.4 Bond Returns

To calculate the return rates of bonds one can write (6.1) as

$$R_t = \frac{P_{t+1} - P_t + c_t}{P_t},\tag{7.1}$$

where R_t is the rate of return of the bond during period t, P_{t+1} is the market price of the bond at the end of period t, P_t is the market price of the bond at the beginning of period t, c_t denotes the coupon payments of the bond during period t.

---

[1]LIBOR (London Inter Bank Offered Rate) is the average interest rate estimated by leading banks in London that they would be charged if borrowing from other banks. It is the primary benchmark, along with the Euribor, for short term interest rates around the world.

[2]Tier 1 capital is the core measure of a bank's financial strength from a regulator's point of view. It is composed of common stock and disclosed reserves (or retained earnings), but may also include non-redeemable, non-cumulative preferred stock [10].

Note that the only variable known in advance in (7.1) is the coupon payments that are stipulated at the issuing of the bond (except for variable coupon bonds). Since the bond prices are subject to market conditions (specifically the prevailing interest rates), the returns are random and subject to fluctuations during the lifetime of a bond. Consequently, it is possible to make or lose money by buying and selling bonds. Therefore, a common strategy is to buy a bond at time of issue and keep it until maturity. This way the investor knows that (assuming that the issuer does not default) she will get her principal at maturity and the coupon payments during the lifetime of the bond. The drawback of this strategy is the opportunity cost[3] incurred by the investor if the current interest rates are above the coupon rate of the bond.

7.5 Bond Valuation

The reader might wonder how the price of a bond can be determined at anytime during the lifetime of a bond. Like any other investment, bonds are valued on the basis of their future stream of income, discounted for the risk borne. Intuitively, it should be clear that the price of a bond is determined by its coupon rate, by the time left to its maturity (or potential call date), the prevailing market indexInterest!rateinterest rates and the risk premium.

The price of a bond can be seen as the present value of its expected cash flows as given by the formula

$$P = \sum_{t=1}^{n} c_t \frac{1}{(1+i)^t} \, , \tag{7.2}$$

where n is the number of periods in the investment horizon (term to maturity), c_t is the cash flow (periodic interest income and principal) received in period t and i is the rate of discount (market yield) for the issue. In most practical cases, the rate of discount is the promised yield to maturity that can be earned by purchasing the issue and holding it until expiry. Aggressive bond investors buy and sell bonds. For them, realized yield is a more relevant performance indicator. For such investors (7.2) represents an expected yield.

One speaks of five different types of yield in bond market terminology: nominal yield, current yield, promised yield, yield to call and realized yield. *Nominal yield* is the coupon rate. *Current yield* is similar to dividend yield of stocks and is computed as

$$CY = c_t / P_m \, , \tag{7.3}$$

---

[3]Opportunity cost refers to a benefit that a person could have received, but gave up, to take another course of action.

where c_t is the annual coupon payment of the issue and P_m is the current market price of the issue.[4] As one can see, current yield excludes capital recovery (the potential for capital gain or loss).

Perhaps the most important bond valuation model is the *promised yield*. It indicates the fully compounded rate of return (coupon payments being reinvested in the issue) offered to the investor at prevailing prices, assuming that the investor holds the issue until its maturity. Therefore, it is also known as *yield to maturity* (YTM). If coupons are not reinvested or if the future investment rates during the lifetime of the issue are less than the promised yield at purchase, then the realized yield will be less than the promised yield. This effect of earning interest on interest (compounding) is often overlooked, but becomes more important as time to maturity and as coupon rates increase. To visualize this effect, imagine that an investor has 25 year US Treasury Bonds with a face value of \$1'000'000 and 8% yearly coupon. At the end of 25 years he will get his \$1'000'000 principal back and he will have received (25 x 1'000'000 x 0.08 =) \$2'000'000 in coupon payments, assuming that he reinvested the coupons in the same bond. The interest he will receive on the coupon payments will be about \$4'100'000 (the interested reader can check this assuming there will be 50 periods at the end of which 4% interest is received).

To calculate the YTM of a bond at any time, one can take (7.2) and modify it slightly to

$$P_m = \sum_{t=1}^{n} \frac{c_t}{(1+i)^t} + \frac{P_p}{(1+i)^n}, \tag{7.4}$$

where P_p is the par value of the obligation, n is the number of years to maturity, c_t is the annual coupon of the issue, and P_m is the current market price of the bond; i then denotes the YTM of the bond. Clearly, there is no general analytical way to solve this equation for YTM but the Excel function YIELD solves the problem for the user numerically. The calculation of the YTM of a \$1'000 8% bond with 20 years remaining to maturity and a current price of \$900 using that function results in 8.17% assuming yearly coupon payments.

Example: Let us calculate the YTM of a USD bond with par value 100, market value 108, 5% annual coupon and four years left to maturity. To be able to do that one would have to solve the following equation for i:

$$P_m = \sum_{t=1}^{4} \frac{c_t}{(1+i)^t} + \frac{P_p}{(1+i)^4}$$

$$= \underbrace{\frac{c_1}{(1+i)} + \frac{c_2}{(1+i)^2} + \frac{c_3}{(1+i)^3} + \frac{c_4}{(1+i)^4}}_{\text{discounted coupon payments}} + \underbrace{\frac{P_p}{(1+i)^4}}_{\text{capital pay back}}.$$

---

[4]Caveat emptor: many inexperienced investors make their investment decisions by just looking at the coupon of a bond and forgetting to look at its price or call date.

One can see that it is not easy to solve this equation for i analytically.[5] Therefore, using the YIELD function in YIELD with the appropriate parameters as
YIELD(today(); today()+4*365; 5; 108; 100; 2) results in 4.6%. The first parameter
is the settlement date, the second parameter is the maturity in (exactly) four years,
the third parameter is annual coupon as percentage, the fourth parameter is the market price of the bond, the fifth parameter is the par value of the bond, and the last
parameter is the number of coupons per year.   □
    For increased accuracy, one uses the formula

$$P_m = \sum_{t=1}^{2n} \frac{c_t/2}{(1+i/2)^t} + \frac{P_p}{(1+i/2)^{2n}}, \tag{7.5}$$

for issues which pay semiannual coupons like bonds denominated in USD. *Yield to
call* can be calculated using formulas (7.4) or (7.5) and taking the next possible call
date as the date of maturity.
    Bond prices are quoted in two different ways. The so-called *dirty price* is the
actual amount one has to pay to purchase a bond. However, the most-quoted price
in the bond market is the *clean price*, which is an artificial price. It is equal to the
dirty price minus accrued interest. The accrued interest is the prorated amount of the
next coupon payment. Clean price is popular, because it enables the comparison of
bonds with different coupon payment dates and also because it does not show a step
decrease as the result of a coupon payment.
    There are many books [44, 57, 98], special hand held calculators (*e.g.,* HP 17BII
Financial Calculator, TI BAII Plus Professional) or even on-line calculators which
can help bond analysts with their calculations.

7.6 Fundamental Determinants of Interest Rates and Bond Yields

As already pointed out, the value of a bond is the present value of its future cash flow
stream. However, the prevailing interest rates are bound to change with time and so
does the value of a bond. Figure 7.2 shows the historical change of interest rates for
USD.
    Conceptually, the interest rates i can be modeled as

$$i = RFR + RP + I_e, \tag{7.6}$$

where *RFR* indicates the risk-free rate of interest (taken as the interest the government
pays on its debt, namely government bonds), *RP* the risk premium and I_e the expected

---

[5]The analytical solution of such quartic equations was published already in 1545 by Gerolamo
Cardano in his book *Ars Magna*. The general solution of a quartic equation, requiring the solution
of a cubic equation (also credited to Cardano), is so unwieldy that it is seldom used, if at all.

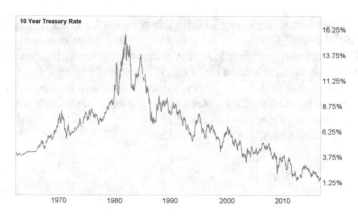

Fig. 7.2 Interest rates of ten year Treasury Notes (US Government bonds). Note the contrast between the peak rates above 15% at the beginning of 1980s and the rates less than 4% in the 2010s. Source: Federal Reserve

Table 7.1 Yields of various bonds on October 20, 2016. ISIN stands for International Securities Identification Number and uniquely identifies any internationally traded stock, bond, fund, etc.S&P stands for Standard and Poor which is an internationally recognized independent credit rating agency which classifies issuers of bonds according to strictly defined criteria

ISIN code	Issuer	Maturity date	YTM	S&P rating
XS0559237952	Lebanese Government	Nov. 12, 2018	5.36%	B-
US71647NAB55	Petrobras	Jan. 15, 2019	3.53%	B+
USP14996AG02	Banco Panamericano	Apr. 23, 2020	8.92%	CCC
USG2370YAB23	Consolidated Minerals	May 15, 2020	32.18%	CCC+
USP97475AG56	Republic of Venezuela	Dec. 9, 2020	29.09%	CCC
US9128282A70	US Treasury	Oct. 17, 2026	1.79%	AA+

inflation. Whereas both *RFR* and I_e are the consequence of global market forces, *RP* is a characteristic of the issuer and the specific issue. With the help of Equation (7.6) one can see that the differences in the yields of corporate and government issues are not caused by economic forces but by differential issuer and issue characteristics. Table 7.1 shows yields of selected USD denominated bonds as a snapshot taken on a particular date.

The three major components of the risk premium of any issue are quality considerations which reflect risk of default as captured by rating agencies, term to maturity as it affects the level of uncertainty (longer terms mean a higher level of risk) and indenture provisions like the type and amount of collateral and, naturally, the call feature.

7.7 Rating Agencies

A credit rating agency is a company that ascribes ratings that assess a debtor's ability to pay back debt (both interest payments and the principal) on time. This way, a credit rating facilitates the trading of such securities on a secondary market. The rating by a credit rating agency affects the interest rate that a bond has to pay out. Higher ratings lead to lower interest rates and vice versa. The debt instruments rated by rating agencies include government and corporate bonds.

Most importantly, a debtor is rated as being investment grade or non-investment grade. Many of the largest institutional investors like mutual funds, insurance companies or pension funds are, by law, regulation or their internal by-laws, not allowed to invest in non-investment grade bonds. Considering that these institutional investors alone manage close to a USD 100 trillion (10^{14}), ratings pronounced by these agencies carry enormous weight. Finer grades within these two groups also exist, resulting in significant differences in how much a debtor has to pay in interest in order to be able to borrow money in the international capital markets.

Historically, US's sovereign long-term credit rating AAA was the highest possible rating. This changed in 2011 when S%P lowered it to AA+. In long term studies prior to that change, it has been found that depending on its credit rating, the debtor had to pay a spread above US Treasuries. Table 7.2 shows the historical default rates and additional interest a borrower has to pay for various ratings.

However, one has to take the ratings of such agencies with a healthy pinch of salt: just remember that during the financial crisis of 2008, bonds worth hundreds of billions were downgraded from the best rating to junk level in a very short period of time, suggesting that the initial ratings were faulty [88]. Similarly, in 2012, rapid downgrading of countries (sovereign debt) precipitated a financial crisis in Europe [28].

There are dozens of credit rating agencies. However, the "Big Three" credit rating agencies (Moody's, Standard and Poor's, Fitch) control about 95% of the ratings business.

7.8 The Yield Curve

The yield curve is a curve which plots the yields of public debt with similar features as a function of their maturities. This curve shows the relation between the level of interest and the time to maturity of the debt for a given borrower in a given currency. This dependency is significant, because several empirical studies *e.g.,* [82], have found that the yield curve has different shapes at different times and that the shape of the yield curve has a lot to say about the economic cycle, because it represents the market players' expectations of the future economic developments. Figure 7.3 shows qualitatively different types of yield curves.

Table 7.2 The second column indicates this average spread in percentage points. This means, for instance, if the 5 year US Treasuries at a given time paid an interest of 5%, a borrower with the BBB rating would have to pay for the same maturity an interest of 6.66%. The third column indicates the default rate for investment grade and non-investment grade bonds during last five years averaged over 30 years prior to 2014.　Source: S&P's 2014 Annual Global Corporate Default Study and Rating Transitions Report

Rating	Spread	Default rate
AAA	43	
AA	73	1.03%
A	99	
BBB	166	
BB	299	
B	404	15.49%
CCC	724	

There are a number of economic theories attempting to explain how yields vary with maturity. According to the so-called expectations hypothesis (which is intuitive, relatively simple and well documented, hence widely accepted both in academia and the market), the shape of the yield curve is explained by how market players expect the interest rates to change over time. In other words, any long-term interest rate is the mean of current and future one year rates expected to prevail during the remaining life time of an issue, which can be expressed with the formula

$$(1 +_t i_N) = \sqrt[N]{(1 +_t i_1)(1 +_{t+1} i_1) \cdots (1 +_{t+N-1} i_1)}, \tag{7.7}$$

where i_N is the actual long term rate, N is the number of years left to maturity, $_t i_1$ is the current one-year rate and $_{t+k} i_1$ is the expected one-year rate during a future period $t + k$. For small interest rates one can simplify (7.7) by using the arithmetic mean instead of the geometric mean. If short-term rates are expected to rise in the future, then the yield curve will be ascending. If short-term rates are expected to fall, then the long-term rates will be lower than short-term rates and the yield curve will be descending.

The slope of the yield curve is an important predictor of future economic growth, inflation, and recessions. An inverted yield curve is often the harbinger of an imminent recession and a positively sloped yield curve often precedes a period of inflationary growth. Empirically, it is observed that all of the recessions in the US during the last 50 years have been preceded by a declining (inverted) yield curve [40]. Figure 7.4 shows the yield curve for Turkish government debt on October 20, 2016.

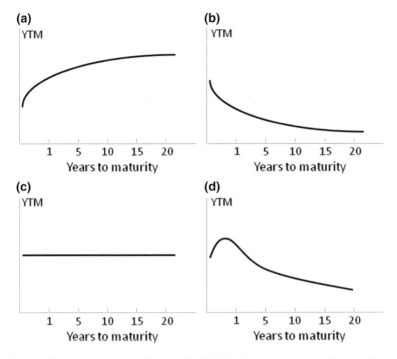

Fig. 7.3 A yield curve plots the yield to maturity (YTM) of similar quality bonds versus their terms left to maturity. There are qualitatively four different types of yield curves. **a** Rising, **b** Declining (inverted), **c** Flat, and **d** Humped

7.9 Duration

Some empirical truths (which are both intuitive and proven mathematically [81]) that should be kept in mind are (1) Bond prices move inversely to bond yields, (2) For a given change in market yield, changes in bond prices are greater for longer-term maturities, (3) Price volatility of a bond varies inversely with the coupon rate and (4) Price volatility varies directly with its term to maturity.

The last two of these facts are difficult to balance when selecting bonds. A measure developed by Macaulay [80] which combines those two factors is called *duration*. This measure takes into account not only the ultimate recovery of capital at maturity, but also the size and timing of coupon payments before maturity. Technically, duration is defined as the weighted average time to full recovery of principal and interest payments, or in other words, the weighted average maturity of cash flows.[6] Formally,

---

[6]There are several different definitions of durations which are known under the names *modified duration, effective duration, key-rate duration* or *Macaulay duration*. This Book uses only the Macaulay duration.

$$D = \frac{\left(\sum_{t=1}^{n} \frac{tc_t}{(1+i)^t} \right)}{\left(\sum_{t=1}^{n} \frac{c_t}{(1+i)^t} \right)}, \tag{7.8}$$

where t is the period in which the coupon and/or principal payment is made, c_t is the coupon and/or principal payment made in period t, i is the yield to maturity and n is the number of periods left to maturity. As can be seen from this definition, the duration of a zero coupon bond is equal to the maturity of the bond. Coupon paying bonds always have durations less than their maturities.

Example: What is the Macaulay duration of a five-year, $100 bond with a 10% coupon paid annually?

$$D = \frac{\left(\frac{1 \cdot 10}{(1.1)^1} + \frac{2 \cdot 10}{(1.1)^2} + \frac{3 \cdot 10}{(1.1)^3} + \frac{4 \cdot 10}{(1.1)^4} + \frac{5 \cdot 10}{(1.1)^5} + \frac{5 \cdot 100}{(1.1)^5} \right)}{\left(\frac{10}{(1.1)^1} + \frac{10}{(1.1)^2} + \frac{10}{(1.1)^3} + \frac{10}{(1.1)^4} + \frac{10}{(1.1)^5} + \frac{100}{(1.1)^5} \right)}$$

$$= \frac{416.99}{100} = 4.17 \text{years}.$$

$\square$

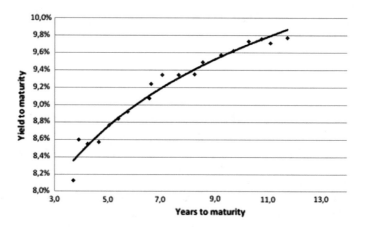

Fig. 7.4 The yield curve for outstanding Turkish government debt on October 20, 2016 is constructed from single actual yields which are fitted to a logarithmic curve. Based on this curve, a period of inflationary growth can be expected

7.10   Exercises

A1: What is a bond?

A2: Which types of bonds do you know?

A3: What are the major parameters of bonds?

A4: What are treasury bonds?

A5: How is the value of a bond calculated?

A6: What is yield to maturity?

A7: Why do yields of different bonds differ?

A8: How are the credit ratings of different issuers determined?

A9: What is the yield curve?

A10: What does the qualitative shape of the yield curve say about the future of the economy?

A11: What is the duration of a bond? Why is it an important metric?

B1: Calculate the yield to maturity of a bond with $100 par value which trades for $96. It matures in two years and has a 5% annual coupon.

B2: A bond with EUR 100 par value which matures in two years and which has a 3% annual coupon trades for EUR 102. Calculate its yield to maturity.

B3: IBM has a 2% coupon, 5 year bond and a 2% coupon 10 year bond. Which one do you expect to be more expensive and why? In one year time the market interest rates go up. Which bond's price will be affected more, in what way and why?

C1: Develop an App which allows you to compare the yields of different bonds. Begin with the basics and enhance the App with sophisticated features.

C2: Develop an App which allows you to compare the durations of different bonds analogous to C1.

C3: Develop a software tool which can make recommendations on which bonds to buy from a given list based on their yields, durations and investor characteristics like her cash needs over time and risk appetite. Make sure you take the issuer risk into account when you make recommendations.

C4: There are different definitions of duration. Study these definitions in detail and describe their nuances. Which definition is best suited for which type of problems? Use these definitions to determine the interest rate sensitivity or exposure of various types of bonds like zero bonds annual coupon bonds, annual annuity bonds and fixed-floating swaps.

Chapter 8
Portfolio Management

Everything existing in the universe is the fruit of chance.

— Democritus

I don't believe in providence and fate, as a technologist
I am used to reckoning with the formulae of probability.

— Max Frisch

8.1 Introduction

This Chapter first looks at an approach for methodically diversifying among different assets, the so-called Modern Portfolio Theory (MPT), as applied to equity stocks. The control engineering problem of selecting the sampling frequency is brought in connection with the question how frequently to re-balance a portfolio by selling and buying individual stocks. Empirical studies with examples from Swiss, Turkish and American stock markets are presented. Smart algorithms for managing stock portfolios based on these empirical studies are proposed. Management of bond portfolios using the Vašíček model and static bond portfolio management methods finish off this chapter.

8.2 Measuring Risk and Return of Investments

In the presence of risk, the outcome of any action is not known with certainty. As discussed in Chapter 3, the outcomes are usually represented by their frequency distribution functions. Let us remember that a frequency distribution function is a listing of all possible outcomes along with the probability of the occurrence of each.

© Springer International Publishing AG 2018   215
S. S. Hacısalihzade, *Control Engineering and Finance*, Lecture Notes in Control
and Information Sciences 467, https://doi.org/10.1007/978-3-319-64492-9_8

Table 8.1   A frequency distribution function example

Event	Probability	Return
1	0.20	12%
2	0.50	9%
3	0.30	6%

Table 8.1 shows an example of a frequency distribution function. The investment has three possible returns. If event 1 occurs, the investor receives a return of 12%; if event 2 occurs, he receives 9% and if event 3 occurs, he receives 6%.

The possibilities for real assets are sufficiently numerous that developing a table like Table 8.1 for each asset becomes practically impossible. In general, two basic measures are used for capturing the characteristics of a frequency function. The first one is the expected value (weighted average value) of the frequency function which is the expected return. The second one is the dispersion around that expected value, *i.e.,* the standard deviation, which is used as a risk measure.

The expected value of the frequency distribution function is calculated as:

$$E[R_i] = \sum_{j=1}^{M} P_{ij} R_{ij} , \qquad (8.1)$$

where R_{ij} denotes the j-th possible outcome for the return on asset i. P_{ij} is the probability of the j-th return on the i-th asset and M is the total number of possible events.

The standard deviation from the expected value is calculated as:

$$\sigma_i = \sqrt{\sum_{j=1}^{M} \left[P_{ij}(R_{ij} - E[R_i])^2 \right]} . \qquad (8.2)$$

In general, just the mean and the standard deviation (or equivalently the mean and the variance) are used rather than the frequency distribution function itself. This makes perfect sense if R_{ij} are assumed to be normally distributed random variables.

8.3   Modern Portfolio Theory

In a seminal paper [83] Markowitz[1] introduced the concept that maximize expected return for a given level of risk, reaping the benefits of diversification by investing in more than one stock. The Modern Portfolio Theory (MPT), as this theory is famously

---

[1]Harry Markowitz, American economist, (1927–).

known, aims to reduce the total risk on the return of the portfolio by combining different assets whose returns are not perfectly correlated. Key components of this theory include

- defining the rate of return of a portfolio of assets as the sum of the returns of its components,
- defining the risk as the standard deviation of the return,
- determining the set of portfolios that has the maximum return for every given level of risk, or the minimum risk for every level of return—formally called the Efficient Frontier,
- assuming that the asset returns are elliptically (for instance normally) distributed random variables, that investors are rational (they aim to maximize their gains), and that the markets are efficient (asset prices fully reflect all available information).

Efficient Markets

One of the corner stones of finance is the so-called Efficient Market Hypothesis (EMH) which can be traced back to Friedrich Hayek (1899–1992), an Austrian-British economist and philosopher best known for his defense of classical liberalism, and perhaps as the most notable intellectual champion of the free market economy as opposed to planned economy in the 20th century [53]. EMH is based on the argument that markets are the most effective way of combining the information distributed amongst individuals within a society. Rational market players with the ability to profit from private information, are motivated to act on their private information. That way they contribute to increasingly more efficient market prices. This process converges to the situation where market prices reflect all available information and prices can only move in response to new information.

Although EMH is intuitive and generally accepted, some empirical studies have found results which are at odds with it: stocks with low book-market ratios (often called value stocks) tend to achieve much higher returns than what would be expected using another pillar of finance, namely the Capital Asset Pricing Model (CAPM) [42]. CAPM is a model that describes the relationship between systematic risk and expected return for assets, particularly stocks. This model is widely used throughout finance for the pricing of risky securities, generating expected returns for assets given the risk of those assets and calculating cost of capital.

According to MPT, investing is a trade-off between the expected return and the risk. For a given amount of risk, MPT indicates how to select a portfolio with the highest possible expected return. Alternatively, for a given expected return, MPT describes how to select a portfolio with the lowest possible risk. MPT explains mathematically how to find the best possible diversification strategy.

8.3.1 Measuring Expected Return and Risk for a Portfolio of Assets

The return of a portfolio of assets is simply a weighted average of the return on the individual assets. The weight applied to each return is the fraction of the portfolio invested in the related asset.

Let R_{Pj} represent the j-th possible return on the portfolio. w_i is the fraction of the investor's funds invested in the i-th asset and the total number of assets is N. The expression for the return is given by (8.3).

$$R_{Pj} = \sum_{i=1}^{N} w_i R_{ij} \,.$$ (8.3)

The expected return of the portfolio becomes:

$$E[R_P] = E\left[\sum_{i=1}^{N} (w_i R_{ij})\right] = \sum_{i=1}^{N} E[w_i R_{ij}] = \sum_{i=1}^{N} w_i E[R_{ij}] \,.$$ (8.4)

It can be seen in (8.4) that the expected return of a portfolio is a weighted average of the expected returns of individual assets. This is hardly surprising when one remembers that the expected value operator is a linear operator.

However, the risk of a portfolio is *not* a simple average of the risk of the individual assets in the portfolio. Let σ_P represent the standard deviation of the portfolio return which is given by (8.5):

$$\sigma_P = \sqrt{\sum_{j=1}^{N} (w_j \sigma_j)^2 + \sum_{j=1}^{N} \sum_{\substack{k=1 \\ k \neq j}}^{N} (w_j w_k \sigma_{jk})} \,.$$ (8.5)

The formula for the variance of the portfolio return consists of two parts:

- Variance Part:

$$\sum_{j=1}^{N} (w_j \sigma_j)^2 \,,$$ (8.6)

- Covariance Part:

$$\sum_{j=1}^{N} \sum_{\substack{k=1 \\ k \neq j}}^{N} w_j w_k \sigma_{jk} \,.$$ (8.7)

The contribution of the covariance part to the total variance, thus the risk of the portfolio return can be adjusted by careful choice of the assets and their weights.

Many empirical studies have shown that distribution of returns deviate from normal distribution over time [119]. Therefore, rather than using volatility defined by the formula (8.5), it has been suggested to use an exponentially weighted volatility definition [55]. It has been rumored that the hedge fund called Long-Term Capital Management (LTCM) lost more than $4.6 billion in less than four months in 1998 precisely because they failed to see that their asset returns were not normally distributed. What makes the story so spicy is that two Nobel laureates Scholes and Merton[2] were the managers of LTCM.

Investing versus Speculating

There is a wide spread confusion in the public mind about the difference between investing and speculation. As the investment guru Graham pointed out already in 1934

> An investment operation is one which, upon thorough analysis promises safety of principle and an adequate return. Operations not meeting these requirements are speculative [47].

It is noteworthy that after almost a century and several major global financial crises later, this pithy statement holds its validity. Therefore, the reader should be weary of accepting the common jargon which applies the term "investor" to everybody buying and selling stocks. It is not uncommon to read even in the most reputable of journals such contradictions in terms like "reckless investors". Indeed, there were times when buying stocks was generally considered risky, especially after major sustained declines in the stock markets.

It should be kept in mind that speculation is not illegal. It is not even immoral. But it is, in Graham's words, "...[not] ...fattening to the pocketbook". Furthermore, speculation is even necessary and unavoidable in some cases. Just think of Amazon.com: speculators bought the shares of this loss making start-up company on the bet that it would at some time begin making money and its value, hence the value of its shares, would skyrocket. And they did. Amazon.com shares were worth $1.54 on May 31, 1997. On May 31, 2017 they were worth $715.62, a 465-fold increase over 20 years!

One has to recognize that speculation is fascinating, thrilling and—when one is ahead—fun! It is in the human nature. Therefore, one bit of timeless advice is always to be aware when one is speculating and not investing. A practical consequence of this advice is to put a small portion of one's capital in a separate account dedicated to speculating and never to add more money to this fund just because the market has gone up.

---

[2]Robert Merton, American economist (1944–).

8.3.2 *Efficient Frontier*

In theory, an investor can consider all possible combinations of assets, one by one, as her investment portfolio. Such a strategy would require practically an infinite number of possibilities to be considered. If one could find a set of portfolios that offered a bigger return for the same risk or offered a lower risk for the same return, one would have identified the portfolios that are interesting for the rational investor and all other portfolios could be ignored.

If all possible portfolios were plotted in the risk-return plane, the resulting diagram might look like Figure 8.1.

Examine the portfolios A and B in Figure 8.1. Portfolio B would be preferred by a rational investor to portfolio A, because it offers a higher return with the same level of risk. One can also see that portfolio E would be preferable to Portfolio F because it offers less risk at the same level of return. At this point in the analysis, there are no portfolios that are better than portfolio B or portfolio C. One can see that a rational set of portfolios cannot lie in the interior region of the shaded area.

Portfolio D can be eliminated from consideration by the existence of portfolio E, because portfolio E has a higher return for the same risk. By the same token, every other portfolio on the edge of the shaded region from point D to point C is not preferable. There is no portfolio in the region of possible portfolios that has less risk for the same return or more return for the same risk. Portfolio C represents the global minimum risk portfolio in the investment universe.

Further, there is no need to consider portfolio F because portfolio E, has less risk for the same return. On the edge of the shaded region between portfolio F and portfolio B no portfolio is preferable over portfolio B, which has the highest return. Actually, portfolio B represents the portfolio that offers the highest expected return in the investment universe. The portfolio with the highest expected return usually consists of a single asset.

The rational set of portfolios all lie on the edge of the shaded region between the minimum risk portfolio (point C) and the maximum return portfolio (point B). This set of points is called the *efficient frontier*. Figure 8.2 illustrates an example efficient frontier graph.

Fig. 8.1 Possible risks versus returns for various portfolios. Each point in the risk-return space represents a portfolio (After [38])

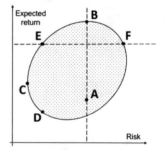

Fig. 8.2 The efficient
frontier (after [139]). Each
dot represents a portfolio.
The dots closest to the
efficient frontier represent
portfolios that are expected
to show the best performance
for a given risk level. No
portfolios are possible above
the efficient frontier

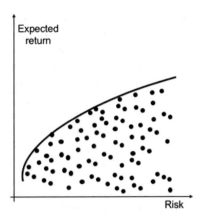

8.4   Portfolio Optimization as a Mathematical Problem

The construction of the efficient frontier requires the optimal portfolios to be sorted
out from all possible portfolio combinations. The portfolios on the efficient frontier
can be obtained by solving a quadratic optimization problem.

The mathematical representation of the problem of finding the portfolio with
the minimum risk for a given expected return results in a constrained optimization
problem which can be stated as

$$\text{minimize} \quad \sum_{i=1}^{N} \sum_{j=1}^{N} w_i w_j \sigma_{ij} \, , \tag{8.8}$$

subject to the constraints

$$\sum_{i=1}^{N} w_i \mu_i = R^* \, , \tag{8.9}$$

$$\sum_{i=1}^{N} w_i = 1 \, , \tag{8.10}$$

$$0 \le w_i \le 1, \qquad i = 1, \dots, N \tag{8.11}$$

where,
N is the number of assets available.
μ_i is the expected return of asset i ($i = 1, \dots, N$).
σ_{ij} is the covariance between assets i and j ($i = 1, \dots, N; \ j = 1, \dots, N$).
R^* is the desired expected return.
w_i is the decision variable that represents the proportion of asset i in the portfolio
($i = 1, \dots, N$).

(8.8) minimizes the total risk associated with the portfolio; (8.9) ensures that the portfolio has the desired expected return, R^*; (8.10) makes sure that the weights of the assets add to one, which means that all the funds are invested; and finally (8.11) ensures that there is no short selling of any stocks.[3]

Because of the product $w_i w_j$ in (8.8), the mathematical formulation above is a quadratic (non-linear) optimization problem. There are many computationally effective algorithms available in the literature for solving quadratic programming (QP) problems [48], [52] and [49]. Thanks to these algorithms, the generation of the efficient frontier for any particular portfolio set becomes a straight forward process.[4] The efficient frontier can be obtained by solving the described QP for varying values of R^*.

An alternative to numerical solution of this constrained quadratic optimization problem is to solve it analytically. To do so, the condition (8.11) about not allowing short sales is left out for the moment and the problem is rewritten using vector notation as

$$\text{Minimize } f(\mathbf{w}) = \mathbf{w}^T \boldsymbol{\Sigma} \mathbf{w} \tag{8.12}$$

$$\text{subject to } g_1(\mathbf{w}) = \mathbf{w}^T \boldsymbol{\mu} - R^* = 0 \tag{8.13}$$

$$g_2(\mathbf{w}) = \mathbf{w}^T \mathbf{1} - 1 = 0, \tag{8.14}$$

with $\mathbf{w} = [w_1, w_2, \ldots, w_N]^T$ being the weights of the assets in the portfolio, $\boldsymbol{\mu} = [\mu_1, \mu_2, \ldots, \mu_N]^T$ being the expected returns of the assets, $\mathbf{1} = [1, 1, \ldots, 1]^T$ being a vector of length N with N being the number of assets in the portfolio and $\boldsymbol{\Sigma}$ being the covariance matrix. R^* is again the desired return of the portfolio.

Solving this problem involves the following steps:

- Build the Lagrangian L with the Lagrange multipliers λ_1 and λ_2.
- Compute the derivatives of L (see Appendix B for relevant rules for matrix derivation).
- Equate the derivatives of L to zero for optimality as seen in Section 2.5 and solve the resulting algebraic equations to find the optimal portfolio weights as shown below.

$$L(\mathbf{w}, \lambda_1, \lambda_2) = \underbrace{\mathbf{w}^T \boldsymbol{\Sigma} \mathbf{w}}_{L_1} + \underbrace{\lambda_1 (\mathbf{w}^T \boldsymbol{\mu} - R^*)}_{L_2} + \underbrace{\lambda_2 (\mathbf{w}^T \mathbf{1} - 1)}_{L_3} . \tag{8.15}$$

---

[3] Short selling is the selling of securities, especially stocks, that are not currently owned by the investor with the promise to repurchase ("cover") them sometime in the future. If the price of the stock declines, the short seller will profit, because the cost of covering will be less than the proceeds received from the initial short sale (see Chapter 9). This practice dates back to the beginning of the 17th century [126]. Because short selling can exert downward pressure on the underlying stocks and thus drives down the stock prices, many countries do not allow short selling. A recent example is China in 2015 [13].

[4] MATLAB's Optimization Toolbox offers a very efficient algorithm based on trust region [22] which can be used very effectively for this purpose.

$$\frac{\partial L}{\partial \mathbf{w}} = 2\boldsymbol{\Sigma}\mathbf{w} + \lambda_1\boldsymbol{\mu} + \lambda_2\mathbf{1} := \mathbf{0} \tag{8.16}$$

$$\frac{\partial L}{\partial \lambda_1} = \mathbf{w}^T\boldsymbol{\mu} - R^* := 0 \tag{8.17}$$

$$\frac{\partial L}{\partial \lambda_2} = \mathbf{w}^T\mathbf{1} - 1 := 0 \tag{8.18}$$

Solving (8.16) for $\mathbf{w}$ yields

$$\mathbf{w}(\lambda_1, \lambda_2) = -\frac{1}{2}(\lambda_1\boldsymbol{\Sigma}^{-1}\boldsymbol{\mu} + \lambda_2\boldsymbol{\Sigma}^{-1}\mathbf{1}). \tag{8.19}$$

Substituting (8.19) in (8.15) yields $L_1(\lambda_1, \lambda_2)$, $L_2(\lambda_1, \lambda_2)$, $L3(\lambda_1, \lambda_2)$. One has to remember that $\boldsymbol{\Sigma}$ is a covariance matrix and is therefore symmetrical. Consequently, its inverse is also symmetrical. Furthermore, since $\boldsymbol{\mu}^T\boldsymbol{\Sigma}^{-1}\mathbf{1}$ is a scalar α, its transpose $\mathbf{1}^T(\boldsymbol{\Sigma}^{-1})^T\boldsymbol{\mu}$ is equal to itself. Therefore, with some matrix algebra,

$$L_1 = \frac{1}{4}(\lambda_1\boldsymbol{\Sigma}^{-1}\boldsymbol{\mu} + \lambda_2\boldsymbol{\Sigma}^{-1}\mathbf{1})^T\boldsymbol{\Sigma}(\lambda_1\boldsymbol{\Sigma}^{-1}\boldsymbol{\mu} + \lambda_2\boldsymbol{\Sigma}^{-1}\mathbf{1})$$

$$= \frac{1}{4}(\lambda_1\boldsymbol{\mu}^T\boldsymbol{\Sigma}^{-1}\boldsymbol{\Sigma} + \lambda_2\mathbf{1}^T\boldsymbol{\Sigma}^{-1}\boldsymbol{\Sigma})(\lambda_1\boldsymbol{\Sigma}^{-1}\boldsymbol{\mu} + \lambda_2\boldsymbol{\Sigma}^{-1}\mathbf{1})$$

$$= \frac{1}{4}(\lambda_1^2\boldsymbol{\mu}^T\boldsymbol{\Sigma}^{-1}\boldsymbol{\Sigma}\boldsymbol{\Sigma}^{-1}\boldsymbol{\mu} + \lambda_1\lambda_2\boldsymbol{\mu}^T\boldsymbol{\Sigma}^{-1}\boldsymbol{\Sigma}\boldsymbol{\Sigma}^{-1}\mathbf{1}$$
$$+ \lambda_1\lambda_2\mathbf{1}^T\boldsymbol{\Sigma}^{-1}\boldsymbol{\Sigma}\boldsymbol{\Sigma}^{-1}\boldsymbol{\mu} + \lambda_2^2\mathbf{1}^T\boldsymbol{\Sigma}^{-1}\boldsymbol{\Sigma}\boldsymbol{\Sigma}^{-1}\mathbf{1})$$

$$= \frac{1}{4}(\lambda_1^2\boldsymbol{\mu}^T\boldsymbol{\Sigma}^{-1}\boldsymbol{\mu} + \lambda_1\lambda_2\boldsymbol{\mu}^T\boldsymbol{\Sigma}^{-1}\mathbf{1} + \lambda_1\lambda_2\mathbf{1}^T\boldsymbol{\Sigma}^{-1}\boldsymbol{\mu} + \lambda_2^2\mathbf{1}^T\boldsymbol{\Sigma}^{-1}\mathbf{1})$$

$$= \frac{1}{4}(\lambda_1^2\beta + 2\alpha\lambda_1\lambda_2 + \lambda_2^2\gamma).$$

$$L_2 = -\frac{\lambda_1}{2}(\lambda_1\boldsymbol{\Sigma}^{-1}\boldsymbol{\mu} + \lambda_2\boldsymbol{\Sigma}^{-1}\mathbf{1})^T\boldsymbol{\mu} - \lambda_1 R^*$$

$$= -\frac{\lambda_1}{2}(\lambda_1\boldsymbol{\mu}^T\boldsymbol{\Sigma}^{-1}\boldsymbol{\mu} + \lambda_2\mathbf{1}^T\boldsymbol{\Sigma}^{-1}\boldsymbol{\mu}) - \lambda_1 R^*$$

$$= -\frac{1}{2}(\lambda_1^2\beta) - \frac{1}{2}(\lambda_1\lambda_2\alpha) - \lambda_1 R^*.$$

$$L_3 = -\frac{\lambda_2}{2}(\lambda_1\boldsymbol{\Sigma}^{-1}\boldsymbol{\mu} + \lambda_2\boldsymbol{\Sigma}^{-1}\mathbf{1})^T\mathbf{1} - \lambda_2$$

$$= -\frac{\lambda_2}{2}(\lambda_1\boldsymbol{\mu}^T\boldsymbol{\Sigma}^{-1}\mathbf{1} + \lambda_2\mathbf{1}^T\boldsymbol{\Sigma}^{-1}\mathbf{1}) - \lambda_2$$

$$= -\frac{1}{2}(\lambda_1\lambda_2\alpha) - \frac{1}{2}(\lambda_2^2\gamma) - \lambda_2.$$

Substituting $L_1(\lambda_1, \lambda_2)$, $L_2(\lambda_1, \lambda_2)$, $L_3(\lambda_1, \lambda_2)$ in (8.15) results in

$$L(\lambda_1, \lambda_2) = -\frac{1}{4}(\lambda_1^2 \beta) - \frac{1}{2}(\lambda_1 \lambda_2 \alpha) - \frac{1}{4}(\lambda_2^2 \gamma) - \lambda_1 R^* - \lambda_2, \qquad (8.20)$$

where

$$\alpha = \mu^T \Sigma^{-1} \mathbf{1}$$
$$\beta = \mu^T \Sigma^{-1} \mu$$
$$\gamma = \mathbf{1}^T \Sigma^{-1} \mathbf{1}.$$

Solving for λ_1 that optimizes (8.20) yields:

$$\frac{\partial L}{\partial \lambda_1} = -\frac{1}{2}(\lambda_1 \beta) - \frac{1}{2}(\lambda_2 \alpha) - R^* := 0. \qquad (8.21)$$

$$\Rightarrow \lambda_1 = -\frac{(2R^* + \lambda_2 \alpha)}{\beta}. \qquad (8.22)$$

Substituting this value to find λ_2 that optimizes (8.20) yields

$$L(\lambda_2) = -\left(\frac{(2R^* + \lambda_2 \alpha)}{2\beta}\right)^2 \beta + \frac{(2R^* + \lambda_2 \alpha)}{2\beta} \lambda_2 \alpha - \frac{\lambda_2^2 \gamma}{4} + \frac{(2R^* + \lambda_2 \alpha)}{\beta} R^* - \lambda_2 \quad (8.23)$$

$$= \lambda_2^2 \left[\frac{\alpha^2 - \beta\gamma}{4\beta}\right] + \lambda_2 \left[\frac{R^*\alpha - \beta}{\beta}\right] + \left[\frac{R^{*2}}{\beta}\right]. \qquad (8.24)$$

$$\frac{\partial L}{\partial \lambda_2} = \lambda_2 \left[\frac{\alpha^2 - \beta\gamma}{2\beta}\right] + \left[\frac{R^*\alpha - \beta}{\beta}\right] := 0. \qquad (8.25)$$

$$\lambda_2^* = \frac{2(\beta - R^*\alpha)}{\alpha^2 - \beta\gamma}. \qquad (8.26)$$

Substituting λ_2^* in (8.22) results in

$$\lambda_1^* = \frac{2(R^*\gamma - \alpha)}{\alpha^2 - \beta\gamma}. \qquad (8.27)$$

Finally, substituting λ_1^* and λ_2^* in (8.19) yields the optimal weights of the stocks in the portfolio as

$$\mathbf{w}^* = -\frac{1}{2}\left[\left(\frac{2(R^*\gamma - \alpha)}{\alpha^2 - \beta\gamma}\right)\Sigma^{-1}\mu + \left(\frac{2(\beta - R^*\alpha)}{\alpha^2 - \beta\gamma}\right)\Sigma^{-1}\mathbf{1}\right]. \qquad (8.28)$$

This "analytical" solution entails a matrix inversion which needs to be done numerically. In other words, this solution exchanges the numerical solution of the quadratic optimization problem with the numerical inversion of a matrix.

So far it was assumed that unrestricted short selling is allowed. If short selling is disallowed one has to consider (8.11) in the optimization problem, which further complicates the analytical solution. In this case, the so-called Karush-Kuhn-Tucker conditions, often abbreviated as KKT [73], need to be considered.

Engineers often face the task of finding solutions to problems with contradictory requirements. As already seen in Chapter 4, the solution to this quandary is through combining those opposing aims in a single objective function, assigning each different weights. One can apply the same trick here to the problem of defining a portfolio which achieves the maximum return with the minimum risk. This can be done by formulating the problem with the weighting parameter ρ $(0 \le \rho \le 1)$:

$$\text{minimize} \quad \rho\left[\sum_{i=1}^{N}\sum_{j=1}^{N}w_i w_j \sigma_{ij}\right] - (1 - \rho)\left[\sum_{i=1}^{N}w_i \mu_i\right], \qquad (8.29)$$

subject to the constraints given by Eqs. (8.10) and (8.11). It is easy to verify that for (8.29), the case $\rho = 0$ implies maximization of the expected return (irrespective of the involved risk). In this case the optimal portfolio will consist of only a single asset, the asset with the highest expected return. The case $\rho = 1$ implies the minimization of the portfolio risk. The optimal portfolio for the minimum risk solution will in general contain a number of assets. The values of ρ that satisfy the condition $0 < \rho < 1$ represent the trade-off between the risk and the return. Hence ρ can be seen as a parameter that quantifies risk aversion.

8.5 Portfolio Optimization as a Practical Problem

The theoretical considerations so far assume that the expected returns of the assets μ_i and their covariances σ_{ij} in Eqs. (8.9) and (8.8) are known. Alas, in reality, those values can be known only *a posteriori*. So, how on earth can one build a portfolio maximizing returns for a desired risk level or minimizing risk for a desired level of return, given that those critical parameters are not known?

All the chapters so far in this Book has equipped the reader with the necessary tools to attack this problem as a Control Engineering problem. Past values of these parameters are known. Therefore, one can apply the certainty equivalence principle (as discussed in Chapter 4 on Optimal Control) and first *estimate* those parameters using past data and then use these estimates as if they were the actual values. For estimating the parameters of the overall return distribution one needs a model. That model assumes that the daily closing prices of the assets follow a geometric random walk process[5] as discussed on the Chapter 3 on Stochastic Processes and Chapter 5 on Stochastic Calculus.

Let $\mathbf{S}^d$ be the vector containing the daily closing prices of N assets on date d,[6] let $\mathbf{R}$ represent the multivariate random vector of the asset returns, and let $\mathbf{L}$ represent the multivariate random vector of the asset log-returns. Then, the geometric random walk process of the prices can be described by (8.30) and (8.31), where the latter indicates that the asset returns follow a multivariate log-normal distribution[7]:

$$\mathbf{S}^d = \mathbf{S}^{d-1} \otimes \mathbf{R}_{daily}, \tag{8.30}$$

$$\mathbf{R}_{daily} \sim \log \mathcal{N}(\mathbf{A}_{daily}, \mathbf{B}_{daily}), \tag{8.31}$$

where $\otimes$ denotes element-wise multiplication. The logarithms of the daily closing prices follow a simple normal random walk which can be described by Eqs. (8.32), (8.33) and (8.34). The last equation indicates that the asset log-returns follow a multivariate normal distribution.

$$\mathbf{R}_{daily} = \exp(\mathbf{L}_{daily}), \tag{8.32}$$

$$\log(\mathbf{S}^d) = \log(\mathbf{S}^{d-1}) + \mathbf{L}_{daily}, \tag{8.33}$$

$$\mathbf{L}_{daily} \sim \mathcal{N}(\boldsymbol{\mu}_{daily}, \boldsymbol{\Sigma}_{daily}). \tag{8.34}$$

There is, however, a catch: the parameters of the distributions in (8.31) or (8.34) are not directly observable by the investor and have to be estimated, for instance, using historical data. Calling the estimated returns and covariances $\hat{\boldsymbol{\mu}}$ and $\hat{\boldsymbol{\Sigma}}$, respectively, the estimated parameters of the regular multivariate return distribution can be calculated as

$$\hat{\mathbf{A}} = \exp\left(\hat{\boldsymbol{\mu}} + \frac{1}{2} Diag\left(\hat{\boldsymbol{\Sigma}}\right)\right), \tag{8.35}$$

---

[5] A random walk is a mathematical formalization of a path that consists of a succession of random steps.

[6] d is used as an index and not an exponent in the formulas of this section.

[7] Note that the multi-dimensional geometric Brownian motion model used here is based on the stochastic differential equation in Section 5.7.2.

and

$$\hat{b}_{ij} = \exp\left(\hat{\sigma}_{ij} - 1\right)\exp\left(\hat{\mu}_i + \hat{\mu}_j + \frac{\hat{\sigma}_{ii} + \hat{\sigma}_{jj}}{2}\right), \tag{8.36}$$

where $Diag\left(\hat{\Sigma}\right)$ represents the vector that is composed of the diagonal entries of the covariance matrix of log-returns; $\hat{b}_{ij}$ represents the entry of the estimated covariance matrix of returns at the j-th column and i-th row; $\hat{\sigma}_{ij}$ represents the entry of the estimated covariance matrix of log-returns at the j-th column and i-th row. The derivation of the last two equations is left to the reader as an exercise.

Let $\mathbf{L}_{daily}^d$ be the $1 \times N$ vector of the daily log-returns on date d. The average vector of the daily log-returns is used as an estimation for the expected value of the vector of daily log-returns according to

$$\hat{\mu}_{daily} = \frac{1}{D}\sum_{d=1}^{D}\mathbf{L}_{daily}^d. \tag{8.37}$$

D is the total number of considered trading days. Remember that the general formula for the estimation of the expected log-return over d trading days is

$$\hat{\mu}_{ddays} = \hat{\mu}_{daily} \times d. \tag{8.38}$$

For the estimation of the covariance matrix of daily log-returns, the matrix of the daily log-returns is formed. The matrix of the daily log-returns is given by (8.39). l_i^d represents the daily log-return of the i-th asset on date d.

$$\mathbf{M}_{daily\,logreturns} = \begin{bmatrix} \vdots \\ \mathbf{L}_{daily}^{d-1} \\ \mathbf{L}_{daily}^{d} \\ \mathbf{L}_{daily}^{d+1} \\ \vdots \end{bmatrix} = \begin{bmatrix} \ddots & \vdots & \vdots & \vdots & \cdots \\ \cdots & l_{i-1}^{d-1} & l_i^{d-1} & l_{i+1}^{d-1} & \vdots \\ \cdots & l_{i-1}^{d} & l_i^{d} & l_{i+1}^{d} & \vdots \\ \cdots & l_{i-1}^{d+1} & l_i^{d+1} & l_{i+1}^{d+1} & \vdots \\ \cdots & \cdots & \cdots & \cdots & \ddots \end{bmatrix}. \tag{8.39}$$

The covariance matrix of the daily log-returns is estimated according to the definition in Chapter 3 as

$$\hat{\Sigma}_{daily} = \mathrm{E}\left[\left(\mathbf{M}_{daily\,logreturns} - \mathrm{E}[\mathbf{M}_{daily\,logreturns}]\right)\left(\mathbf{M}_{daily\,logreturns} - \mathrm{E}[\mathbf{M}_{daily\,logreturns}]\right)^T\right], \tag{8.40}$$

$$\hat{\Sigma}_{daily} = \begin{bmatrix} \ddots & \vdots & \vdots & \vdots & \cdots \\ \cdots & \hat{\sigma}^d_{(i-1)(j-1)} & \hat{\sigma}^d_{(i-1)(j)} & \hat{\sigma}^d_{(i-1)(j+1)} & \vdots \\ \cdots & \hat{\sigma}^d_{(i)(j-1)} & \hat{\sigma}^d_{(i)(j)} & \hat{\sigma}^d_{(i)(j+1)} & \vdots \\ \cdots & \hat{\sigma}^d_{(i+1)(j-1)} & \hat{\sigma}^d_{(i+1)(j)} & \hat{\sigma}^d_{(i+1)(j+1)} & \vdots \\ \cdots & \cdots & \cdots & \cdots & \ddots \end{bmatrix}. \tag{8.41}$$

Also here, $\hat{\sigma}^d_{ij}$ represents the entry of the estimated covariance matrix of daily log-returns at the i-th column and j-th row. The general formula for the estimation of the covariance of the log-returns of the assets i and j over d trading days is, again according to Chapter 3,

$$\hat{\sigma}^{d\,days}_{ij} = \hat{\sigma}^{daily}_{ij} \times \sqrt{d}. \tag{8.42}$$

The remaining question is—actually this is literally a million-dollar question—how to choose the parameter D in (8.37). In other words, how far back in the past should one look to make an optimal estimate for the current values of μ and Σ? Take, say, five days and the values are current but too few to make proper estimates for all the N elements of μ and $N \cdot N/2$ elements of Σ. Take say, five hundred days and the number of points to make estimates is ample but they include very old, and likely irrelevant, data.

There is yet another million-dollar question which also needs to be answered: how frequently should the portfolios be updated? One extreme is to build a portfolio and keep the composition fixed forever. This, obviously, cannot be the best solution since there is no way to incorporate new data into this solution. The other extreme is to update every day. This would be fantastic in terms of incorporating new data daily into the portfolio optimization problem. However, because of the transaction costs[8] which are in the order of 0.5%, this is not a very good idea either.

8.6   Empirical Examples of Portfolio Optimization Using Historical Data

The two critical questions introduced in the previous section were investigated in a series of empirical studies applying the approach introduced in the previous section, using historical stock market data to model low risk ($\rho = 0.2$), medium risk ($\rho = 0.5$), and high risk ($\rho = 0.8$) optimal portfolios and compare their performances to market indexes over time. By construction, higher risk portfolios contained fewer stocks, and hence were less well diversified. The studies used data from the New York

---

[8]Transaction cost is composed of the difference between the ask and prices on the stock exchange (spread) and the commission the bank or the stock broker charges on each trade.

(market index DJIA[9]), Istanbul (market index IMKB30[10]) and Zurich (market index SMI[11]) stock exchanges, covering the period January 2010 to April 2012. All three portfolios of varying risk levels modeled for each market were composed of stocks included in the respective market indexes. Portfolios were updated using closing prices of previous days. Transaction costs, dividends and stock splits were all taken into account.[12]

The first multi-million dollar question, namely that of determining the optimal length of the time for estimating the expected returns and covariances, was addressed separately for each market using historical data prior to the 2010–2012 study period. It turned out that for all markets best results were obtained by using data from the last 24 months for estimating the covariances and 6 months for estimating the expected returns.

The second big question, namely that of how frequently to update the portfolios, was investigated by comparing the performances of portfolios updated with varying frequencies with each other and the index for the respective market. In all markets, it was demonstrated that the cost of frequent updates offsets the benefits of including the most recent data in the optimization of the portfolios. This was markedly so in the high risk portfolios, which contained fewer number of assets to be updated. Best performances were achieved for portfolios updated every five weeks for New York and Istanbul and 7 weeks for Zurich as shown in Figure 8.3. A further finding was that low risk portfolios are less sensitive to how frequently they were updated.

It is important to recall at this stage that MPT assumes the expected returns and volatilities of individual stocks to be known. However, there is no way to know them. Therefore, they have to be estimated. These estimates, on the other hand, are necessarily imperfect. Also, it cannot be stressed enough that the results of all empirical studies as the ones presented above are valid *only* for the given time frame and given market and that they cannot be generalized.

The length of data, D in equation (8.37), which is used to estimate the expected returns and volatilities of individual stocks in the empirical studies described above, was optimized before the back-tests and then kept constant. But it can be expected that the optimal value for D changes during the study. Therefore, moving the time frame one day ahead every day during the back-testing and finding the new optimal value for D might result in better performances of the portfolios. This was simply not possible, because such optimizations took several weeks of computational effort and could not have been done on a daily basis with the computing power available at that time.

---

[9]The Dow Jones Industrial Average is an index that shows how 30 largest publicly owned companies based in the United States have traded during a standard trading session in the stock market.

[10]IMKB30 (now called BIST30) is an index that shows 30 publicly owned companies with the biggest market capitalization based in Turkey.

[11]Swiss Market Index comprises the 20 largest publicly owned companies based in Switzerland.

[12]This Chapter makes extensive use of Efe Doğan Yılmaz's Master Thesis [139] and Bachelor Theses of Yasin Çotur [31], Ceren Sevinç [119] and Mehmet Hilmi Elihoş [37], all at the Boğaziçi University.

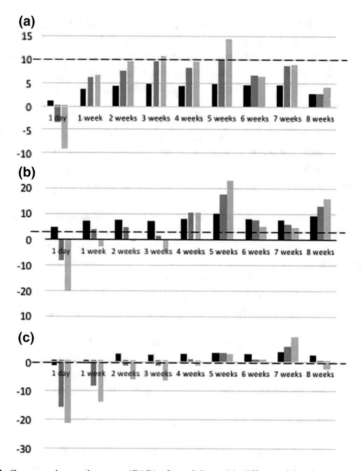

Fig. 8.3 Compound annual returns (CAR) of portfolios with different risk tolerances (black low risk, dark gray medium risk, light gray high risk) between January 2010–April 2012. The horizontal axis indicates the update rate. **a** Portfolios composed of the DJIA component stocks, **b** Portfolios composed of the IMKB30 component stocks, **c** Portfolios composed of the SMI component stocks. The dashed lines are the performances of the respective indexes

A subsequent study was undertaken to refine the solution of the second question—about when to update the portfolios [37]. The approach was to use the volatility of the market index to trigger an update signal for the portfolio. Extensive empirical analyses of the Istanbul Stock Exchange were conducted and it was found that by choosing the threshold value properly and by triggering the update mechanism only when the index volatility exceeds this threshold, one can have a return which is significantly better than the market index. Figure 8.4 shows the index variance during the years 2005–2012. The first million dollar question above throws its shadow here as well: while calculating the variance, how far back should one look into the data? It was found empirically that using a past time frame of 13 trading days to calculate

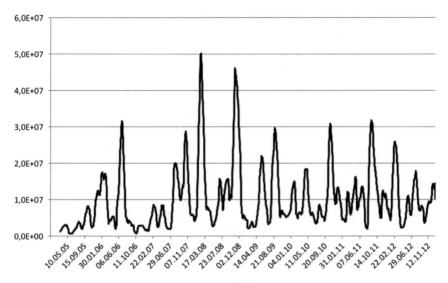

Fig. 8.4 The volatility of BIST30 calculated with data from the previous 50 days. Note the high volatility during the global financial crisis of 2008

the volatility and to update the portfolio every time the volatility exceeds 6 million points gave the best results. Figure 8.5 shows the performance of a portfolio updated with this strategy.

However, it became quickly apparent that this way of updating the portfolio could be improved by using the relative index variance. It was found that updating on days when the relative index volatility calculated with data from the previous 48 days exceeded 10% resulted in a far better performance although there were long stretches of time when the portfolio was not updated at all. After many simulations, it was concluded that, after a while, the portfolio needed to be updated anyway even if the volatility was low. It also became apparent that once the volatility was high it stayed high for some days and this caused unnecessary and costly updates on consecutive days. Therefore, the final "intelligent" algorithm can be summarized as (a) update when the relative index volatility threshold is exceeded, (b) update if 20 days pass without any updates (irrespective of the volatility) and (c) do not update for five days after the last update. The impressive results are shown in Figure 8.6.

Now, a cardinal rule in back-testing—as empirical analysis is often called—is never to use data already used for optimizing parameters. This would be similar to betting on which horses will win the races after the races have been run! Therefore, one has to strictly separate data for parameter optimization purposes and for testing purposes. Elihoş used market data from the years 2005–2012 for analysis and parameter optimization purposes and then back-tested for the period May 2012—May 2014 while using the optimal parameters of the previous six months. Again impressive results are shown in Figure 8.7.

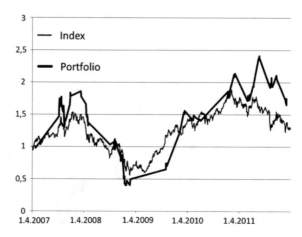

Fig. 8.5 The performance of a portfolio consisting of the BIST30 stocks updated when the volatility calculated with data from the previous 13 days exceeded 6 million points

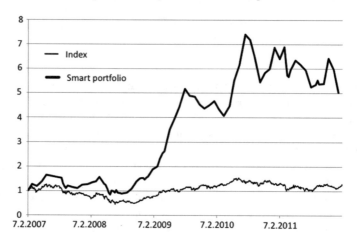

Fig. 8.6 The performance of a portfolio consisting of the BIST30 stocks updated according to an intelligent algorithm

To summarize, Modern Portfolio Theory was derived first heuristically and then mathematically. A practical way for applying MPT using estimated values of the expected returns and covariances of the stocks which are considered for inclusion in the portfolio was presented. Those estimates were based on historical data of the stocks. Empirical studies using this method in different stock markets have resulted in very different performances. Finally, an "intelligent" algorithm was proposed and tested with impressive results.

Fig. 8.7 The graph shows the portfolio value in comparison with updating the portfolio in constant 20 days and updating the portfolio with the intelligent algorithm explained above between May 2012 and May 2014. The parameters were chosen as the parameters that resulted in the highest return in the preceding 6 months

8.7 Bond Portfolios

As discussed in the previous sections, diversification is an excellent tool to reduce risk. Just as one can invest in several stocks to reduce the risk of an investment portfolio, one can also reduce the risk by investing in several bonds simultaneously. In other words, one can build a bond portfolio composed of several bonds. For that purpose, let us look at bonds from a slightly different angle.[13]

The spot interest rate of a bond, $i_S(t, T)$, is the interest during the period from t to T, with $t \leq T$. Therefore, $P(t, T)$, the price of a zero-coupon bond at time t with maturity T is calculated as

$$P(t, T) = e^{(-(T-t)i_S(t,T))} . \qquad (8.43)$$

Consequently,

$$i_S(t, T) = -\frac{1}{T - t} \log(P(t, T)) . \qquad (8.44)$$

The short rate is defined as

$$i_r(t) = \lim_{T \longrightarrow t} i_S(t, T) . \qquad (8.45)$$

---

[13] This section draws on [133] and replicates parts of it.

Similarly, the forward interest rate is defined as the interest rate of a bond determined at time t for a period that begins at a future time T and is repaid at time τ, with $t \le T \le \tau$. Formally,

$$_t i_S(T, \tau) = \frac{1}{\tau - T} \log(\frac{P(t, T)}{P(t, \tau)}) . \tag{8.46}$$

Another group of interest rate models is based on instantaneous forward rates, $f(t, T)$. These are interest rates determined at time t for loans that begin at time T and paid back at time $T + dt$. These forward rates are very important, because every interest rate can be expressed in terms of instantaneous forward rates. Formally,

$$\begin{aligned} f(t, T) &= \lim_{\tau \to T} \frac{1}{\tau - T} \log(\frac{P(t, T)}{P(t, \tau)}) \\ &= -\frac{\partial}{\partial T} \log(P(t, T)) \\ &= -\frac{1}{P(t, T)} \frac{\partial P(t, T)}{\partial T} . \end{aligned} \tag{8.47}$$

The *Vašíček model*[14] is a one-factor short rate model used to describe the momentary interest rates using an Ornstein-Uhlenbeck mean-reverting process defined by the stochastic differential equation

$$di_r(t) = \kappa(\theta - i_r(t))dt + \sigma_r dW(t), \qquad r(0) = r_0, \tag{8.48}$$

where $\kappa > 0$ is the mean reversion speed, θ is the constant reversion level, σ_r is the constant volatility of the short rate, and $W(t)$ is a Brownian motion.

In the Vašíček model, mean reversion is achieved by the drift term: when the short rate $i_r(t)$ is above θ, the drift term tends to pull $r(t)$ downwards and when $i_r(t)$ is below θ, the drift term tends to push $i_r(t)$ upwards (see Chapter 5).

The analytic solution for zero bond prices in the Vašíček model is

$$P(t, T) = e^{[A(t, T) - B(t, T)i_r(t)]}, \tag{8.49}$$

where

$$A(t, T) = i_S(\infty) \left(\frac{1}{\kappa}(1 - e^{-\kappa(t-T)}) - (T - t) \right) - \frac{\sigma_r^2}{4\kappa^3}(1 - e^{-\kappa(t-T)})^2, \tag{8.50}$$

with

$$i_S(\infty) = \left(\theta + \lambda \frac{\sigma_r}{\kappa} - \frac{1}{2} \frac{\sigma_r^2}{\kappa^2} \right), \tag{8.51}$$

---

[14]Oldřich Vašíček, Czech mathematician, (1942–).

where λ denotes the constant market price of interest rate risk, and

$$B(t, T) = \frac{1}{\kappa} \left(1 - e^{-\kappa(t-T)} \right) .$$
(8.52)

The spot interest rate is, therefore,

$$i_S(t, T) = \frac{i_r(t)B(t, T) - A(t, T)}{T - t} .$$
(8.53)

For a complete derivation see e.g., [107].

Having already developed the main ideas behind stock portfolio optimization in the previous sections, let us now discuss how they need to be adapted to the selection of bond portfolios before applying them to the Vašíček model described above.

In principle, MPT can be applied to all financial instruments with known expected returns and volatilities. However, applying it to bond portfolios is not straightforward, because the terminal value of a bond, W_T, has to be taken into account. Perhaps the most significant difference between stocks and bonds is the finite maturity of the bonds.[15] This needs to be addressed in the formulation of the optimization problem. Bonds which have maturities shorter than the investment horizon, T, will no longer exist at the investment horizon. Therefore, an assumption needs to be made about the reinvestment of cash flows received before the investment horizon. It makes sense to assume that all cash flows received before T shall be invested at the current spot interest rate $i_S(t, T)$ until the investment horizon. For simplicity and without loss of generality one can also assume that all the bonds in the portfolio are zero-coupon bonds.

Consider the case where the portfolio consists of zero-coupon bonds with different maturities, with the longest maturity being τ. There is one bond for each maturity date $1, 2, \ldots, \tau - 1, \tau$. Thus, at time $t = 0$, the investor allocates her initial wealth W_0 to τ zero-coupon bonds as

$$W_0 = \sum_{t=1}^{\tau} w_t P(0, t) ,$$
(8.54)

where w_t denotes the purchased quantity of the zero-coupon bond with maturity date t at the current price $P(0, t)$. Further, assuming that there is one "risk-free" bond with maturity T, this leaves $\tau - 1$ risky assets which are combined in the weights vector $\mathbf{w}$ and the price vector $\mathbf{P}_0$ such that

$$W_0 = \mathbf{w}^T \mathbf{P}_0 + w_T P(0, T) ,$$
$$\mathbf{w}^T = (w_1, \ldots, w_{T-1}, w_{T+1}, \ldots, w_\tau) ,$$
$$\mathbf{P}_0^T = (P(0, 1), \ldots, P(0, T - 1), P(0, T + 1), \ldots, P(0, \tau)) ,$$
(8.55)

---

[15] One notable exception is the perpetual bond described in Chapter 7.

where w_T denotes the quantity of bonds with maturity T the investor purchases at time zero.

Because of (8.43) and (8.44), the terminal wealth of the investor can be calculated as

$$W_T = \sum_{t=1}^{T-1} w_t \frac{1}{P(t,T)} + w_T + \sum_{t=T+1}^{T} w_t P(T,t), \qquad (8.56)$$

and with

$$\mathbf{P}_T = \left(\frac{1}{P(1,T)}, \ldots, \frac{1}{P(T-1,T)}, P(T,T+1), \ldots, P(T,\tau) \right)^T, \qquad (8.57)$$

the terminal wealth becomes

$$W_T = \mathbf{w}^T \mathbf{P}_T + w_T . \qquad (8.58)$$

Let us remember that only the expected value and the variance of terminal wealth are needed to apply Markowitz portfolio optimization. These can be calculated in a straight forward manner as

$$E[W_T] = \sum_{t=1}^{T-1} w_t E\left[\frac{1}{P(t,T)} \right] + w_T + \sum_{t=T+1}^{\tau} w_t E[P(T,t)] \qquad (8.59)$$

$$= \mathbf{w}^T E[\mathbf{P}_T] + w_T , \qquad (8.60)$$

$$\begin{aligned}
\text{Var}[W_T] = & \sum_{t=1}^{T-1} \sum_{s=1}^{T-1} w_t w_s \text{Cov} \left(\frac{1}{P(t,T)}, \frac{1}{P(s,T)} \right) \\
& + \sum_{t=T+1}^{\tau} \sum_{s=T+1}^{\tau} w_t w_s \text{Cov}(P(T,t), P(T,s)) \\
& + 2 \sum_{t=1}^{T-1} \sum_{s=T+1}^{\tau} w_t w_s \text{Cov} \left(\frac{1}{P(t,T)}, P(T,s) \right) \\
= & \; \mathbf{w}^T \boldsymbol{\Sigma} \mathbf{w} .
\end{aligned} \qquad (8.61)$$

where $\boldsymbol{\Sigma}$ is the covariance matrix with the elements $\sigma_{i,j}$ with $i, j = 1, ..., \tau - 1$ which are defined as

$$\sigma_{i,j} = \begin{cases} \text{Cov}\left(\frac{1}{P(i,T)}, \frac{1}{P(j,T)}\right) & \text{for } i = 1, 2, \ldots, T-1 \\ & \qquad j = 1, 2, \ldots, T-1 \\ \text{Cov}\left(\frac{1}{P(i,T)}, P(T, j+1)\right) & \text{for } i = 1, 2, \ldots, T-1 \\ & \qquad j = T, T+1, \ldots, \tau-1 \\ \text{Cov}\left(P(T, i+1), \frac{1}{P(j,T)}\right) & \text{for } i = T, T+1, \ldots, \tau-1 \\ & \qquad j = 1, 2, \ldots, T-1 \\ \text{Cov}\left(P(T, i+1), P(T, j+1)\right) & \text{for } i = T, T+1, \ldots, \tau-1 \\ & \qquad j = T, T+1, \ldots, \tau-1 \end{cases}$$

One can thus state the portfolio optimization problem as

$$\min_{\mathbf{w}} \frac{1}{2} \mathbf{w}^T \Sigma \mathbf{w} \tag{8.62}$$

subject to $\mathbf{w}^T \text{E}[\mathbf{P}_T] + w_T = \text{E}[W_T]$ and $\mathbf{w}^T \mathbf{P}_0 + w_T P(0, T) = W_0$. (8.63)

These two constraints can be written as a single constraint as

$$\underbrace{\frac{W_0}{P(0, T)}}_{\alpha} + \mathbf{w}^T \underbrace{\left(\text{E}[\mathbf{P}_T] - \frac{1}{P(o, T)} \mathbf{P}_0\right)}_{\beta} = \text{E}[W_T], \tag{8.64}$$

where α denotes the risk-free compounded initial wealth and β denotes the vector of risk premiums.

The optimization with constraints problem (8.62) subject to the constraints (8.64) can now be solved using Lagrange's method (see Section 2.5). The problem of finding the optimal bond portfolio composition is thus formulated in the same way as that for the stock portfolio composition covered earlier in this chapter. In order to calculate optimal bond portfolios, the following information is required:

- $\text{E}[P(T, t)]$, the expected discount factors at time T for all maturities greater than T, i.e., $t = T+1, T+2, \ldots, \tau$,
- $\text{E}[1/P(t, T)]$, the expected accrual factors from t to T, i.e., $t = 1, 2, \ldots, T-1$,
- $\text{Cov}[(1/P(t, T)), (1/P(s, T))]$, covariances between accrual factors for $t, s = 1, 2, \ldots, T-1$,
- $\text{Cov}[P(t, T), P(s, T)]$, covariances between different discount factors for $t, s = T+1, T+2, \ldots, \tau$,
- $\text{Cov}[(1/P(t, T)), P(s, T)]$, covariances between different accrual factors and discount factors for $t = 1, 2, \ldots, T-1$; $s = T+1, T+2, \ldots, \tau$.

As with the stock portfolio optimization, these expected values and covariances are usually not known and need to be estimated. Then, employing the time honored certainty equivalence principle of control engineering, one must proceed with these estimates as if they were the actual values.

However, although MPT is well accepted both in academia and in the equity markets, it is hardly ever used in bond markets. This is, in addition to the additional complexities of using bonds in portfolios as described above, believed to be caused by the low volatility in interest rates compared to equity prices [137]. So how are bond portfolios managed?

There are four different strategies of managing diversified bond portfolios. These can be classified under four headings [84]: immunization strategies, duration strategies, yield curve strategies, and riding/rollover strategies.

Immunization strategies: These strategies require practically no expectational input and try to build a portfolio such that the portfolio value at the investment horizon is not affected by changing interest rates. In other words, it is immune to changes in interest rates. This can be achieved, assuming that the term structure is flat, by constructing a bond portfolio with a Macaulay duration equal to the investment horizon. The minimum variance portfolio described above should be conceptually very close to an immunized portfolio.

Duration strategies: Here, if the investor expects the interest rates to fall during the investment horizon of the portfolio, she selects a portfolio with a duration longer than the investment horizon. Conversely, if her expectations are about increasing interest rates she chooses a portfolio with a duration shorter than the investment horizon. It was found empirically that, for such strategies to be successful, the investment horizon of the portfolio should be at least five years.

Yield curve strategies: These strategies involve building a portfolio to benefit from expected changes in the yield curve [41]. In practice there are three possible strategies. In a so-called *bullet strategy*, the portfolio is made up of bonds with maturities concentrated around the intermediate point of the yield curve. A *barbell portfolio*, on the other hand, is made up of bonds with maturities concentrated at the short-term and the long-term ends of the yield curve. In a *ladder strategy*, the portfolio is composed of bonds with equal amounts of each maturity. Figure 8.8 illustrates the three different yield curve strategies.

Riding/rollover strategies Riding the yield curve strategy requires buying bonds with maturities longer than the investment horizon and selling them before maturity. Rollover strategy, on the other hand entails buying bonds with maturities shorter than the investment horizon, holding them until maturity and reinvesting the proceeds. Both strategies rely on the current yield curve remaining relatively stable. In practice, when the yield curve is upward sloping, riding strategy is preferred and when the term structure of interests is downward sloping the rollover strategy is more profitable.

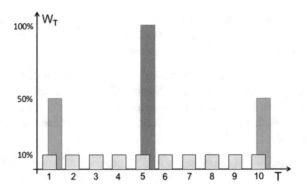

Fig. 8.8 Different yield curve strategies in bond portfolio management are characterized by the distribution of the funds as a function of maturities. Here the lightest gray distribution illustrates the ladder strategy, the darkest gray distribution illustrates the bullet strategy and the in between gray illustrates the dumb bell strategy

8.8 Exercises

A1: What does portfolio diversification mean? Why is it a good idea to diversify one's investments?

A2: How does the Modern Portfolio Theory (MPT) help with portfolio diversification?

A3: What are the key components of MPT?

A4: What does"efficient markets" mean?

A5: How would you distinguish investing from speculating?

A6: Explain how portfolio management relates to control engineering.

A7: MPT requires the expected returns and volatilities to be known. How would you go about determining those values?

A8: What is the "efficient frontier" in MPT?

A9: The analytical solution presented in Section 8.4 allows for unrestricted short selling. How would the mathematical problem change if short selling were not allowed?

B1: Find an analytical solution of the portfolio optimization problem where short selling is not allowed.

B2: A method for estimating the expected returns and volatilities of the stocks in a portfolio based on past data was presented in Section 8.5. Suggest and compare different estimation algorithms.

C1: Use the concept of implied volatilities (see Chapter 9) to estimate the volatilities of the stocks in a portfolio and use these estimated volatilities to perform empirical tests similar to the ones presented in Section 8.6.

C2: An "intelligent" algorithm was presented in Section 8.6 to determine when to update a portfolio. This algorithm uses the volatility of the market index to update a portfolio. Improve this algorithm by using volatilities of individual stocks and volatility of the existing portfolio in addition to the volatility of the market index to update a portfolio.

C3: Develop an on-line high frequency trading program which makes use of MPT.

Chapter 9
Derivatives and Structured Financial Instruments

Derivatives are financial weapons of mass destruction.

— Warren Buffett

Investing should be more like watching paint dry or watching grass grow. If you want excitement, take $800 and go to Las Vegas.

— Paul Samuelson

9.1 Introduction

This Chapter reviews forward contracts, futures and margin accounts. Options are then discussed in detail. Black-Scholes equation is derived and the calculation of option prices using this equation is shown for European and American options. Some popular structured products including swaps and how they might be used to enhance returns or reduce risks of diversified investment portfolios are discussed.

A derivative financial instrument (or simply derivative) is a security where the value depends explicitly on other variables called underlying instruments or simply underlyings.[1] Derivatives comprise instruments like forwards, futures, swaps and options.

Structured products combine classic investment instruments such as shares with derivatives. They are issued as stand-alone products and securitized in a commercial paper. The advantages of structured products for investors are that they provide for

- every market expectation, rising, falling or sideways,
- every risk profile, from low-risk capital protection products to high-risk leverage products,

---

[1] Parts of this Chapter is reproduced by kind permission from [46].

© Springer International Publishing AG 2018
S. S. Hacısalihzade, *Control Engineering and Finance*, Lecture Notes in Control and Information Sciences 467, https://doi.org/10.1007/978-3-319-64492-9_9

- every investment class, including those usually not accessible to many investors, including precious metals, commodities and emerging markets,
- high liquidity in the secondary market as provided by the issuer [128].

Derivatives are traded in a standardized fashion on financial markets (*e.g.*, Chicago Board of Trade CBOT, VIOP Istanbul) and also "over-the-counter", where tailored contracts are sold to investors. Actually, the investor can create practically any structured product with any underlying(s) she wishes. Some of the most popular structured products are listed at the end of the Chapter.

In general, derivatives and structured products have two contradictory powers. On the one hand, they can be efficient tools for reducing risk. At the same time, they have an astounding capacity to amplify returns—and in so doing the risk—through leverage. This Janus faced nature of derivatives can be viewed in terms of two human emotions: fear and greed. The reason why an investor wants to reduce risk is the fear of loss. On the other hand, the motivation to take on large amounts of risk for higher profits is based on greed. Derivatives provide an efficient way to construct a strategy that is consistent with either of these attitudes [17].

The global derivatives market is gigantic. Some market analysts estimate the size of the derivatives market at more that \$1.2 quadrillion (that is 1.2×10^{15}) which is more than 10 times the size of the total world gross domestic product. This is a truly frightening figure, especially when one considers that the derivatives have no "physical" value.

The pricing of derivatives makes use of two basic concepts: arbitrage and risk-free rate of interest. The interest rate at which money can be loaned to the state is called the risk-free rate of interest.[2] Arbitrage is the simultaneous purchase and sale of an asset with a price differential exceeding the risk-free rate of interest. It is a trade that profits by exploiting the price differences of identical or similar financial instruments on different markets or in different forms. Arbitrage exists as a result of market inefficiencies. This would theoretically lead to infinite gains by investing infinite borrowed money in an arbitrage opportunity.

9.2 Forward Contracts

A forward contract is a commitment to purchase or sell an asset at a specific price and at a specific date. Usually, this is a contract with a financial institution like a bank. The spot price is the price of the asset for immediate delivery. The party of the contract who agrees to buy the underlying owns the long position and the party who agrees to sell the underlying owns the short position. The price specified in the

---

[2]For investments in U.S. Dollars the current Treasury bill rates are considered as risk-free rate of interest. T-bills are considered free of default risk because they are fully backed by the U.S. government. However, given the most recent defaults of countries like Ukraine (2016), Argentina (2014), Greece (2012), and Côte d'Ivoire (2011), the phrase "risk-free" needs to be taken with a healthy pinch of salt.

forward contract is called delivery price K. The forward price F_t of a contract is the delivery price that would apply if the contract would be established at that time.

Denoting the price of the underlying security at maturity as S_T and the price of the forward position as K, the pay-off of the long position is

$$P_L = S_T - K .$$

This means that the long party can sell the security at a price of S_T which she bought for K. Obviously, the gain of the long position is the loss of the short position and vice versa. Therefore, the pay-off of the short position is

$$P_S = K - S_T .$$

Figure 9.1 shows the pay-off diagram of a forward contract as a function of underlying security at maturity. The long position gains on rising price of the underlying and the short position loses. Conversely, if the underlying's price falls the short position gains and the long position loses.

The question now is how to set the delivery price without creating arbitrage opportunities? Suppose that the current or spot price of an asset is S_0 and assume that the asset can be stored at zero cost (this is clearly not the case for physical assets like commodities). Calling the risk-free interest rate r, the delivery price in a forward contract with time to delivery T is

$$K = \begin{cases} e^{rT} S_0 & \text{for the continuous case} \\ (1+r)^T S_0 & \text{for the discrete case.} \end{cases}$$

The equation eliminates arbitrage opportunities. If K were greater than $e^{rT} S_0$, an arbitrage portfolio could be constructed as follows: borrow S_0 at rate r and buy the security. At maturity the security can be sold for K and the loan has to be repaid for $e^{rT} S_0$. The guaranteed gain is $K - e^{rT} S_0$. Therefore, K must not be greater than $e^{rT} S_0$. If, on the other hand, were K less than $e^{rT} S_0$, an arbitrage portfolio could

Fig. 9.1 Pay-off of a forward contract at maturity where K is the delivery price and S_T the price of the security at maturity. The long position gains on rising price of the underlying whereas the short position gains on falling price of the underlying

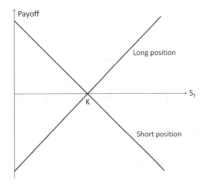

be constructed as follows: sell the security short and invest the proceeds S_0 with the risk-free rate r. At maturity, repurchase the security for K. The gain at maturity is $e^{rT} S_0 - K$. Therefore, K must not be less than $e^{rT} S_0$. Since K cannot be greater or less than $e^{rT} S_0$ it must be equal to $e^{rT} S_0$.

Suppose that the forward price of security S after time t is F_t and that the delivery price is K. Assuming again that the security can be stored at zero cost, the risk-free interest rate is r and the forward contract must be delivered at T, the current value of the forward contract for the long position to prevent any arbitrage is

$$F_t = d(F_t - K),$$

where d is the discount factor which can be calculated according to

$$d = \begin{cases} e^{-r(T-t)} & \text{for the continuous case} \\ \frac{1}{(1+r)^{T-t}} & \text{for the discrete case.} \end{cases}$$

Example: An American industrial company has a contract with a Swiss customer to sell machinery for 1 million Swiss Francs in one year's time. If the exchange rate USD/CHF rises during that time the American company will earn less in dollars. Therefore, wanting to eliminate this risk, it enters a forward contract to sell 1 million

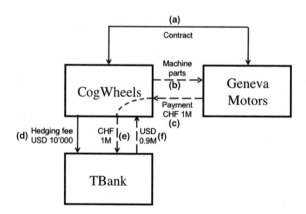

Fig. 9.2 CogWheels, an American company, agrees to deliver some sensitive machine parts to Geneva Motors, a Swiss company, which will integrate it in machinery it produces. They sign a contract on January 23, 2017 (**a**). According to that contract CogWheels will deliver the machine parts on January 23, 2018 (**b**) and Geneva Motors will pay CHF 1 million upon receipt of the elements (**c**); this amount corresponds to USD 0.9 million on the date the contract is signed. Since the CFO of CogWheels does not want her company to be subject to the risk of exchange rate fluctuations, on the same day, she signs a forward contract with TBank and pays a hedging fee of USD 10'000 (**d**) (this is calculated using the difference of interest rates between CHF and USD taken as 2% and 1% respectively). After receiving CHF 1 million from Geneva Motors, she pays that amount to the bank (**e**) and receives USD 0.9 million in return (**f**)

Swiss Francs for a fixed amount of dollars in one year. Clearly, this cannot be free of charge. Otherwise, it would constitute an arbitrage opportunity. The cost of the contract can be seen as an insurance premium which the company is willing to pay to avoid a potential loss (also eliminating the chance of earning more dollars, should USD/CHF fall—but, that would be speculating, not the business of making and selling machinery). Such operations, colloquially known as hedging, are very common for companies which export their products and whose future earnings are in foreign currencies. Figure 9.2 shows how this is done.   ☐

9.3 Futures

A major drawback of forwards is that they are not securitized and cannot be traded on an exchange. To make trading possible, the market has to specify certain standards. Once that is done, the two parties of a future contract do not need to know each other and the stock exchange takes care of the risk of default of the counterpart. A future contract has to specify the following:

- The underlying asset (often commodities, shares or market indexes).
- The contract size.
- Delivery arrangements (location, FOB/COD,[3] ...).
- Delivery date.
- Quality of delivered goods (specifications, *e.g.*, there are about 100 different grades of oil!).
- Price quotes.
- Daily price movement limits, *i.e.*, prevention of large movements.
- Position limits, *i.e.*, maximum of contracts a speculator may hold so as not to be able to manipulate prices.

However, even with standardization, two problems remain unsolved: how to avoid defaults and how to specify the forward price. The specification of the forward price is not obvious since the exchange cannot issue new contracts every time the price of the underlying changes (this happens practically every moment the exchange is open). So-called margin accounts can be used to deal with these two problems.

Imagine that the contract has a forward price of F_0 at the beginning. The next day the forward price will be F_1. The long party then receives $F_1 - F_0$ and the short party pays $F_0 - F_1$. The situation is equivalent to the delivery price in the contract being changed from F_0 to F_1. One can say that the contract is 'renewed' each day like this. The margin account serves as a guarantee that neither party of the contract defaults on

---

[3] Free on board (FOB) shipping is a trade term published by the International Chamber of Commerce (ICC), indicating which party assumes the cost of delivering goods. Some costs of FOB shipping include transportation and insurance. Cash on delivery (COD) is a type of transaction in which the recipient makes payment for a good at the time of delivery. If the purchaser does not make payment when the good is delivered, then the good is returned to the seller.

Fig. 9.3 Convergence of
futures and spot prices as
time approaches maturity.
The future price is always
more than the spot price

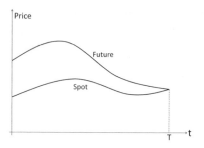

its obligation. On writing the contract, a deposit, known as the initial margin, has to be
paid (usually about 5–10% of the current value of the underlying in the contract). To
make sure that the margin account never has a negative balance, a minimum margin,
called the maintenance margin, is set to about 75–90% of the initial margin. If the
value of the margin account drops below the maintenance margin, a margin call is
issued to the trader and he is requested to top up the margin account to the initial
margin. If the margin account is not topped up by the trader, the institution holding
the margin account is entitled to liquidate a fraction of the contract to account for
the missing margin.

Since the margin account is balanced at the end of every day, the value of the
contract is always zero. The only variable price in the contract is the delivery price.
As the contract reaches its maturity, the futures price must converge to the spot price
(see Figure 9.3). Were this not the case, arbitrage opportunities would arise.

If the future's price at maturity were above the spot price, one could buy the
underlying asset at the spot price, take a short position in the future and then settle
the contract making a profit. Conversely, if the future's price were to be below the
spot price at maturity one could take a long position in the future and immediately sell
the asset in the spot market for a profit. Therefore, the future's price has to converge
to the spot price at maturity.

The question of how to set the future's price is still unanswered at this point and
is dealt with in the next Section within the context of the Black[4]-Scholes[5] equation.
A highly readable book which explains financial derivatives in layman's terms is
[17]. It features an introduction to the entire realm of derivatives with many real life
examples to provide a broad outlook on the subject matter. It also presents a lucid
conceptual background to derivatives by avoiding unnecessary technical details. A
very important statement stressed in the book is that *a derivative is a contract that
is used to transfer risk*. It is also pointed out that *a derivative is a contract where the
ultimate pay-off depends on future events*.

---

[4]Fischer Black, American economist (1938–1995); famous for developing the Black-Scholes formula for pricing derivative investment instruments like options together with Myron Scholes.

[5]Myron Scholes, Canadian-American financial economist, (1941–); famous for developing the
Black-Scholes formula for pricing derivative investment instruments like options together with
Robert Black.

If the risk-free rate is constant and deterministic, the corresponding futures and forward prices are identical. In practice, there are numerous reasons which cause differences between futures and forward prices such as transaction costs, margin accounts, taxes etc. For a more detailed discussion of forwards and futures see [58, 79].

Example: On a given day, the price of gold for delivery on a specific future date is $1'000. One future contract corresponds to 100 oz. gold as specified by New York Mercantile Exchange (NYMEX) and is worth $100'000. A trader buys two contracts and pays $12'000 (just 6% of $200'000 the trader would have to bring up to actually buy 200 oz. gold, giving this operation a leverage of almost 17) as initial margin as specified by NYMEX in a margin account. Further, NYMEX sets $4'000 per contract as the maintenance margin based on the current levels of volatility in gold prices. On Day 1 the price of gold future drops to $990. This has no consequence for the trader other than a paper loss of $2'000. On Day 2, the price of gold future drops to $980. The trader now has a cumulative paper loss of $4'000. On Day 3 the price of gold future drops yet again, this time to $970. The trader has so far suffered a paper loss of $6'000. More significant is the fact that the value of his margin account has dropped to $6'000 which is below the maintenance level. Therefore, he gets a call from NYMEX saying that he has to top up his margin account to the initial level. This means that the trader has to pay $6'000 in his margin account. If he is not able or willing to do so, NYMEX will liquidate one contract to protect itself. The trader grudgingly pays the required amount. On Day 4 the price of gold future drops yet again, this time to $960 bringing the trader's cumulative loss up to $8'000. On Day 5 things look a bit better and the price of gold goes up to $970 reducing the trader's cumulative loss to $6'000. On Day 6 the price of gold goes up to $980 reducing the trader's cumulative loss to $4'000. Saying enough is enough, the trader sells his futures contracts and liquidates the margin account. So, the saga is finished for the trader with a net loss of $4'000, corresponding to a return of -33.3% in six days. Table 9.1 summarizes the operations. A happier scenario would be increasing gold prices as shown in Table 9.2 which would net the trader $6'000, a return of 50% in six days! One can thus see the effect of leveraging in derivative instruments on gains or losses.   □

Table 9.1 Buying futures on margin might incur steep losses in falling markets

Day	Future's price	Daily gain (loss)	Cum'l gain (loss)	Margin account balance	Margin call
0	$1'000			$12'000	
1	$990	($2'000)	($2'000)	$10'000	$0
2	$980	($2'000)	($4'000)	$8'000	$0
3	$970	($2'000)	($6'000)	$6'000	$6'000
4	$960	($2'000)	($8'000)	$10'000	$0
5	$970	$2'000	($6'000)	$12'000	$0
6	$980	$2'000	($4'000)	$14'000	$0

Table 9.2 Buying futures on margin might result in huge gains in bull markets

Day	Future's price	Daily gain (loss)	Cum'l gain (loss)	Margin account balance	Margin call
0	$1'000			$12'000	
1	$1'005	$1'000	$1'000	$13'000	$0
2	$1'010	$1'000	$2'000	$14'000	$0
3	$1'020	$2'000	$4'000	$16'000	$0
4	$1'020	$0	$4'000	$16'000	$0
5	$1'025	$1'000	$5'000	$17'000	$0
6	$1'030	$ 1'000	$6'000	$18'000	$0

9.4   Options

9.4.1   Basics

An option is the right, but not the obligation, to buy (or sell) an asset under specified terms. An option which gives the right to purchase something is called a *call option*, whereas an option which gives the right to sell something is called a *put option*. An option is a derivative security whose underlying asset is the asset that can be bought or sold, such as a commodity, an index or simply a share. The options on common stocks, like all other options, need specifications in order to classify and price them. The specification of an option are

- A clear description of what can be bought (for a call) or sold (for a put); here the name of the underlying stock and the market where it is traded suffices.
- The exercise price, or strike price (K) at which the asset can be purchased upon exercising the option.
- The period of time for which the option is valid; here the expiration date (T) is required.
- Exercise type; an *American option* allows exercising it any time before and including the expiration date. A *European option*, however, allows exercising it only on the expiration date (the terms European and American option classify the different options and have no bearing on the location where they are issued).

In issuing an option there are two sides involved: the party who grants the option and the party who purchases the option. The party which grants the option is said to write an option. The party which purchases an option faces no risk of loss other than the original purchase price. However, the party which writes the option, usually a financial intermediary such as a bank, may face a large loss, since this party must buy or sell this asset at specified terms if the option is exercised. In the case of an exercised call option, if the writer does not already own the asset, she must purchase it in order to deliver it at the specified strike price, which may be much lower than the current market price. The market regulations require that the financial institutions writing options must always cover their positions, for instance, by writing call and put options at the same time on the same underlying.

9.4.2 Some Properties of Options

Suppose a trader owns a European call option on a given stock with the strike price K and suppose the option has reached the expiration time T. What is the value of the option at this time as a function of the underlying stock price S? If the stock price S is larger than the strike price K, the value of the call option is $C = S - K$. If, on the other hand, $S < K$ the trader would not exercise the option and thus the option has no value, $C = 0$. Therefore, the option pay-off at expiration time is

$$C(S, T) = \max(0, S(T) - K).$$

The argument for put options is identical and thus

$$P(S, T) = \max(K - S(T), 0).$$

The value of an option at expiration is shown graphically in Figure 9.4. Before the expiration time, the call option value is above its pay-off function at expiration time, since the time remaining to expiration offers the possibility that the stock price might further increase. In other words, the option has a time value. For a call option where $S > K$ the option is said to be *in the money*, where $S = K$ *at the money*, and where $S < K$ *out of the money*.

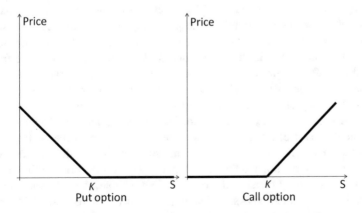

Fig. 9.4 Value of an option C at the expiration date as a function of the price of the underlying S. K is the strike price

Fig. 9.5 Stock S_1 and stock S_2 have the same value at time t_0. Imagine a trader buys put options on those two stocks with the same strike price and the same strike date T. Which option would cost more on $t_0 < T$?

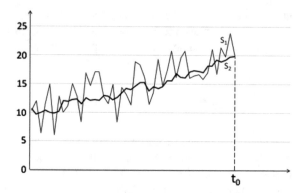

In addition to its time value an option also has a volatility value. Consider two stocks at the same price and two put options with the same strike price as shown in Figure 9.5. Stock one is more volatile than stock two. The put option on stock one has a higher value, because the probability that stock one falls below the strike price is higher than for stock two. Since a put option is similar to an insurance which pays in case of a loss, the insurance premium must go up, as the likelihood of this case gets larger just like a driver who is accident prone has to pay a higher insurance premium.

9.4.3 Economics of Options

In order to derive a way of pricing for options, one needs to think about the underlying economics of options and financial markets. Option pricing centers on the concept of arbitrage-free markets. Once again, the notion of arbitrage is, loosely put, the possibility to make a profit with no risks involved. In an arbitrage-free market, one can only obtain a risk-free return at the prevailing risk-free interest rate, such as paid by government bonds. Let us form a portfolio G of a stock at the current price S_0 and a put option P with the strike price K. If the market is arbitrage-free, the expected pay-off at expiration r_e is exactly the risk-free interest rate r, because the combination of a put option and the same underlying stock removes all possibility of a loss. If the pay-off is larger than the risk-free rate, one could borrow money (L) at the risk-free rate, and buy the portfolio. Consequently, one would achieve a profit larger than the risk-free rate at no risk without any capital investments. Because this would be such a wonderful return generating scheme, one would start to buy huge quantities of the portfolio, and this would increase the option price until the pay-off at expiration is exactly equal to the risk-free interest rate. A similar argument can be made if the expected pay-off is below the risk-free interest rate. The possible scenarios are shown in the table below.

Initial Investment	Return	Pay-off
$P + S - L = 0$	$r_e > r$	$E[r_e(P + S) - rL] > 0$
$-P - S + L = 0$	$r_e < r$	$E[-r_e(P + S) + rL] > 0$
$P + S - L = 0$	$r_e = r$	$E[r_e(P + S) - rL] = 0$

The same argument can be made to explain the value of options at any other time than expiration. Take a call option some time before expiration which is out of the money (meaning the underlying stock is below the strike price). Assuming that the value of the option is zero, one could now obtain infinitely large quantities of this option and face no risk of loss and need no initial capital outlay. The pay-off at expiration is either zero or positive and can be larger than the prevailing risk-free interest rate. This would clearly be an arbitrage possibility. Therefore, even the out of the money options must a have a positive time value.

9.4.4 Black-Scholes Equation

Let us now derive the *Black-Scholes Equation*, arguably the most famous equation in finance, to price options correctly. The derivation for call options is based on the no arbitrage argument above. Detailed derivations of the Black-Scholes Equation can be found in [12, 79, 89, 99], or [35].

A model of the dynamics of stock prices is required to price stock options. As mentioned in Chapter 5, the geometric Brownian motion model is used to model stock prices for three main reasons. Firstly, asset prices are always positive, secondly, stocks behave like bonds with continuous compounding and stochastic returns, and thirdly, stock prices have returns with distributions resembling log-normal distributions. The stochastic differential equation governing a stock price is thus given by

$$dS(t) = \mu S(t)dt + \sigma S(t)dW , \qquad (9.1)$$

where μ and σ are assumed to be constant and $W(\cdot)$ is a Brownian motion. One could, however, also use the more general stock price dynamics

$$dS(t) = \mu(S, t)S(t)dt + \sigma S(t)dW , \qquad (9.2)$$

where σ is constant but $\mu(S, t)$ is a function of S and t.

The derivation of the Black-Scholes equation is based on the two central arguments that in a perfect market no arbitrage possibilities exist and that, to be rational, the writer of an option should not be taking any speculative risks by writing the option.

Let $C(t, S)$ denote the value of the call option on the underlying stock S at time t, and r denote the current risk-free interest rate. The stock price dynamics are assumed to be as in Equation (9.1). The call option is a derivative security of the underlying stock S. Therefore, as shown in Chapter 5, it changes its value dC according to Itô's chain rule (5.12) as

$$dC = \left(\frac{\partial C}{\partial t} + \frac{\partial C}{\partial S}\mu S + \frac{1}{2}\frac{\partial^2 C}{\partial S^2}\sigma^2 S^2\right)dt + \frac{\partial C}{\partial S}\sigma S dW .$$ (9.3)

The writer of the option does not want to take any risks by writing the option and therefore forms a portfolio G which is comprised of the amount x of the underlying stock and of the amount y of a bond B. The value of the portfolio $G = xS + yB$ should match the option value at all times. Therefore, this portfolio is called the replication portfolio.

The differential equation of the bond dynamics is derived from $B = B_0 e^{rt}$ as

$$dB = rBdt ,$$ (9.4)

where B_0 is the initial value of the bond.

The replication portfolio should be self-financing, *i.e.*, no money should be needed except for the initial capital outlay. Using the chain rule, the differential equation governing the portfolio dynamics is thus given by

$$\frac{dG}{dt} = x\frac{dS}{dt} + S\frac{dx}{dt} + y\frac{dB}{dt} + B\frac{dy}{dt} ,$$ (9.5)

or in the differential form as

$$dG = xdS + ydB + \underbrace{dxS + dyB}_{0} = xdS + ydB .$$ (9.6)

The term $dxS + dyB$ equals zero, because a change in the amount of stocks and bonds held in the portfolio at constant stock and bond prices equals an inflow or outflow of money. Since the portfolio should be self-financing, the positions in the stocks and bonds are only allowed to change in such a way that the change in one asset finances the change in the other asset.

Let us now substitute dS from (9.1) and dB from (9.4) into (9.6). This yields

$$dG = x(\mu S(t)dt + \sigma S(t)dW) + y(rB)dt = (x\mu S + yrB)dt + x\sigma S dW. (9.7)$$

Since the portfolio dynamics should match the dynamics of the option, the coefficients of dt and dW in (9.7) and (9.3) must be matched. To do this, first the coefficient of dW must be matched by setting

$$x = \frac{\partial C}{\partial S} .$$ (9.8)

The value of the replication portfolio should always be equal to the value of the option. This means $G = C$. Because $G = xS + yB$, it follows that

$$G = yB + \frac{\partial C}{\partial S} S = C$$

$$\Rightarrow y = \frac{1}{B}\left(C - \frac{\partial C}{\partial S} S\right). \tag{9.9}$$

Substituting the (9.8) and (9.9) into (9.7) results in

$$dG = \left[\frac{\partial C}{\partial S}\mu S + r\left(C - \frac{\partial C}{\partial S} S\right)\right]dt + \frac{\partial C}{\partial S}\sigma S dW .$$

Matching the coefficient of dt in (9.3) gives

$$\frac{\partial C}{\partial S}\mu S + \frac{1}{B}\left(C - \frac{\partial C}{\partial S} S\right)r B = \frac{\partial C}{\partial t} + \frac{\partial C}{\partial S}\mu S + \frac{1}{2}\frac{\partial^2 C}{\partial S^2}\sigma^2 S^2.$$

Thus, finally, the (in)famous Black-Scholes equation emerges as

$$rC = \frac{\partial C}{\partial t} + \frac{\partial C}{\partial S}rS + \frac{1}{2}\frac{\partial^2 C}{\partial S^2}\sigma^2 S^2. \tag{9.10}$$

The solution of this PDE together with the appropriate boundary conditions gives the pricing of the option.

But is (9.10) really arbitrage-free? Consider the strategy of writing one call option and investing the yield in the replication portfolio. Let us call the initial market price of the option $C^*(S_0, 0)$. The theoretical Black-Scholes price is $C(S_0, 0)$. By construction, the replication portfolio always matches the option until the expiration date. If one could convince somebody to pay more than the theoretical price for the option, in other words $C^*(S_0, 0) > C(S_0, 0)$, then one could make money with no risk, because one would invest $C(S_0, 0)$ in the replication portfolio and pocket the initial difference $C^*(S_0, 0) - C(S_0, 0)$ as profit. Since there is no initial capital required and the replication portfolio procedure removes all risks, one would have found an arbitrage possibility. An analogous argument can be made for the case if the market price were below the theoretical price, again resulting in an arbitrage opportunity. Therefore, any arbitrage possibility is removed only if the market price and the theoretical price were identical, i.e., $C^*(S_0, 0) = C(S_0, 0)$.

Having derived the Black-Scholes PDE, let us now look at the Black-Scholes formula for European call options.

The underlying stock, in a trivial way, is a derivative of the stock itself. Therefore, $C(t, S) = S(t)$ should satisfy the PDE. One can easily check

$$\frac{\partial C}{\partial S} = 1, \qquad \frac{\partial^2 C}{\partial S^2} = 0, \qquad \frac{\partial C}{\partial t} = 0 \qquad \Rightarrow \qquad rS(t) = rS(t),$$

and see that $C(t, S) = S(t)$ is indeed a possible solution.

The Black-Scholes formula for a European call option with strike price K, expiration time T, a stock with the current price S which pays no dividends, and a constant risk-free interest rate r, is given by

$$C(t, S) = S(t)N(d_1) - Ke^{-r(T-t)}N(d_2), \qquad (9.11)$$

where $N(d)$ is the cumulative normal probability distribution

$$N(d) = \frac{1}{\sqrt{2\pi}} \int_{-\infty}^{d} e^{-\frac{y^2}{2}} dy,$$

and d_1 and d_2 are defined as

$$d_1 = \frac{\log(\frac{S}{K}) + (r + \frac{\sigma^2}{2})(T - t)}{\sigma \sqrt{T - t}}$$

$$d_2 = d_1 - \sigma \sqrt{T - t}.$$

The formula for the pricing of European call options (9.11) is simply stated here. For its derivation see, e.g., [12, 79] or [89].

Let us now show that (9.11) satisfies the Black-Scholes equation in (9.10) together with the boundary conditions $C(T, S) = \max(S(T) - K, 0)$. For $t = T$ (expiration time)

$$d_1(S, T) = d_2(S, T) = \begin{cases} +\infty & S(T) > K \\ -\infty & S(T) < K \end{cases}.$$

One needs to distinguish these two cases, because $S > K \Rightarrow \log(S/K) > 0$ and $S < K \Rightarrow \log(S/K) < 0$.

Since $N(\infty) = 1$ and $N(-\infty) = 0$ Equation (9.11) results in

$$C(S, T) = \begin{cases} S - K & S(T) > K \\ 0 & S(T) < K \end{cases}.$$

This confirms that (9.11) satisfies the boundary condition. The derivatives of (9.11) can be calculated with some algebra as

$$\frac{\partial C}{\partial S} = N(d_1)$$

$$\frac{\partial^2 C}{\partial S^2} = \frac{e^{-\frac{d_1^2}{2}}}{S\sigma\sqrt{2\pi(T-t)}}$$

$$\frac{\partial C}{\partial t} = -\frac{e^{-\frac{d_1^2}{2}}S\sigma}{2\sqrt{2\pi(T-t)}} - rKe^{-rT}N(d_2).$$

Thus,

$$\frac{\partial C}{\partial t} + \frac{\partial C}{\partial S}rS + \frac{1}{2}\frac{\partial^2 C}{\partial S^2}\sigma^2 S^2 = -\frac{e^{-\frac{d_1^2}{2}}S\sigma}{2\sqrt{2\pi(T-t)}} - rKe^{-rT}N(d_2)$$

$$+N(d_1)rS + \frac{e^{-\frac{d_1^2}{2}}}{2S\sigma\sqrt{2\pi(T-t)}}\sigma^2 S^2$$

$$= r\left(S(t)N(d_1) - Ke^{-r(T-t)}N(d_2)\right)$$

$$= rC(t,S).$$

This shows that (9.11) satisfies the Black-Scholes equation.

A very important and perhaps surprising result is that the Black-Scholes equation and the call option formula are valid not only when the stock price dynamics is modeled with a geometric Brownian motion (9.1), but also with the more general dynamics shown in (9.2). Since the drift term does not enter the PDE, it does not affect the pricing of options. The replication procedure remains the same as well. See [23] for a detailed discussion of this property.

The formula for European put options can be easily derived from the formula for European call options. To derive the formula, one needs to use a simple theoretical relationship between the prices of the corresponding puts and calls. If one buys a call option and sells a put option with the same strike price, the portfolio behaves almost like the stock itself. The pay-off at expiration time is $\max(S(T) - K, 0) - \max(K - S(T), 0) = S(T) - K$. In other words, the difference to the stock price is the strike price K. By lending $e^{-r(T-t)}K$, one obtains a pay-off of K at the expiration time and the portfolio composed of the put and the call options and the credit resembles exactly the stock. Thus the call-put-parity is $C - P + e^{-r(T-t)}K = S$.

One can now use the so-called put-call-parity and the Black-Scholes call option formula to derive the European put option formula.

$$C - P + e^{-r(T-t)}K = S$$

$$S(t)N(d_1) - Ke^{-r(T-t)}N(d_2) - P(t,S) + e^{-r(T-t)}K = S(t)$$

$$P(t,S) = S(t)(N(d_1) - 1) - Ke^{-r(T-t)}(1 - N(d_2)).$$

Since $N(x) - 1 = N(-x)$ (see Chapter 3 and Appendix C), the formula for European put options can be stated as

$$P(t, S) = Ke^{-r(T-t)}N(-d_2) - S(t)N(-d_1). \tag{9.12}$$

In the option contracts discussed so far, the holder may exercise the option at a certain time specified in the contract. However, the holder is often given the opportunity to exercise early, which means that the option may be exercised not only on a given date but at any at time before expiration date. Such contracts are known as American options.[6]

American options or any other "exotic" options, where the pay-off depends not only of the value of the underlying security at expiration date but on how it gets there, are a lot more difficult to evaluate. The valuation of American options usually are based on a worst case principle, where it is assumed that the option will be exercised at a time at which this is most profitable to the holder, which represents a worst case for the writer. Let us now use the example of an American put option to illustrate the pricing of American options.

For American options the Black-Scholes PDE is replaced by an inequality

$$\frac{\partial P}{\partial t} + \frac{\partial P}{\partial S}rS + \frac{1}{2}\frac{\partial^2 P}{\partial S^2}\sigma^2 S^2 - rP \leq 0, \tag{9.13}$$

in which the equality holds if the option is not exercised, *i.e.*, if its value exceeds the revenue of exercising. For the put option, this is expressed by the inequality

$$P(t, S) > \max(K - S(t), 0).$$

For the inequality (9.13) the condition on expiration is

$$P(T, S) = \max(K - S(T), 0).$$

In addition, two boundary conditions are imposed on the PDE. The first boundary condition is that the put option has no value for arbitrarily large stock prices

$$\lim_{S\to\infty} P(T, S) = 0.$$

and the second boundary condition is that the value of the put option for $S(t) = 0$ is equal to the discounted strike price

$$P(T, 0) = e^{-r(T-t)}K.$$

---

[6]A not so uncommon mixture between American and European options are options which can be exercised at a fixed number of dates before the expiration date. Such options are called "Bermuda" options.

The inequality (9.13), together with these boundary conditions, the terminal value, and the early exercise inequality, yield no analytical solution and one must seek numerical methods for its solution as already explained in Section 5.9.2. A detailed discussion of the numerical solution procedure can be found in [116].

9.4.5 General Option Pricing

A PDE similar to the Black-Scholes equation for the pricing of any single factor stock price process can also be calculated. If one assumes that the underlying asset follows an SDE as given by

$$dS(t) = \mu(t, S)dt + \sigma(t, S)dW,$$

the PDE for pricing any possible option $C(t, S)$ is given by

$$rC = \frac{\partial C}{\partial t} + \frac{\partial C}{\partial S}rS + \frac{1}{2}\frac{\partial^2 C}{\partial S^2}\sigma(t, S)^2. \tag{9.14}$$

It is noteworthy, that the drift does not enter the PDE and that the diffusion term which is a function of the price of the underlying stock and the time changes the PDE. The derivation is analogous to the derivation of the Black-Scholes PDE. Note that the Equation (9.14) applies only when the underlying stock price is modeled by a scalar SDE. If the price of the underlying stock is modeled by a system of SDE, a multi dimensional form of (9.14) can be derived as shown in [89].

9.5 Swaps

One of the most common derivative products is the so called swap. As the name suggests, a swap is a contract between two parties to exchange any financial instruments. Most swaps involve cash flows based on a notional principal amount that both parties agree to. As with the other derivatives, the principal does not have to change hands. Each cash flow is one leg of the swap. One cash flow is fixed, while the other is variable based on an interest rate (most common swap), a currency exchange rate and so on. Swaps are over the counter (OTC) agreements between financial institutions and big investors. They are not traded on exchanges.

Interest rate swaps are useful instruments for hedging as described in Section 9.2 but they can also be used as highly leveraged speculation instruments. Let us see how an interest rate swap works.

Example: CogWheels Company requires a large amount of money over the next year to develop new materials to be used in its products. As described in Chapter 7, it issues a five-year, $10million bond with a variable coupon of '12-month LIBOR + 240 basis points'. This is the most favorable deal CogWheels can get under the circumstances. The current 12-month LIBOR is 1.1% which makes the current yearly coupon of the bond 3.5%. However, LIBOR is close to its historical lows. Therefore, the CFO of CogWheels is concerned about increasing interest rates in the coming years. What to do?

She, therefore, finds a pet food company, FatCats Inc. to help CogWheels by agreeing to pay the interest payments on the whole bond issue over its life time. This, of course, does not come free: CogWheels agrees to pay FatCats 5.5% on the notional value of $10million for five years. This way, CogWheels benefits from this swap deal if the interest rates go up significantly over the next five years. FatCats benefit if the rates fall or go up moderately. Let us look at two such scenarios in detail.

Scenario A: LIBOR rises 0.5% per year over the next five years.
This means the interest payments of CogWheels to its bond holders will be (3.5% + 4% + 4.5% + 5% + 5.5%)× $10'000'000 = $2'250'000. CogWheels pays FatCats 5 × 5.5%× $10'000'000 = $2'750'000 as part of their swap deal and FatCats pays CogWheels $2'250'000, the interest to the bond holders. As a result CogWheels makes a loss of $2'750'000 - $2'250'000= $500'000 on this swap deal.

Scenario B: LIBOR rises 1.5% per year over the next five years.
This means the interest payments of CogWheels to its bond holders will be (3.5% + 5% + 6.5% + 8% + 9.5%)× $10'000'000 = $3'250'000. CogWheels pays FatCats 5 × 5.5%× $10'000'000 = $2'750'000 as part of their swap deal and FatCats pays CogWheels $3'250'000, the interest to the bond holders. As a result CogWheels makes a gain of $3'250'000 - $2'750'000=$500'000 on this swap deal.

Figure 9.6 illustrates the two scenarios. In most cases the CogWheels and FatCats would act through a bank, which would take a fee for its services. The numbers in this example are chosen to be illustrative. In reality, the increase in interest rates would not be so dramatic. Therefore, the swap fee of the bank would not be negligible. □

Fig. 9.6 Swaps are used to reduce financial risks (hedging) as in this example. They can, however, be used to increase leverage for speculating

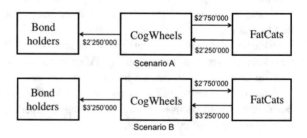

9.6 Structured Products

Switzerland is by far the largest market globally in derivatives issued and sold. Common structured products can be classified in several categories suggested by the Swiss Structured Products Association as shown in Table 9.3.

According to Swiss Structured Products Association, structured products are innovative, flexible investment instruments and an attractive alternative to direct financial investments such as shares, bonds, currencies and the like. They are sufficiently flexible to accommodate any risk profile, no matter how challenging the markets. Basically, structured products are bearer bonds with the issuer liable to the extent of all his assets. Thus, structured product quality is directly linked to the debtor's, or issuer's, creditworthiness. Bearer bonds (bonds and structured products) are subject to issuer risk. In case of issuer bankruptcy, traditional bonds as well as structured products form part of assets under bankruptcy law. To keep issuer risk to a minimum, investors should stick to top-quality issuers, and spread their investments among several issuers.Diversification is, of course, a wise strategy for all forms of investment including time deposits, bonds and structured products. Issuer creditworthiness over time should also be monitored. A frightening reminder is Lehman Brothers, one of the world's largest financial services firm. On September 15, 2008, it filed for the largest bankruptcy in history...

Example: Let us assume that a trader expects the price of a share (underlying) to rise during the next year. However, she also expects rising volatility in its price and further thinks that a sharp drop in the price of the underlying is possible. She might then be interested in a structured product called Capital Protection Certificate with Participation. Figure 9.7 shows how her trade fares under different circumstances.

Minimum redemption at expiry is equivalent to the capital protection which is defined as a percentage of the nominal (it can be 100%, or less as shown here). Capital protection refers to the nominal only, and not to the purchase price. Value of the product may fall below its capital protection during its lifetime. Participation in underlying's price increase above the strike is less than 100%. Any pay-outs like dividends attributable to the underlying are used in favor of the strategy and is not handed out to the owner.   □

Fig. 9.7 The horizontal axis denotes the price of the underlying. The vertical axis denotes the return of the Capital Protection Certificate with Participation product

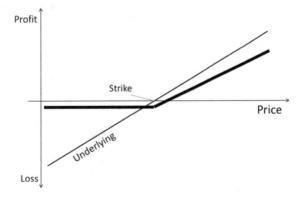

Table 9.3 Different classes of derivative products can be employed for different investment needs and market expectations

Purpose	Market expectation	Products
Capital protection	Rising underlying, rising volatility, sharply falling underlying possible	Capital protection certificate with participation
		Convertible certificate
		Capital protection certificate with coupon
	Rising underlying, sharply falling underlying possible, underlying is not going to touch or go above the barrier during product lifetime	Barrier capital protection certificate
Yield enhancement	Underlying moving sideways or slightly rising, falling volatility.	Discount certificate, Reverse convertible
	Underlying moving sideways or slightly rising, falling volatility, underlying will not breach barrier during product lifetime	Barrier discount certificate, Barrier reverse convertible, Express certificate
Participation	Rising underlying	Tracker certificate
	Rising underlying, rising volatility	Out-performance certificate
	Underlying moving sideways or rising, underlying will not breach barrier during product lifetime	Bonus certificate
	Rising underlying, underlying will not breach barrier during product lifetime	Bonus out-performance certificate
	Rising or slightly falling underlying, underlying will not breach barrier during product lifetime	Twin-win certificate
Leverage	Call: Rising underlying, rising volatility, Put: Falling underlying, rising volatility	Warrant
	Bull: Rising underlying, Bear: Falling underlying	Spread Warrant
	Call: Rising underlying, Put: Falling underlying	Warrant with knock-out
	Long: Rising underlying, Short: Falling underlying	Mini-future
	Long: Rising underlying, Short: Falling underlying	Constant leverage certificate

9.7 Exercises

A1: What is a financial derivative?

A2: What are the contradictory powers of derivatives and structured products?

A3: What is the estimated size of the global derivatives market?

A4: What is arbitrage?

A5: What are the main concepts that lead to the pricing of derivatives?

A6: What is a forward contract?

A7: What is hedging? How does it work?

A8: How are futures different from forward contracts?

A9: What is an option? What kind of options do you know? How are they different?

A10: Explain the concepts of time value and volatility value of an option.

A11: What is the difference between an American option and a European option?

B1: An investor can chose among n risk-bearing assets for his portfolio. In order to maximize his utility, the investor has to solve the following optimal control problem:

$$\max_{\mathbf{u}} \ \mathrm{E}\left[-\frac{1}{\gamma} e^{-\gamma X(T)} \right],$$

subject to

$$dX = X\left[r + \mathbf{u}^T (\boldsymbol{\mu} - \mathbf{1}r) \right]dt + X\mathbf{u}^T \boldsymbol{\Sigma} d\mathbf{W},$$

with

$$t \in [0, T], r \in \mathbb{R}, \boldsymbol{\mu} \in \mathbb{R}^{n \times 1}, \mathbf{e} \in \mathbb{R}^{n \times 1}, \mathbf{u} \in \mathbb{R}^{n \times 1}, \boldsymbol{\Sigma} \in \mathbb{R}^{n \times m}, \mathbf{W} \in \mathbb{R}^{m \times 1}.$$

The SDE is the so-called "Wealth equation", where X is the wealth of an investor. The aim of optimal control is to maximize the utility of his final wealth. A utility function measures the inclination for increased wealth and also governs the investor's risk aversion. $\gamma > 0$ is the risk aversion. The larger $\gamma > 0$, the more risk averse is the investor.

$\mathbf{W}$ denotes an m-dimensional Brownian Motion, r is the risk-free interest rate, $\mathbf{1}$ denotes the unity vector. The control vector $\mathbf{u}$ denotes the percentage of the wealth that is invested in different "assets", $\boldsymbol{\mu}$ denotes the expected return of the risky assets and $\boldsymbol{\Sigma}$ is the covariance matrix.

Write down the Hamilton-Jacobi equation for the problem.

B2: Solve the embedded maximization for $\mathbf{u}$ which results in $\mathbf{u}^*$.

B3: Use **u*** for **u** to get the PDE which needs to be solved.

B4: Solve the resulting PDE.

B5: Study some particular solutions of the PDE and discuss them.

C1: Design a software environment which would allow the user to design various structured products. The environment should have a user interface for defining the characteristics of the desired product. It should also be able to solve the related Black-Scholes equations numerically. Finally, the output should include the composition of the replication portfolio (bonds, options) and its statistical parameters.

C2: Derivative products are based on certain assumptions like the Brownian price dynamics of the underlying stocks. How would the theory change when you generalize these assumptions? Will you be able to achieve more realistic pricing mechanisms for the derivatives?

Appendix A
Dynamic Systems

A.1 Introduction

The concepts of "system" and "signal" are very general and encompass a broad range of applications. Therefore, it is not easy to define them precisely without making compromises in terms of generality. However, by means of descriptions and by giving some examples, it is possible to create a common and acceptable understanding of these concepts[1]

The term *system* is used here for a bounded and ordered collection of elements that fulfill a common purpose in unison. A system interacts with its surroundings through *signals* at its inputs and outputs. These signals can be in the form of energy, matter or information. A system converts its input signals to its output signals. Accordingly, a system can also be seen as a mapping which maps its input variables to its output variables.

This description of a system brings with itself the difficulty of defining the border where a system ends and its surrounding begins. Since this border is arbitrary and application dependent, it makes sense to define the border such that the interface between a system and its outputs are as clearly defined as possible. The inputs which cannot be influenced are conveniently labeled as disturbances.

A common method of depicting systems is by means of *block diagrams* in which systems are symbolized as rectangular blocks and signals as arrows entering (inputs) or leaving (outputs) a block (see Figure A.1).

---

[1]This Appendix is an abridged version of the opening Chapter of [51].

© Springer International Publishing AG 2018
S. S. Hacısalihzade, *Control Engineering and Finance*, Lecture Notes in Control and Information Sciences 467, https://doi.org/10.1007/978-3-319-64492-9

Fig. A.1 The block diagram

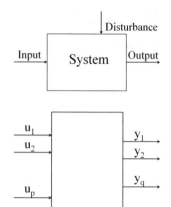

Fig. A.2 A multivariable
system with p inputs and q
outputs

A.2   Some Special Types of Systems

In general, a system might have several inputs and several outputs. Such systems are also called multivariable systems or multi-input-multi-output (MIMO) systems. Figure A.2 shows such a system with p inputs $u_1, u_2, \ldots, u_p$ and q outputs $y_1, y_2, \ldots, y_q$.

For multivariable systems, it is convenient to aggregate all inputs $u_1, u_2, \ldots, u_p$ in a p-dimensional input vector $\mathbf{u}$ and all the outputs $y_1, y_2, \ldots, y_q$ in a q-dimensional output vector $\mathbf{y}$. The system is thus defined by the functional relationship

$$\mathbf{y}(t) = \mathbf{f}\big(\mathbf{u}(t), t\big). \tag{A.1}$$

In other words, any output of a multivariable system is a time function of a combination of its inputs.

Let us now define several important special cases of (A.1):

A system for which the superposition of the inputs results in the superposition of the outputs is called a *linear system*. Formally,

$$\mathbf{f}\big(\alpha_1\mathbf{u}_1(t) + \alpha_2\mathbf{u}_2(t)\big) = \alpha_1\mathbf{f}\big(\mathbf{u}_1(t)\big) + \alpha_2\mathbf{f}\big(\mathbf{u}_2(t)\big). \tag{A.2}$$

A system for which the choice of the initial time t_0 has no effect on the output is called a *time invariant system*. Formally,

$$\mathbf{y}(t - t_0) = \mathbf{f}\big(\mathbf{u}(t - t_0)\big) \quad \forall t_0. \tag{A.3}$$

A system for which two identical input signals (starting with identical initial conditions) result in two identical output signals is called a *causal system*. Formally,

$$\mathbf{u}_1(t) = \mathbf{u}_2(t) \quad \Rightarrow \quad \mathbf{y}_1(t) = \mathbf{y}_2(t). \tag{A.4}$$

A system whose output depends only on the momentary value of the input is called a *static system*. Formally,

$$\mathbf{y}(t_1) = \mathbf{f}\big(\mathbf{u}(t_1)\big), \qquad -\infty < t_1 < \infty. \tag{A.5}$$

A system whose output depends on the previous as well as the momentary values of its input is called a *dynamic system*. Formally,

$$\mathbf{y}(t_1) = \mathbf{f}\big(\mathbf{u}(t)\big), \qquad t_0 \le t \le t_1. \tag{A.6}$$

A.3 Mathematical Description and Analysis of Systems

A.3.1 Input-Output Description of Systems in Time Domain

In general, a system defined as in (A.1) cannot easily be handled analytically. Therefore, most of the analysis methods assume the system in question to be linear, causal and time invariant. A possible way of describing such a system is by means of its impulse response.

Let us first look at the concept of the impulse. The impulse in Figure A.3 can be described as

$$\delta_\Delta(t, t_1) = \begin{cases} 1/\Delta & t_1 \le t \le t_1 + \Delta \\ 0 & \text{else}. \end{cases} \tag{A.7}$$

The shaded area is 1. Now let $\Delta \to 0$ while keeping the area as 1. This results in the so-called Dirac impulse function[2]

$$\delta(t, t_1) = \lim_{\Delta \to 0} \delta_\Delta(t, t_1). \tag{A.8}$$

This definition implies

$$\int_{-\infty}^{\infty} \delta(t, t_1)dt = 1, \tag{A.9}$$

Fig. A.3 An impulse function

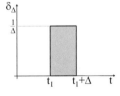

---

[2]Mathematicians prefer the term "distribution", because this limit case is not strictly a function.

and

$$\int_{-\infty}^{\infty} u(t)\delta(t, t_1)dt = u(t_1).$$ (A.10)

Why is this relevant? Well, as shown in Figure A.4, the input $u(t)$ of any SISO system can be approximated as a sum of impulses

$$u(t) = \sum_{i=0}^{\infty} u(t_i)\delta_\Delta(t, t_i)\Delta.$$ (A.11)

The output of a linear, causal and time invariant system with the input in (A.11) is

$$y(t_k) = f\big(u(t), t_k\big)$$ (A.12)

$$= f\big(u(0, \ldots, t_k), t_k\big)$$ (A.13)

$$= f\Big(\sum_{i=0}^{k-1} u(t_i)\delta_\Delta(t, t_i)\Delta, t_k\Big)$$ (A.14)

$$= \sum_{i=0}^{k-1} u(t_i)f\big(\delta_\Delta(t, t_i), t_k\big)\Delta$$ (A.15)

$$= \sum_{i=0}^{k-1} u(t_i)f\big(\delta_\Delta(t, 0), t_k - t_i\big)\Delta.$$ (A.16)

Letting $\Delta \to 0$ results in

$$y(t) = \int_0^t f\big(\delta(t, 0), t - \tau\big)u(\tau)d\tau.$$ (A.17)

$f\big(\delta(t, 0), t\big)$ is called the impulse response of the system and is often symbolized as $g(t)$. Thus, the output of a system with vanishing initial conditions is the convolution of the system's impulse response with its input

Fig. A.4 Approximation of a system's input as a sum of impulses

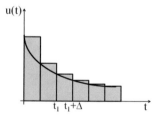

$$y(t) = g(t) \circ u(t) \tag{A.18}$$

$$= \int_0^t g(t - \tau)u(\tau)d\tau . \tag{A.19}$$

It is important to note that knowing the response of a linear, causal and time invariant system to an impulse is sufficient to know what the system's output will be for an arbitrary input function since the latter can be decomposed to a sum of impulses. The output is, according to (A.2), computable as the sum of the single impulse responses.

For multivariable systems (A.18) simply becomes

$$\mathbf{y}(t) = \int_0^t \mathbf{G}(t - \tau)\mathbf{u}(\tau)d\tau , \tag{A.20}$$

with

$$\mathbf{G}(t) = \begin{pmatrix} g_{1,1} & g_{1,2} & \cdots & g_{1,p} \\ g_{2,1} & g_{2,2} & \cdots & g_{2,p} \\ \vdots & \vdots & \ddots & \vdots \\ g_{q,1} & g_{q,2} & \cdots & g_{q,p} \end{pmatrix} ,$$

where $g_{i,j}$ denotes the response of the i-th output to an impulse at the j-th input.

A.3.2 Input-Output Description of Systems in Frequency Domain

Taking the Laplace[3] transform of (A.18) results in the much simpler relationship between the input and the output of a system with vanishing initial conditions, namely

$$Y(s) = G(s)U(s) , \tag{A.21}$$

where $G(s)$ is called the transfer function of the system. Similarly, the Laplace transform of (A.20) results in

$$\mathbf{Y}(s) = \mathbf{G}(s)\mathbf{U}(s) , \tag{A.22}$$

where $\mathbf{G}(s)$ is called the transfer function matrix of the system which has the elements $G_{i,j}(s)$ which are the transfer functions between the i-th output and the j-th input.

---

[3]Pierre-Simon Laplace, French mathematician and astronomer (1749–1827); famous for his work on celestial mechanics and the transform named after him for solving differential equations.

The transfer function can be determined as the Laplace transform of the impulse response or as the quotient of the system output and the system input in the Laplace domain. The transfer functions of a linear, causal and time invariant system with lumped parameters is in the form of a proper rational function or in other words, $G(s)$ is a fraction with polynomials as its nominator and denominator with the order of the nominator being less than the order of the denominator.

The differential equation of such a system is

$$y^{(n)} + a_{n-1}y^{(n-1)} + \cdots + a_1\dot{y} = b_0 u + b_1\dot{u} + \cdots + b_{n-1}u^{(n-1)}. \qquad (A.23)$$

The Laplace transforms of $y^{(n)}$ and $u^{(n)}$ (the n-th derivative of $y(t)$ and $u(t)$) are

$$\mathscr{L}\{y^{(n)}\} = s^n Y(s) - s^{n-1}y(0) - s^{n-2}\dot{y}(0) - \cdots sy^{(n-2)}(0) - y^{(n-1)}(0) \quad (A.24)$$
$$\mathscr{L}\{u^{(n)}\} = s^n U(s) - s^{n-1}u(0) - s^{n-2}\dot{u}(0) - \cdots su^{(n-2)}(0) - u^{(n-1)}(0).$$

Substituting (A.24) in (A.23) and rearranging results in

$$Y(s) = G(s)U(s) + G_0(s), \qquad (A.25)$$

with

$$G(s) = \frac{b_{n-1}s^{n-1} + \cdots + b_1 s + b_0}{s^n + a_{n-1}s^{n-1} + \cdots + a_1 s + a_0},$$
$$G_0(s) = \frac{\beta_{n-1}s^{n-1} + \cdots + \beta_1 s + \beta_0}{s^n + a_{n-1}s^{n-1} + \cdots + a_1 s + a_0},$$

where

$$\beta_{n-1} = y(0) \qquad\qquad\qquad\qquad\qquad\qquad\qquad (A.26)$$
$$\beta_{n-2} = \dot{y}(0) + a_{n-1}y(0) - b_{n-1}u(0)$$
$$\beta_{n-3} = \ddot{y}(0) + a_{n-2}y(0) + a_{n-1}\dot{y}(0) - b_{n-2}u(0) - b_{n-1}\dot{u}(0)$$
$$\vdots$$
$$\beta_1 = y^{(n-2)}(0) + \sum_{i=2}^{n-1} a_i y^{(i-2)}(0) - \sum_{i=2}^{n-1} b_i u^{(i-2)}(0)$$
$$\beta_0 = y^{(n-1)}(0) + \sum_{i=2}^{n-1} a_i y^{(i-1)}(0) - \sum_{i=2}^{n-1} b_i u^{(i-1)}(0).$$

Clearly, $G(s)$ corresponds to the transfer function in (A.21) and is independent of the initial conditions. $G_0(s)$ on the other hand depends on the initial conditions. Since the system is linear, both parts can be handled separately and their responses can ultimately be superimposed as Figure A.5 shows.

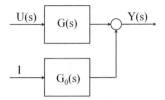

Fig. A.5 A linear system's output is the result of the superimposition of a part dependent on the input and a part dependent on the initial conditions

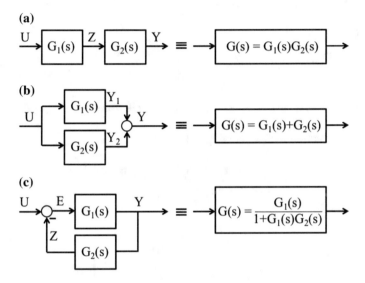

Fig. A.6 Block diagrams and equivalent transfer functions of systems **a** in series, **b** in parallel and **c** with feedback. A circle block builds the sum of its inputs. For instance, in **b** $Y_1 + Y_2$, in **c** $E = U - Z$

An interesting consequence of (A.21) is that systems and signals can be "multiplied" in the frequency domain. Figure A.6 shows some basic combinations of systems and their equivalent transfer functions.

A.3.3 State Space Description of Systems

The description of a linear system with a differential equation as in (A.23) is not suitable for the solution of many problems like numerical integration etc. For some purposes it is more convenient to replace this single n-th order differential equation by n first order differential equations. A possible way of achieving this is by defining

$$y = x_1 \tag{A.27}$$

$$\frac{dy}{dt} = x_2$$

$$\vdots$$

$$\frac{d^{n-1}y}{dt^{n-1}} = x_n \, ,$$

and by substituting $\dot{x}_i$ for dx_i/dt as

$$\dot{x}_1 = x_2 \tag{A.28}$$

$$\dot{x}_2 = x_3$$

$$\vdots$$

$$\dot{x}_n = -a_0 x_1 - a_1 x_2 - \cdots - a_{n-1} x_n + b_0 u + b_1 \dot{u} + \cdots + b_{n-1} u^{(n)} \, .$$

Or in matrix notation with $\mathbf{x}$ denoting the state vector

$$\dot{\mathbf{x}} = \mathbf{A}\mathbf{x} + \mathbf{b}_0 u + \mathbf{b}_1 \dot{u} + \cdots + \mathbf{b}_{n-1} u^{(n)}$$

$$y = \mathbf{c}^T \mathbf{x} + du \, ,$$

with

$$\mathbf{A} = \begin{pmatrix} 0 & 1 & 0 & \cdots & & 0 \\ 0 & 0 & 1 & 0 & & \cdots \\ 0 & 0 & 0 & 1 & & \cdots \\ \vdots & \vdots & \vdots & \vdots & 1 & \\ -a_0 & -a_1 & \cdots & -a_{n-2} & -a_{n-1} \end{pmatrix}$$

$$\mathbf{b}_0 = \mathbf{b}_1 = \cdots = \mathbf{b}_{n-1} = \begin{pmatrix} 0 \\ 0 \\ \vdots \\ 0 \\ 1 \end{pmatrix}$$

$$\mathbf{c}^T = (1 \ 0 \ 0 \ \cdots \ 0), \qquad d = 0 \, .$$

In most practical cases the derivatives of the input $u(t)$ do not appear explicitly, therefore $\mathbf{b}_i$ are zero except for $\mathbf{b}_0$ which is frequently abbreviated simply as $\mathbf{b}$. Thus, the general state space description for a linear multivariable system is

$$\dot{\mathbf{x}} = \mathbf{Ax} + \mathbf{Bu}, \tag{A.29}$$
$$\mathbf{y} = \mathbf{Cx} + \mathbf{Du}.$$

(The direct coupling between the input and the output, $\mathbf{D}$, can be left out in most cases without loss of generality.) For the names and the dimensions of the matrices $\mathbf{A}, \mathbf{B}, \mathbf{C}$ and $\mathbf{D}$ see Table A.1. Figure A.7 shows the block diagram of a linear system in state space description.

For a non-linear time variant system (A.29) generalizes to

$$\dot{\mathbf{x}} = \mathbf{f}\big(\mathbf{x}(t), \mathbf{u}(t), t\big), \tag{A.30}$$
$$\mathbf{y} = \mathbf{g}\big(\mathbf{x}(t), \mathbf{u}(t), t\big).$$

The concept of a system's state is very important and plays a central role in system theory. The number of states of a system is equal to the number of energy storage elements of the system. This number can be finite or infinite. Any system with lumped parameters which can be described by an ordinary differential equation of the order n can be equivalently described in the state space with n equations. A system with distributed parameters which can only be described by a partial differential equation needs an infinite number of states to be described in the state space. For practical purposes, even a distributed parameter system is often approximated by a finite number of states.

Table A.1 Common names and dimensions of the matrices $\mathbf{A}, \mathbf{B}, \mathbf{C}$ and $\mathbf{D}$ in (A.29). n is the system order, p is the number of inputs and q is the number of outputs

Symbol	Name	Dimension
A	System matrix	[n,n]
B	Input matrix	[n,p]
C	Output matrix	[q,n]
D	Direct input-output coupling matrix	[p,q]

Fig. A.7 The general block diagram of a linear system in the state space. $\mathbf{u}$ is the input vector, $\mathbf{y}$ is the output vector, and $\mathbf{x}$ is the state vector

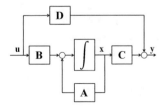

Table A.2 Some properties
of the transition matrix $\boldsymbol{\Phi}(t)$

$\frac{d\boldsymbol{\Phi}}{dt} = A\boldsymbol{\Phi}$
$\boldsymbol{\Phi}(0) = \mathbf{I}$
$\boldsymbol{\Phi}(t-p)\boldsymbol{\Phi}(p-q) = \boldsymbol{\Phi}(t-q)$
$\boldsymbol{\Phi}(t-p)\boldsymbol{\Phi}(p-t) = \boldsymbol{\Phi}(t-t) = \mathbf{I}$

The choice of the states for a given system is *not* unique. In other words, any set of system states $x_1, x_2, \ldots, x_n$ can be transformed to any other set of states $\hat{x}_1, \hat{x}_2, \ldots, \hat{x}_n$ by means of a non-singular affine transformation

$$\hat{\mathbf{x}} = \mathbf{Px} \qquad \det(\mathbf{P}) \neq 0.$$

without changing the input-output behavior of the system. Of course, the matrices $\mathbf{A}, \mathbf{B}, \mathbf{C}$ and $\mathbf{D}$ must also be transformed accordingly.

If a system is presented in a state space description, the corresponding n-th order differential equation can be reconstructed by computing the transfer function matrix of the system according to

$$\mathbf{G}(s) = \mathbf{C}(s\mathbf{I} - \mathbf{A})^{-1}\mathbf{B} + \mathbf{D}. \tag{A.31}$$

($\mathbf{I}$ is the $[n, n]$ identity matrix) and subsequently by substituting s in (A.21) by the differential operator d/dt.

In order to solve the vector differential equation in (A.29) one has to introduce the matrix exponential:

$$e^{\mathbf{A}t} = \mathbf{I} + \mathbf{A}t + \mathbf{A}^2\frac{t^2}{2!} + \ldots + \mathbf{A}^n\frac{t^n}{n!} + \ldots. \tag{A.32}$$

Most of the rules for handling scalar exponentials hold in the matrix case as well. Specifically,

$$e^{\mathbf{A}t_1}e^{\mathbf{A}t_2} = e^{\mathbf{A}(t_1+t_2)}, \tag{A.33}$$

and

$$\frac{d}{dt}e^{\mathbf{A}t} = \mathbf{A}e^{\mathbf{A}t} = e^{\mathbf{A}t}\mathbf{A}. \tag{A.34}$$

The matrix $e^{\mathbf{A}t}$ is called the transition matrix or the fundamental matrix of the system and is often written as $\boldsymbol{\Phi}(t)$. Different ways of computing it can be found, for instance in [96]. Some of the interesting properties of the transition matrix are shown in Table A.2.

To solve

$$\dot{\mathbf{x}} = \mathbf{Ax} + \mathbf{b}u, \tag{A.35}$$

first find the solution of the homogeneous differential equation

$$\frac{d}{dt}\mathbf{x}(t) = \mathbf{A}\mathbf{x}(t),$$

as

$$\mathbf{x}(t) = e^{\mathbf{A}t}\mathbf{v}, \tag{A.36}$$

and use this solution as an *Ansatz* for the solution of the inhomogeneous equation. Substituting (A.36) in (A.35) results with (A.34) in

$$\dot{\mathbf{x}}(t) = e^{\mathbf{A}t}\dot{\mathbf{v}}(t) + \mathbf{A}e^{\mathbf{A}t}\mathbf{v}(t) = \mathbf{A}e^{\mathbf{A}t}\mathbf{v}(t) + \mathbf{b}u,$$
$$\dot{\mathbf{v}}(t) = e^{-\mathbf{A}t}\mathbf{b}u,$$

and integrating $\dot{\mathbf{v}}(t)$ yields

$$\mathbf{v}(t) = \mathbf{v}_0 + \int_{t_0}^{\tau} e^{-\mathbf{A}t}\mathbf{b}u(\tau)d\tau.$$

Thus the general solution of (A.35) is

$$\mathbf{x}(t) = e^{\mathbf{A}t}\mathbf{v}_0 + \int_{t_0}^{t} e^{\mathbf{A}(t-\tau)}\mathbf{b}u(\tau)d\tau, \tag{A.37}$$

and the particular solution for the initial conditions $\mathbf{x}(t = t_0) = \mathbf{x}_0$ is

$$\mathbf{x}(t) = e^{\mathbf{A}(t-t_0)}\mathbf{x}_0 + \int_{t_0}^{t} e^{\mathbf{A}(t-\tau)}\mathbf{b}u(\tau)d\tau. \tag{A.38}$$

This solution is easily transferable to multivariable systems by substituting $\mathbf{b}$ by $\mathbf{B}$ and u by $\mathbf{u}$. Also, this result should not depend on the choice of t_0 for a time invariant system. Therefore, it can be simplified by taking $t_0 = 0$. Obviously, the solution has two terms, one depending on the initial conditions but not on the input and the other depending on the input but not on the initial conditions. This is another manifestation of the system's output being the sum of the system's response to the input and to the initial conditions as already stated in (A.25) and depicted in Figure A.5.

The output of the system can now be calculated from (A.38) to be

$$y(t) = \underbrace{\mathbf{c}^T e^{\mathbf{A}t}x_0}_{(a)} + \underbrace{\int_{t_0}^{t} \mathbf{c}^T e^{\mathbf{A}(t-\tau)}\mathbf{b}u(\tau)d\tau}_{(b)}. \tag{A.39}$$

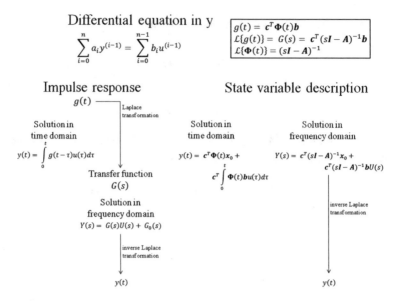

Fig. A.8 Different ways of solving linear, causal and time invariant systems and relationships between different forms of describing them

For vanishing initial conditions

$$y(t) = \int_{t_0}^{t} \mathbf{c}^T e^{\mathbf{A}(t-\tau)} \mathbf{b} u(\tau) d\tau \ . \tag{A.40}$$

As can be seen, this equation has the same form as (A.18)! This observation gives us a way of computing a system's impulse response from the state space description. Namely,

$$g(t) = \mathbf{c}^T e^{\mathbf{A}t} \mathbf{b} = \mathbf{c}^T \boldsymbol{\Phi}(t) \mathbf{b} \ .$$

To sum up this Section, the relationships between different forms of describing linear systems are shown in Figure A.8.

A.4   Some Important System Properties

There are several qualitative properties of a system which decide on its applicability for a given problem. Those are *controllability*, *observability* and *stability*.

A system of order n is called *controllable* if a control variable $\mathbf{u}(t)$ exists such that the system can be brought from an arbitrary starting point in the state space (initial condition) $\mathbf{x}_0$ to an arbitrary end point $\mathbf{x}_e$ in finite time.

Fig. A.9 Decomposition of a system in a controllable and observable part (S_1), a controllable but not observable part (S_2), a not controllable but observable part (S_3), and a not controllable and not observable part (S_4)

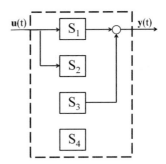

A system is called *observable* if the initial conditions $\mathbf{x}_0$ can be determined unambiguously from the measurement of the output $\mathbf{y}(t)$ for a finite time with known input $\mathbf{u}(t)$.

With these definitions of controllability and observability, any system can be divided into a) a controllable and observable part, b) a controllable but not observable part, c) a not controllable but observable part, and d) a not controllable and not observable part as schematized in Figure A.9 [67]. This is called Kalman[4] decomposition.

Stability is one of the most studied issues in system theory. Consequently, it has several different definitions.

A system is said to be BIBO (bounded-input-bounded-output) stable if and only if its output is bounded for all bounded inputs. For linear systems this statement is equivalent to

$$\int_0^\infty |g(t)|dt \leq p < \infty, \tag{A.41}$$

with $g(t)$ denoting the impulse response of the system and p a real number. However, in general, this does not necessarily imply that $|g(t)| < m$ or that $g(t \to \infty) = 0$, except for systems with proper rational transfer functions. For a multivariable system to be BIBO stable all of the impulse responses $g_{i,j}(t)$ must fulfill (A.41). For $t \to \infty$ a BIBO stable system's output approaches a periodic function with the same period as the input function. If, on the other hand, the input is constant, the output also approaches a constant but not necessarily the same value as the input.

There is a simple way one can tell whether a linear system is BIBO stable: a SISO system with a proper rational transfer function $G(s)$ is BIBO stable if and only if all of the poles of $G(s)$ (zeros of the denominator polynomial) are in the open left half complex plane (to which the imaginary axis does not belong). Figure A.10 shows the pole configuration of some BIBO stable or unstable systems. For a multivariable system with the transfer function matrix $\mathbf{G}(s)$ to be BIBO stable, all of its transfer functions $G_{i,j}$ must be BIBO stable.

---

[4]Rudolf E. Kalman, Hungarian-American electrical engineer and mathematician (1930–2016); famous for his work on system theory and an algorithm named after him used for filtering and estimation.

Fig. A.10 Pole
configurations of five
systems of varying orders.
System 1 and 2 are BIBO
stable, the others are not

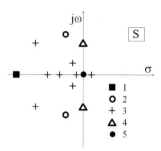

To determine the exact location of the zeros of a polynomial is not always easy; for polynomials of the order 5 or higher it is, in general, even impossible. Nevertheless, it is possible to determine whether a polynomial of any given order has zeros in the right half plane or not. The most common way to solve this problem goes back more than a century to Routh[5] and Hurwitz,[6] who, independently of each other, developed methods to determine the number of zeros a polynomial has in the right half plane.

The polynomial under study is given as

$$P(s) = a_0 s^n + a_1 s^{n-1} + \cdots + a_{n-1}s + a_n . \tag{A.42}$$

A necessary but not sufficient condition for all of the zeros of $P(s)$ to be in the open left half plane is that all coefficients of the polynomial in (A.42) must be positive (this condition is also sufficient for polynomials up to the second degree). In other words, if even a single coefficient is negative or missing this means that $P(s)$ has at least one zero in the right half plane. On the other hand, if all coefficients are positive, it still cannot be concluded that $P(s)$ has no zeros in the right half plane. So, how can be sure that it does not?

Let us decompose $P(s)$ in two polynomials

$$P_1(s) = a_0 s^n + a_2 s^{n-2} + \cdots$$
$$P_2(s) = a_1 s^{n-1} + a_3 s^{n-3} + \cdots$$

The quotient of these polynomials can be expressed as a continuous fraction

$$\frac{P_1(s)}{P_2(s)} = \alpha_1 s + \cfrac{1}{\alpha_2 s + \cfrac{1}{\ddots \cfrac{}{\alpha_{n-1}s + \frac{1}{\alpha_n s}}}} . \tag{A.43}$$

It can be shown that the polynomial in (A.42) has all its roots in the open left half plane if and only if all α_i in (A.43) are positive. See Table A.3 for an easy way of calculating α_i.

---

[5]Edward Routh, English mathematician (1831–1907).
[6]Adolf Hurwitz, German mathematician (1859–1919).

Table A.3 Routh scheme for the calculation of the coefficients in the continuous fraction representation of the quotient of even and odd parts of a polynomial

$\alpha_1 = \frac{a_0}{a_1}$	s^n	$a_0\ a_2\ a_4\ a_6\ \ldots$	
			$b_1 = \frac{a_1 a_2 - a_0 a_3}{a_1}$
$\alpha_2 = \frac{a_1}{b_1}$	s^{n-1}	$a_1\ a_3\ a_5\ \ldots$	
$\alpha_3 = \frac{b_1}{c_1}$	s^{n-2}	$b_1\ b_2\ b_3\ \ldots$	$b_2 = \frac{a_1 a_4 - a_0 a_5}{a_1}$
	s^{n-3}	$c_1\ c_2\ c_3\ \ldots$	
	$\vdots$		$c_1 = \frac{b_1 a_3 - a_1 b_2}{b_1}$
$\alpha_n = \frac{j_1}{k_1}$	s	$j_1\ 0$	$\vdots$
	1	$k_1\ 0$	

The polynomial $P(s)$ has all its roots in the left half plane if all the elements $a_1, b_1, c_1, \ldots, k_1$ in the Routh table are positive. Furthermore, the number of roots in the right half plane is equal to the change of signs in the first column of the Routh table.

So far the stability of a system given by its input-output description in the frequency domain was addressed. How can the results be transferred for a system described in its state space? One has to remember that the solution of the set of equations in (A.29) as given in (A.39) consists of two terms, the first one depending on the initial conditions but not on the input and the second term depending on the input but not on the initial conditions. Therefore, the question of stability should be studied separately for both parts.

One cannot apply the BIBO stability on the first part since it is not dependent on the input. The study of the effect of initial conditions on the system response leads to a new concept: the stability of an equilibrium point. An equilibrium point $\mathbf{x}_e$ of a system $\dot{\mathbf{x}} = \mathbf{f}(\mathbf{x})$ is given by $\dot{\mathbf{x}} = \mathbf{f}(\mathbf{x}_e) = \mathbf{0}$. In other words, a system is at equilibrium if its state does not change.

Let us now define an equilibrium point $\mathbf{x}_e$ of a system $\dot{\mathbf{x}} = \mathbf{f}(\mathbf{x})$ to be Lyapunov stable if and only if for an arbitrarily small ϵ a $\delta(\epsilon)$ exists such that

$$\|\mathbf{x}_e - \mathbf{x}_0\| < \delta \quad \Rightarrow \quad \|\mathbf{x}(t) - \mathbf{x}_e\| < \epsilon \qquad \forall t \geq 0, \qquad (\text{A.44})$$

(see Figure A.11) for an illustration in two dimensions. Further, the equilibrium point is said to be asymptotically stable if it is Lyapunov stable and for any $\gamma > 0, \mu > 0$ a time $T(\mu, \gamma)$ exists, such that

$$\|\mathbf{x}_e - \mathbf{x}_0\| < \gamma \quad \Rightarrow \quad \|\mathbf{x}(t) - \mathbf{x}_e\| < \mu \qquad \forall t \geq T. \qquad (\text{A.45})$$

Fig. A.11 An equilibrium point of a system is Lyapunov stable if and only if for any arbitrary spherical domain Ω_ϵ around it another spherical domain Ω_δ exists such that any system trajectory starting in Ω_δ never leaves Ω_ϵ as shown in the example of a second order system

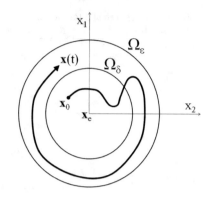

Note that those definitions are also valid for a non-linear system which can have none, one, several or an infinite number of equilibrium points. A linear system, on the other hand, has either an equilibrium point at the origin of the state space (if the system matrix $\mathbf{A}$ is regular) or an infinite number of equilibrium points (if the system matrix $\mathbf{A}$ is singular). Furthermore, the stability of an equilibrium point is a local property of a non-linear system. However, for a linear system, the stability (or instability) is a global property.

The equilibrium point $\mathbf{x}_e$ of a linear system $\dot{\mathbf{x}} = \mathbf{A}\mathbf{x}$, thus the linear system itself is asymptotically stable if and only if all the eigenvalues of $\mathbf{A}$ have negative real parts. (The Routh-Hurwitz scheme can be used to determine the position of the eigenvalues, in other words, the zeros of the characteristic polynomial of $\mathbf{A}$.)

It is important to note that each pole of a system's transfer function $G(s)$ is also an eigenvalue of the system's $\mathbf{A}$ matrix. However, the reversal of this statement is in general not true, because $G(s)$ describes the relationship the input and the output of only the observable and controllable part of the system. Therefore, asymptotic stability of a linear system implies BIBO stability, but a BIBO stable system is not necessarily Lyapunov stable as the following example demonstrates.

Example: The state and output equations of a second order system is given as

$$\dot{x}_1 = x_1$$
$$\dot{x}_2 = -0.5x_2 + u$$
$$y = x_2 .$$

The system matrix of this system is

$$\mathbf{A} = \begin{pmatrix} 1 & 0 \\ 0 & -0.5 \end{pmatrix},$$

which has the eigenvalues $\lambda_1 = 1$ and $\lambda_2 = -0.5$, thus is not Lyapunov stable. The transfer function can be computed with (A.31) to be

$$G(s) = \frac{1}{s + 0.5},$$

which has its only pole in the left half plane, thus is BIBO stable. □

Appendix B
Matrices

B.1 Introduction

A matrix is a rectangular array of numbers, symbols, or expressions, arranged in rows and columns. The size of a matrix is defined by the number of rows and columns that it contains. A matrix with m rows and n columns is called an $m \times n$ matrix. The entry in the i-th row and j-th column of a matrix $\mathbf{A}$ is sometimes referred as the (i, j)-th entry of the matrix, and most commonly denoted as $a_{i,j}$ (Table. B.1).

Table B.1 Some special matrices

Name	Size	Example
Row vector	$1 \times n$	$[3 \ 4 \ -1]$
Column vector	$n \times 1$	$\begin{bmatrix} 3 \\ 2 \\ 0 \end{bmatrix}$
Square matrix	$n \times n$	$\begin{bmatrix} 3 & 2 & 4 \\ 1 & 1 & 0 \\ 1 & 1 & 5 \end{bmatrix}$
Identity matrix	$n \times n$	$\begin{bmatrix} 1 & 0 & 0 \\ 0 & 1 & 0 \\ 0 & 0 & 1 \end{bmatrix}$
Zero matrix	$m \times n$	$\begin{bmatrix} 0 & 0 & 0 & 0 \\ 0 & 0 & 0 & 0 \end{bmatrix}$

© Springer International Publishing AG 2018
S. S. Hacısalihzade, *Control Engineering and Finance*, Lecture Notes in Control and Information Sciences 467, https://doi.org/10.1007/978-3-319-64492-9

Table B.2 Basic matrix operations

Operation	Definition	Example
Addition	The sum of two m by n matrices is calculated element-wise: $(\mathbf{A}+\mathbf{B})_{ij} = \mathbf{A}_{ij} + \mathbf{B}_{ij}$.	$\begin{bmatrix} 3 & 2 \\ 2 & 1 \\ 0 & 7 \end{bmatrix} + \begin{bmatrix} 1 & 1 \\ -1 & 0 \\ 0 & -2 \end{bmatrix} = $ $\begin{bmatrix} 3+1 & 2+1 \\ 2-1 & 1+0 \\ 0+0 & 7-2 \end{bmatrix} = \begin{bmatrix} 4 & 3 \\ 1 & 1 \\ 0 & 5 \end{bmatrix}$
Scalar multiplication	The product of a scalar c and a matrix $\mathbf{A}$ is computed by multiplying every entry of $\mathbf{A}$ by c: $(c\mathbf{A})_{ij} = c \cdot \mathbf{A}_{ij}$	$3 \cdot \begin{bmatrix} 3 & 1 \\ 1 & 0 \end{bmatrix} = \begin{bmatrix} 3\cdot3 & 3\cdot1 \\ 3\cdot1 & 3\cdot0 \end{bmatrix} = \begin{bmatrix} 9 & 3 \\ 3 & 0 \end{bmatrix}$
Transposition	The transpose of an m-by-n matrix $\mathbf{A}$ is the n-by-m matrix $\mathbf{A}^T$ formed by turning rows into columns and vice versa: $\mathbf{A}^T_{ij} = \mathbf{A}_{ji}$	$\begin{bmatrix} 3 & 3 & 1 & 7 \\ 1 & 2 & 3 & -1 \end{bmatrix}^T = \begin{bmatrix} 3 & 1 \\ 3 & 2 \\ 1 & 3 \\ 7 & -1 \end{bmatrix}$
Matrix multiplication	Multiplication of two matrices is defined if and only if the number of columns of the left matrix is the same as the number of rows of the right matrix. If $\mathbf{A}$ is an m-by-n matrix and $\mathbf{B}$ is an n-by-p matrix, then their matrix product $\mathbf{AB}$ is the m-by-p matrix whose entries are given by dot product of the corresponding row of $\mathbf{A}$ and the corresponding column of $\mathbf{B}$: $(AB)_{ij} = \sum_{r=1}^{n} A_{ir} B_{rj}$	$\begin{bmatrix} 1 & 2 & 3 \\ 0 & 1 & 7 \end{bmatrix} \begin{bmatrix} 0 \\ 1 \\ 7 \end{bmatrix} = $ $\begin{bmatrix} 1\cdot0 + 2\cdot1 + 3\cdot7 \\ 0\cdot0 + 1\cdot1 + 7\cdot7 \end{bmatrix} = \begin{bmatrix} 24 \\ 50 \end{bmatrix}$
Trace operator	The trace, $tr(\mathbf{A})$ of a square matrix $\mathbf{A}$ is the sum of its diagonal entries. It is also the sum of the eigenvalues of the matrix.	$tr \begin{bmatrix} 1 & 2 & 3 \\ 4 & 5 & 6 \\ 7 & 8 & 9 \end{bmatrix} = 1+5+9 = 15$

B.2   Basic Matrix Operations

There are several basic operations that can be applied to matrices. These are addition, scalar multiplication, transposition and matrix multiplication (Table B.2).

One has to keep in mind that the matrix addition is commutative, meaning $\mathbf{A}+\mathbf{B} = \mathbf{B}+\mathbf{A}$. However, matrix multiplication is not commutative, meaning, in general, $\mathbf{AB} \neq \mathbf{BA}$. Some useful matrix algebra rules are shown in Table B.3.

Table B.3 Some useful matrix algebra rules

Rule	Remark
$A + B = A + B$	
$(A(B + C)) = AB + AC$	
$(cA)^T = cA^T$	
$(A + B)^T = A^T + B^T$	
$(AB)^T = B^T A^T$	
$(AB)^{-1} = B^{-1}A^{-1}$	Provided that the individual inverses exist
$(A^T)^T = A$	
$AI = A$	Provided $\dim[A] = m \times n$ and $\dim[I] = n \times n$
$IA = A$	Provided $\dim[A] = m \times n$ and $\dim[I] = m \times m$
$AA^{-1} = A^{-1}A = I$	Provided that A is invertible (or non-singular)
$(A^T)^{-1} = (A^{-1})^T$	
$tr(AB) = tr(BA)$	
$tr(A) = tr(A^T)$	

B.3 Matrix Calculus

This Section deals with vector and matrix derivatives. Bold capital letters denote matrices, bold lower case letters vectors and lower case italics denote scalars. The transpose of a vector $\mathbf{x}$ is denoted as $\mathbf{x}^T$. First, the conventions used in derivatives are explained, then some useful derivative rules and identities are listed.

B.3.1 Definitions and Conventions

When taking derivatives with respect to vectors or matrices, there are two different conventions in writing the result. Looking at the derivative of a vector with respect to another vector, *e.g.*, $\frac{\partial \mathbf{y}}{\partial \mathbf{x}}$ where $\mathbf{y} \in \mathbb{R}^m$ and $\mathbf{x} \in \mathbb{R}^n$, the result can be laid out as either an $m \times n$ matrix or $n \times m$ matrix, *i.e.*, the elements of $\mathbf{y}$ laid out in columns and the elements of $\mathbf{x}$ laid out in rows, or vice versa. In other words, there are two equivalent ways of writing the result, namely the numerator layout (according to $\mathbf{y}$ and $\mathbf{x}^T$) and the denominator layout (according to $\mathbf{y}^T$ and $\mathbf{x}$).

Similarly, when taking the derivative of a scalar with respect to a matrix as in $\frac{\partial y}{\partial \mathbf{X}}$ and the derivative of a matrix with respect to a scalar as in $\frac{\partial \mathbf{Y}}{\partial x}$, the consistent numerator layout lays out according to $\mathbf{Y}$ and $\mathbf{X}^T$, while consistent denominator layout lays out according to $\mathbf{Y}^T$ and $\mathbf{X}$.

The results of differentiating various kinds of variables with respect to other kind of variables are illustrated in Table B.4.

Table B.4 Results of differentiation with vectors and matrices using the numerator layout convention

	Scalar $y \in \mathbb{R}$		Vector $\mathbf{y} \in \mathbb{R}^m$		Matrix $\mathbf{Y} \in \mathbb{R}^{m \times n}$	
	Notation	Type	Notation	Type	Notation	Type
Scalar $x \in \mathbb{R}$	$\frac{\partial y}{\partial x}$	scalar	$\frac{\partial \mathbf{y}}{\partial x}$	size-m column vector	$\frac{\partial \mathbf{Y}}{\partial x}$	$m \times n$ matrix
Vector $\mathbf{x} \in \mathbb{R}^n$	$\frac{\partial y}{\partial \mathbf{x}}$	size-n row vector	$\frac{\partial \mathbf{y}}{\partial \mathbf{x}}$	$m \times n$ matrix	$\frac{\partial \mathbf{Y}}{\partial \mathbf{x}}$	
Matrix $\mathbf{X} \in \mathbb{R}^{p \times q}$	$\frac{\partial y}{\partial \mathbf{X}}$	$q \times p$ matrix	$\frac{\partial \mathbf{y}}{\partial \mathbf{X}}$		$\frac{\partial \mathbf{Y}}{\partial \mathbf{X}}$	

The derivative of a vector

$$\mathbf{y} = \begin{bmatrix} y_1 \\ y_2 \\ \vdots \\ y_m \end{bmatrix},$$

with respect to a scalar is known as the *tangent vector* and x is given as

$$\frac{\partial \mathbf{y}}{\partial x} = \begin{bmatrix} \frac{\partial y_1}{\partial x} \\ \frac{\partial y_2}{\partial x} \\ \vdots \\ \frac{\partial y_m}{\partial x} \end{bmatrix}.$$

The derivative of a scalar y with respect to vector

$$\mathbf{x} = \begin{bmatrix} x_1 \\ x_2 \\ \vdots \\ x_n \end{bmatrix},$$

is known as the *gradient* and is given as

$$\frac{\partial y}{\partial \mathbf{x}} = \begin{bmatrix} \frac{\partial y}{\partial x_1} & \frac{\partial y}{\partial x_2} & \cdots & \frac{\partial y}{\partial x_n} \end{bmatrix}.$$

The derivative of an m-dimensional vector $\mathbf{y}$ with respect to an n-dimensional vector $\mathbf{x}$ is known as the *Jacobian* and is given as

$$\frac{\partial \mathbf{y}}{\partial \mathbf{x}} = \begin{bmatrix} \frac{\partial y_1}{\partial x_1} & \frac{\partial y_1}{\partial x_2} & \cdots & \frac{\partial y_1}{\partial x_n} \\ \frac{\partial y_2}{\partial x_1} & \frac{\partial y_2}{\partial x_2} & \cdots & \frac{\partial y_2}{\partial x_n} \\ \vdots & \vdots & \ddots & \vdots \\ \frac{\partial y_m}{\partial x_1} & \frac{\partial y_m}{\partial x_2} & \cdots & \frac{\partial y_m}{\partial x_n} \end{bmatrix}.$$

The derivative of a matrix $\mathbf{Y}$ with respect to a scalar x is known as the *tangent matrix* and is given by

$$\frac{\partial \mathbf{Y}}{\partial x} = \begin{bmatrix} \frac{\partial y_{11}}{\partial x} & \frac{\partial y_{12}}{\partial x} & \cdots & \frac{\partial y_{1n}}{\partial x} \\ \frac{\partial y_{21}}{\partial x} & \frac{\partial y_{22}}{\partial x} & \cdots & \frac{\partial y_{2n}}{\partial x} \\ \vdots & \vdots & \ddots & \vdots \\ \frac{\partial y_{m1}}{\partial x} & \frac{\partial y_{m2}}{\partial x} & \cdots & \frac{\partial y_{mn}}{\partial x} \end{bmatrix}.$$

The derivative of a scalar y with respect to a $p \times q$ matrix $\mathbf{X}$ is given by

$$\frac{\partial y}{\partial \mathbf{X}} = \begin{bmatrix} \frac{\partial y}{\partial x_{11}} & \frac{\partial y}{\partial x_{21}} & \cdots & \frac{\partial y}{\partial x_{p1}} \\ \frac{\partial y}{\partial x_{12}} & \frac{\partial y}{\partial x_{22}} & \cdots & \frac{\partial y}{\partial x_{p2}} \\ \vdots & \vdots & \ddots & \vdots \\ \frac{\partial y}{\partial x_{1q}} & \frac{\partial y}{\partial x_{2q}} & \cdots & \frac{\partial y}{\partial x_{pq}} \end{bmatrix}.$$

B.3.2 Matrix Derivatives

Some commonly used identities for derivatives of expressions involving vectors and matrices are listed in this Section. It is assumed that all of the expressions and the multipliers in the tables are in numerator layout (Tables B.5 and B.6).

Table B.5 Some scalar by vector differentiation identities of the form $\frac{\partial y}{\partial \mathbf{x}}$

Condition	Expression	Identity
a is not a function of $\mathbf{x}$	$\frac{\partial a}{\partial \mathbf{x}}$	$\mathbf{0}^T$
a is not a function of $\mathbf{x}$, $u = u(\mathbf{x})$	$\frac{\partial au}{\partial \mathbf{x}}$	$a\frac{\partial u}{\partial \mathbf{x}}$
$u = u(\mathbf{x})$, $v = v(\mathbf{x})$	$\frac{\partial(u+v)}{\partial \mathbf{x}}$	$\frac{\partial u}{\partial \mathbf{x}} + \frac{\partial v}{\partial \mathbf{x}}$
$u = u(\mathbf{x})$, $v = v(\mathbf{x})$	$\frac{\partial(uv)}{\partial \mathbf{x}}$	$u\frac{\partial v}{\partial \mathbf{x}} + v\frac{\partial u}{\partial \mathbf{x}}$
$u = u(\mathbf{x})$	$\frac{\partial g(u)}{\partial \mathbf{x}}$	$\frac{\partial g(u)}{\partial u} \cdot \frac{\partial u}{\partial \mathbf{x}}$
$u = u(\mathbf{x})$	$\frac{\partial f(g(u))}{\partial \mathbf{x}}$	$\frac{\partial f(g)}{\partial g} \cdot \frac{\partial g(u)}{\partial u} \cdot \frac{\partial u}{\partial \mathbf{x}}$
$\mathbf{u} = \mathbf{u}(\mathbf{x})$, $\mathbf{v} = \mathbf{v}(\mathbf{x})$	$\frac{\partial(\mathbf{u} \cdot \mathbf{v})}{\partial \mathbf{x}}$	$\mathbf{u}^T\frac{\partial \mathbf{v}}{\partial \mathbf{x}} + \mathbf{v}^T\frac{\partial \mathbf{u}}{\partial \mathbf{x}}$
$\mathbf{a}$ is not a function of $\mathbf{x}$	$\frac{\partial(\mathbf{a} \cdot \mathbf{x})}{\partial \mathbf{x}}$	$\mathbf{a}^T$
$\mathbf{A}$ is not a function of $\mathbf{x}$, $\mathbf{b}$ is not a function of $\mathbf{x}$	$\frac{\partial(\mathbf{b}^T \mathbf{A}\mathbf{x})}{\partial \mathbf{x}}$	$\mathbf{b}^T \mathbf{A}$
$\mathbf{A}$ is not a function of $\mathbf{x}$	$\frac{\partial(\mathbf{x}^T \mathbf{A}\mathbf{x})}{\partial \mathbf{x}}$	$\mathbf{x}^T (\mathbf{A} + \mathbf{A}^T)$

Table B.6 Some vector by vector differentiation identities of the form $\frac{\partial \mathbf{y}}{\partial \mathbf{x}}$

Condition	Expression	Identity
	$\frac{\partial \mathbf{x}}{\partial \mathbf{x}}$	$\mathbf{I}$
$\mathbf{y}$ is not a function of $\mathbf{x}$	$\frac{\partial \mathbf{y}}{\partial \mathbf{x}}$	$\mathbf{0}$
$\mathbf{A}$ is not a function of $\mathbf{x}$	$\frac{\partial \mathbf{Ax}}{\partial \mathbf{x}}$	$\mathbf{A}$
$\mathbf{A}$ is not a function of $\mathbf{x}$	$\frac{\partial \mathbf{x}^T \mathbf{A}}{\partial \mathbf{x}}$	$\mathbf{A}^T$
a is not a function of $\mathbf{x}$, $\mathbf{u} = \mathbf{u}(\mathbf{x})$	$\frac{\partial a\mathbf{u}}{\partial \mathbf{x}}$	$a\frac{\partial \mathbf{u}}{\partial \mathbf{x}}$
$a = a(\mathbf{x})$, $\mathbf{u} = \mathbf{u}(\mathbf{x})$	$\frac{\partial a\mathbf{u}}{\partial \mathbf{x}}$	$a\frac{\partial \mathbf{u}}{\partial \mathbf{x}} + \mathbf{u}\frac{\partial a}{\partial \mathbf{x}}$
$\mathbf{A}$ is not a function of $\mathbf{x}$, $\mathbf{u} = \mathbf{u}(\mathbf{x})$	$\frac{\partial \mathbf{Au}}{\partial \mathbf{x}}$	$\mathbf{A}\frac{\partial \mathbf{u}}{\partial \mathbf{x}}$
$\mathbf{u} = \mathbf{u}(\mathbf{x})$, $\mathbf{v} = \mathbf{v}(\mathbf{x})$	$\frac{\partial (\mathbf{u}+\mathbf{v})}{\partial \mathbf{x}}$	$\frac{\partial \mathbf{u}}{\partial \mathbf{x}} + \frac{\partial \mathbf{v}}{\partial \mathbf{x}}$
$\mathbf{u} = \mathbf{u}(\mathbf{x})$	$\frac{\partial \mathbf{g}(\mathbf{u})}{\partial \mathbf{x}}$	$\frac{\partial \mathbf{g}(\mathbf{u})}{\partial \mathbf{u}} \cdot \frac{\partial \mathbf{u}}{\partial \mathbf{x}}$
$\mathbf{u} = \mathbf{u}(\mathbf{x})$	$\frac{\partial \mathbf{f}(\mathbf{g}(\mathbf{u}))}{\partial \mathbf{x}}$	$\frac{\partial \mathbf{f}(\mathbf{g})}{\partial \mathbf{g}} \cdot \frac{\partial \mathbf{g}(\mathbf{u})}{\partial \mathbf{u}} \cdot \frac{\partial \mathbf{u}}{\partial \mathbf{x}}$

For more formulas, the reader is referred to [102].

Appendix C
Normal Distribution Tables

A random variable X is said to be normally distributed if it has the density function

$$f_X(t) = \frac{1}{\sqrt{2\pi}\sigma} e^{-\frac{(t-\mu)^2}{2\sigma^2}} . \tag{C.1}$$

μ (real) and σ (positive) are its parameters. A special case of the normal distribution is obtained when $\mu = 0$ and $\sigma = 1$. This is called the standard normal distribution and has the density

$$\varphi(t) = \frac{1}{\sqrt{2\pi}} e^{-\frac{t^2}{2}} . \tag{C.2}$$

The distribution function

$$\Phi(t) = \frac{1}{\sqrt{2\pi}} \int_{-\infty}^{t} e^{-\frac{\tau^2}{2}} d\tau , \tag{C.3}$$

cannot be determined analytically. Therefore, it is tabulated here for various values of t. Note that $\Phi(0) = 0.5$, $\Phi(-t) = 1 - \Phi(t)$, $D(t) = \Phi(t) - \Phi(-t)$.

Any normal distribution can be transformed to the standard normal distribution by transforming the variable as

$$t = \frac{X - \mu}{\sigma} .$$

© Springer International Publishing AG 2018
S. S. Hacısalihzade, *Control Engineering and Finance*, Lecture Notes in Control and Information Sciences 467, https://doi.org/10.1007/978-3-319-64492-9

z	$\Phi(z)$	$\Phi(-z)$	$D(z)$	z	$\Phi(z)$	$\Phi(-z)$	$D(z)$
0,01	0,5040	0,4960	0,0080	0,31	0,6217	0,3783	0,2434
0,02	0,5080	0,4920	0,0160	0,32	0,6255	0,3745	0,2510
0,03	0,5120	0,4880	0,0239	0,33	0,6293	0,3707	0,2586
0,04	0,5160	0,4840	0,0319	0,34	0,6331	0,3669	0,2661
0,05	0,5199	0,4801	0,0399	0,35	0,6368	0,3632	0,2737
0,06	0,5239	0,4761	0,0478	0,36	0,6406	0,3594	0,2812
0,07	0,5279	0,4721	0,0558	0,37	0,6443	0,3557	0,2886
0,08	0,5319	0,4681	0,0638	0,38	0,6480	0,3520	0,2961
0,09	0,5359	0,4641	0,0717	0,39	0,6517	0,3483	0,3035
0,10	0,5398	0,4602	0,0797	0,40	0,6554	0,3446	0,3108
0,11	0,5438	0,4562	0,0876	0,41	0,6591	0,3409	0,3182
0,12	0,5478	0,4522	0,0955	0,42	0,6628	0,3372	0,3255
0,13	0,5517	0,4483	0,1034	0,43	0,6664	0,3336	0,3328
0,14	0,5557	0,4443	0,1113	0,44	0,6700	0,3300	0,3401
0,15	0,5596	0,4404	0,1192	0,45	0,6736	0,3264	0,3473
0,16	0,5636	0,4364	0,1271	0,46	0,6772	0,3228	0,3545
0,17	0,5675	0,4325	0,1350	0,47	0,6808	0,3192	0,3616
0,18	0,5714	0,4286	0,1428	0,48	0,6844	0,3156	0,3688
0,19	0,5753	0,4247	0,1507	0,49	0,6879	0,3121	0,3759
0,20	0,5793	0,4207	0,1585	0,50	0,6915	0,3085	0,3829
0,21	0,5832	0,4168	0,1663	0,51	0,6950	0,3050	0,3899
0,22	0,5871	0,4129	0,1741	0,52	0,6985	0,3015	0,3969
0,23	0,5910	0,4090	0,1819	0,53	0,7019	0,2981	0,4039
0,24	0,5948	0,4052	0,1897	0,54	0,7054	0,2946	0,4108
0,25	0,5987	0,4013	0,1974	0,55	0,7088	0,2912	0,4177
0,26	0,6026	0,3974	0,2051	0,56	0,7123	0,2877	0,4245
0,27	0,6064	0,3936	0,2128	0,57	0,7157	0,2843	0,4313
0,28	0,6103	0,3897	0,2205	0,58	0,7190	0,2810	0,4381
0,29	0,6141	0,3859	0,2282	0,59	0,7224	0,2776	0,4448
0,30	0,6179	0,3821	0,2358	0,60	0,7257	0,2743	0,4515

z	$\Phi(z)$	$\Phi(-z)$	$D(z)$	z	$\Phi(z)$	$\Phi(-z)$	$D(z)$
0,61	0,7291	0,2709	0,4581	0,91	0,8186	0,1814	0,6372
0,62	0,7324	0,2676	0,4647	0,92	0,8212	0,1788	0,6424
0,63	0,7357	0,2643	0,4713	0,93	0,8238	0,1762	0,6476
0,64	0,7389	0,2611	0,4778	0,94	0,8264	0,1736	0,6528
0,65	0,7422	0,2578	0,4843	0,95	0,8289	0,1711	0,6579
0,66	0,7454	0,2546	0,4907	0,96	0,8315	0,1685	0,6629
0,67	0,7486	0,2514	0,4971	0,97	0,8340	0,1660	0,6680
0,68	0,7517	0,2483	0,5035	0,98	0,8365	0,1635	0,6729
0,69	0,7549	0,2451	0,5098	0,99	0,8389	0,1611	0,6778
0,70	0,7580	0,2420	0,5161	1,00	0,8413	0,1587	0,6827
0,71	0,7611	0,2389	0,5223	1,01	0,8438	0,1562	0,6875
0,72	0,7642	0,2358	0,5285	1,02	0,8461	0,1539	0,6923
0,73	0,7673	0,2327	0,5346	1,03	0,8485	0,1515	0,6970
0,74	0,7704	0,2296	0,5407	1,04	0,8508	0,1492	0,7017
0,75	0,7734	0,2266	0,5467	1,05	0,8531	0,1469	0,7063
0,76	0,7764	0,2236	0,5527	1,06	0,8554	0,1446	0,7109
0,77	0,7794	0,2206	0,5587	1,07	0,8577	0,1423	0,7154
0,78	0,7823	0,2177	0,5646	1,08	0,8599	0,1401	0,7199
0,79	0,7852	0,2148	0,5705	1,09	0,8621	0,1379	0,7243
0,80	0,7881	0,2119	0,5763	1,10	0,8643	0,1357	0,7287
0,81	0,7910	0,2090	0,5821	1,11	0,8665	0,1335	0,7330
0,82	0,7939	0,2061	0,5878	1,12	0,8686	0,1314	0,7373
0,83	0,7967	0,2033	0,5935	1,13	0,8708	0,1292	0,7415
0,84	0,7995	0,2005	0,5991	1,14	0,8729	0,1271	0,7457
0,85	0,8023	0,1977	0,6047	1,15	0,8749	0,1251	0,7499
0,86	0,8051	0,1949	0,6102	1,16	0,8770	0,1230	0,7540
0,87	0,8078	0,1922	0,6157	1,17	0,8790	0,1210	0,7580
0,88	0,8106	0,1894	0,6211	1,18	0,8810	0,1190	0,7620
0,89	0,8133	0,1867	0,6265	1,19	0,8830	0,1170	0,7660
0,90	0,8159	0,1841	0,6319	1,20	0,8849	0,1151	0,7699

z	$\Phi(z)$	$\Phi(-z)$	$D(z)$	z	$\Phi(z)$	$\Phi(-z)$	$D(z)$
1,21	0,8869	0,1131	0,7737	1,51	0,9345	0,0655	0,8690
1,22	0,8888	0,1112	0,7775	1,52	0,9357	0,0643	0,8715
1,23	0,8907	0,1093	0,7813	1,53	0,9370	0,0630	0,8740
1,24	0,8925	0,1075	0,7850	1,54	0,9382	0,0618	0,8764
1,25	0,8944	0,1056	0,7887	1,55	0,9394	0,0606	0,8789
1,26	0,8962	0,1038	0,7923	1,56	0,9406	0,0594	0,8812
1,27	0,8980	0,1020	0,7959	1,57	0,9418	0,0582	0,8836
1,28	0,8997	0,1003	0,7995	1,58	0,9429	0,0571	0,8859
1,29	0,9015	0,0985	0,8029	1,59	0,9441	0,0559	0,8882
1,30	0,9032	0,0968	0,8064	1,60	0,9452	0,0548	0,8904
1,31	0,9049	0,0951	0,8098	1,61	0,9463	0,0537	0,8926
1,32	0,9066	0,0934	0,8132	1,62	0,9474	0,0526	0,8948
1,33	0,9082	0,0918	0,8165	1,63	0,9484	0,0516	0,8969
1,34	0,9099	0,0901	0,8198	1,64	0,9495	0,0505	0,8990
1,35	0,9115	0,0885	0,8230	1,65	0,9505	0,0495	0,9011
1,36	0,9131	0,0869	0,8262	1,66	0,9515	0,0485	0,9031
1,37	0,9147	0,0853	0,8293	1,67	0,9525	0,0475	0,9051
1,38	0,9162	0,0838	0,8324	1,68	0,9535	0,0465	0,9070
1,39	0,9177	0,0823	0,8355	1,69	0,9545	0,0455	0,9090
1,40	0,9192	0,0808	0,8385	1,70	0,9554	0,0446	0,9109
1,41	0,9207	0,0793	0,8415	1,71	0,9564	0,0436	0,9127
1,42	0,9222	0,0778	0,8444	1,72	0,9573	0,0427	0,9146
1,43	0,9236	0,0764	0,8473	1,73	0,9582	0,0418	0,9164
1,44	0,9251	0,0749	0,8501	1,74	0,9591	0,0409	0,9181
1,45	0,9265	0,0735	0,8529	1,75	0,9599	0,0401	0,9199
1,46	0,9279	0,0721	0,8557	1,76	0,9608	0,0392	0,9216
1,47	0,9292	0,0708	0,8584	1,77	0,9616	0,0384	0,9233
1,48	0,9306	0,0694	0,8611	1,78	0,9625	0,0375	0,9249
1,49	0,9319	0,0681	0,8638	1,79	0,9633	0,0367	0,9265
1,50	0,9332	0,0668	0,8664	1,80	0,9641	0,0359	0,9281

z	$\Phi(z)$	$\Phi(-z)$	$D(z)$	z	$\Phi(z)$	$\Phi(-z)$	$D(z)$
1,81	0,9649	0,0351	0,9297	2,11	0,9826	0,0174	0,9651
1,82	0,9656	0,0344	0,9312	2,12	0,9830	0,0170	0,9660
1,83	0,9664	0,0336	0,9328	2,13	0,9834	0,0166	0,9668
1,84	0,9671	0,0329	0,9342	2,14	0,9838	0,0162	0,9676
1,85	0,9678	0,0322	0,9357	2,15	0,9842	0,0158	0,9684
1,86	0,9686	0,0314	0,9371	2,16	0,9846	0,0154	0,9692
1,87	0,9693	0,0307	0,9385	2,17	0,9850	0,0150	0,9700
1,88	0,9699	0,0301	0,9399	2,18	0,9854	0,0146	0,9707
1,89	0,9706	0,0294	0,9412	2,19	0,9857	0,0143	0,9715
1,90	0,9713	0,0287	0,9426	2,20	0,9861	0,0139	0,9722
1,91	0,9719	0,0281	0,9439	2,21	0,9864	0,0136	0,9729
1,92	0,9726	0,0274	0,9451	2,22	0,9868	0,0132	0,9736
1,93	0,9732	0,0268	0,9464	2,23	0,9871	0,0129	0,9743
1,94	0,9738	0,0262	0,9476	2,24	0,9875	0,0125	0,9749
1,95	0,9744	0,0256	0,9488	2,25	0,9878	0,0122	0,9756
1,96	0,9750	0,0250	0,9500	2,26	0,9881	0,0119	0,9762
1,97	0,9756	0,0244	0,9512	2,27	0,9884	0,0116	0,9768
1,98	0,9761	0,0239	0,9523	2,28	0,9887	0,0113	0,9774
1,99	0,9767	0,0233	0,9534	2,29	0,9890	0,0110	0,9780
2,00	0,9772	0,0228	0,9545	2,30	0,9893	0,0107	0,9786
2,01	0,9778	0,0222	0,9556	2,31	0,9896	0,0104	0,9791
2,02	0,9783	0,0217	0,9566	2,32	0,9898	0,0102	0,9797
2,03	0,9788	0,0212	0,9576	2,33	0,9901	0,0099	0,9802
2,04	0,9793	0,0207	0,9586	2,34	0,9904	0,0096	0,9807
2,05	0,9798	0,0202	0,9596	2,35	0,9906	0,0094	0,9812
2,06	0,9803	0,0197	0,9606	2,36	0,9909	0,0091	0,9817
2,07	0,9808	0,0192	0,9615	2,37	0,9911	0,0089	0,9822
2,08	0,9812	0,0188	0,9625	2,38	0,9913	0,0087	0,9827
2,09	0,9817	0,0183	0,9634	2,39	0,9916	0,0084	0,9832
2,10	0,9821	0,0179	0,9643	2,40	0,9918	0,0082	0,9836

z	$\Phi(z)$	$\Phi(-z)$	$D(z)$	z	$\Phi(z)$	$\Phi(-z)$	$D(z)$
2,41	0,9920	0,0080	0,9840	2,71	0,9966	0,0034	0,9933
2,42	0,9922	0,0078	0,9845	2,72	0,9967	0,0033	0,9935
2,43	0,9925	0,0075	0,9849	2,73	0,9968	0,0032	0,9937
2,44	0,9927	0,0073	0,9853	2,74	0,9969	0,0031	0,9939
2,45	0,9929	0,0071	0,9857	2,75	0,9970	0,0030	0,9940
2,46	0,9931	0,0069	0,9861	2,76	0,9971	0,0029	0,9942
2,47	0,9932	0,0068	0,9865	2,77	0,9972	0,0028	0,9944
2,48	0,9934	0,0066	0,9869	2,78	0,9973	0,0027	0,9946
2,49	0,9936	0,0064	0,9872	2,79	0,9974	0,0026	0,9947
2,50	0,9938	0,0062	0,9876	2,80	0,9974	0,0026	0,9949
2,51	0,9940	0,0060	0,9879	2,81	0,9975	0,0025	0,9950
2,52	0,9941	0,0059	0,9883	2,82	0,9976	0,0024	0,9952
2,53	0,9943	0,0057	0,9886	2,83	0,9977	0,0023	0,9953
2,54	0,9945	0,0055	0,9889	2,84	0,9977	0,0023	0,9955
2,55	0,9946	0,0054	0,9892	2,85	0,9978	0,0022	0,9956
2,56	0,9948	0,0052	0,9895	2,86	0,9979	0,0021	0,9958
2,57	0,9949	0,0051	0,9898	2,87	0,9979	0,0021	0,9959
2,58	0,9951	0,0049	0,9901	2,88	0,9980	0,0020	0,9960
2,59	0,9952	0,0048	0,9904	2,89	0,9981	0,0019	0,9961
2,60	0,9953	0,0047	0,9907	2,90	0,9981	0,0019	0,9963
2,61	0,9955	0,0045	0,9909	2,91	0,9982	0,0018	0,9964
2,62	0,9956	0,0044	0,9912	2,92	0,9982	0,0018	0,9965
2,63	0,9957	0,0043	0,9915	2,93	0,9983	0,0017	0,9966
2,64	0,9959	0,0041	0,9917	2,94	0,9984	0,0016	0,9967
2,65	0,9960	0,0040	0,9920	2,95	0,9984	0,0016	0,9968
2,66	0,9961	0,0039	0,9922	2,96	0,9985	0,0015	0,9969
2,67	0,9962	0,0038	0,9924	2,97	0,9985	0,0015	0,9970
2,68	0,9963	0,0037	0,9926	2,98	0,9986	0,0014	0,9971
2,69	0,9964	0,0036	0,9929	2,99	0,9986	0,0014	0,9972
2,70	0,9965	0,0035	0,9931	3,00	0,9987	0,0013	0,9973

References

1. Alexander, C.: Quantitative Methods in Finance. John Wiley & Sons, Chichester (2010)
2. Alexander, C.: Market Models. John Wiley & Sons, Chichester (2011)
3. Anderson, B.D.O., Moore, J.B.: Optimal Control: Linear Quadratic Methods. Prentice-Hall, Englewood Cliffs (1990)
4. Anderson, B.D.O., Moore, J.B.: Linear Quadratic Methods. Dover Books on Engineering, New York (1990)
5. Arnăutu, V., Neittaanmäki, P.: Optimal Control from Theory to Computer Programs. Kluwer Academic Publishers, Dordrecht (2003)
6. Arnold, L.: Stochastische Differentialgleichungen: Theorie und Anwendung. Oldenbourg, München (1973)
7. Arrow, K.J.: The Theory of Risk Aversion," in "Aspects of the Theory of Risk Bearing", by Yrjö Jahnssonin Säätiö, Helsinki. Reprinted. In: "Essays in the Theory of Risk Bearing. Markham, Chicago (1971)
8. Asimov, I.: Foundation. Gnome Press Publishers, New York (1951)
9. Atkinson, A.C., Donev, A.N., Tobias, R.: Optimum Experimental Designs. With SAS. Oxford University Press, Oxford (2007)
10. Bank for International Settlements, "Basel III: A Global Regulatory Framework for More Resilient Banks and Banking Systems". Basel (2010)
11. Bellman, R.E.: Dynamic Programming. Princeton University Press, Princeton (1957)
12. Bjørk, T.: Arbitrage Theory in Continuous Time. Oxford University Press, Oxford (1998)
13. Bloomberg News, August 3, 2015, China Stocks Rise as Brokerages Ban Short Selling to Stem Losses. Accessed 5 Nov 2016
14. Bonnans, J.F., Gilbert, J.C., Lemarechal, C., Sagastzábal, C.A.: Numerical Optimization: Theoretical and Practical Aspects. Springer, Berlin (2009)
15. Bowers, Q.D.: The Expert's Guide to Collecting & Investing in Rare Coins: Secrets of Success. Whitman, Atlanta (2006)
16. Boyd, S., Vandenberghe, L.: Convex Optimization. Cambridge University Press, Cambridge (2004)
17. Boyle, P., Boyle, F.: Derivatives: The Tools that Changed Finance. Risk Waters Group, London (2001)
18. Bressan, A.: Non-cooperative Differential Games: A Tutorial. Penn State University, University Park, Department of Mathematics (2010)
19. Bronstein, I.N., Semendjajew, K.A., Musiol, G., Mühling, H.: Taschenbuch der Mathematik. Verlag Harri Deutsch, Frankfurt (2016)
20. Burrage K., Burrage P.M., Tian T.: Numerical Methods for Strong Solutions of Stochastic Differential Equations: An Overview. In: Proceedings of the Royal Society, London, vol. A, 460 (2004)

© Springer International Publishing AG 2018
S. S. Hacısalihzade, *Control Engineering and Finance*, Lecture Notes in Control and Information Sciences 467, https://doi.org/10.1007/978-3-319-64492-9

21. Butenko, S., Pardalos, P.M.: Numerical Methods and Optimization: An Introduction. Chapman & Hall, Boca Raton (2014)
22. Byrd R., Schnabel R., Schultz G.: A trust region algorithm for non-linearly constrained optimization. SIAM J. Numer. Anal. **24** (1987)
23. Campbell, J.Y., Lo, A.W., MacKinlay, A.C.: The Econometrics of Financial Markets. Princeton University Press, Princeton (1997)
24. Carmona, R.: Lectures on BSDEs, Stochastic Control, and Stochastic Differential Games with Financial Applications. SIAM Series on Financial Mathematics. Cambridge University Press, Cambridge (2016)
25. Cellier F.E.: Prisoner's dilemma revisited: a new strategy based on the general system problem solving framework. Int. J. Gen. Syst. **13** (1987)
26. Cellier, F.E., Greifeneder, J.: Continuous System Modeling. Springer, New York (1991)
27. Cellier, F.E., Kofman, E.: Continuous System Simulation. Springer, New York (2006)
28. CFR.org Staff: The Credit Rating Controversy, Council of Foreign Relations. Accessed 19 Oct 2016
29. Clark M.R., Stark L.: Time optimal behavior of human saccadic eye movement. IEEE Trans. Autom. Control **AC-20** (1975)
30. Cobelli, C., Carson, E.R.: Introduction to Modeling in Physiology and Medicine. Elsevier, London (2008)
31. Çotur Y.: Effect of Different Instrument Classes in Portfolio Optimization, Bachelor Thesis, Boğaziçi University, Electrical and Electronics Engineering Department (2013)
32. Crassidis, J.L., Junkins, J.L.: Optimal Estimation of Dynamic Systems. CRC Press, Boca Raton (2004)
33. Cyganowski, S., Kloeden, P., Ombach, J.: From Elementary Probability to Stochastic Differential Equations with MAPLE. Springer, Berlin (2002)
34. Datz, S.R.: Stamp Investing. General Philatelic Corporation, Loveland (1999)
35. Duffie, D.: Dynamic Asset Pricing Theory. Princeton University Press, Princeton (1996)
36. Einstein, A.: Über die von der molekularkinetischen Theorie der Wärme geforderte Bewegung von in ruhenden Flüssigkeiten suspendierten Teilchen. Annalen der Physik **17** (1905)
37. Elihoş M.H.: Modern Portfolio Optimization Strategies, Bachelor Thesis, Boğaziçi University, Electrical and Electronics Engineering Department (2014)
38. Elton, E., Gruber, M., Brown, S., Goetzmann, W.: Modern Portfolio Theory and Investment Analysis. John Wiley & Sons, Hoboken (1995)
39. Elton E., Gruber, M.: Modern Portfolio Theory, 1950 to Date. J. Bank. Financ. **21** (1997)
40. Estrella, A., Mishkin, F.S.: Predicting U.S. recessions: financial variables as leading indicators. Rev. Econ. Stat. **80** (1998)
41. Fabozzi, F.J.: Bond Markets, Analysis, and Strategies. Prentice Hall, New York (2004)
42. Fama, E., French, K.: The cross-section of expected stock returns. J. Financ. **47** (1992)
43. Fedorov, V.V., Hackl, P.: Model-Oriented Design of Experiments. Springer, New York (1997)
44. Financial Publishing Company: Expanded Bond Values Tables: Coupons 1% to 12%. Financial Publishing Company, Boston (1970)
45. Fleming, W.M., Rishel, R.W.: Deterministic and Stochastic Optimal Control. Springer, New York (1975)
46. Geering, H.P., Dondi, G., Herzog, F., Keel, S.: Stochastic Systems. Measurement and Control Laboratory, ETH Zurich (2011)
47. Graham, B.: The Intelligent Investor. Harper, New York (2006)
48. Graves, R.L., Wolfe, P. (eds.): Recent Advances in Mathematical Programming. McGraw-Hill, New York (1963)
49. Guerard, J.B. (ed.): Handbook of Portfolio Construction: Contemporary Applications of Markowitz Techniques. Springer, New York (2009)
50. Gugercin S., Antoulas A.C.: A survey of model reduction by balanced truncation and some new results. Int. J. Control **77** (2004)
51. Hacısalihzade, S.S.: Biomedical Applications of Control Engineering. Springer, Berlin (2013)
52. Hadley, G.: Non-linear and Dynamic Programming. Addison & Wesley, Boston (1964)

53. Hayek, F.A.: Individualism and Economic Order. University of Chicago Press, Chicago (1948)
54. Heffernan, S.: Modern Banking. John Wiley & Sons, Hoboken (2005)
55. Herzog, F.: Getting in the Smile and Realistic Underlying Market Dynamics. SwissQuant Group, Zurich (2012)
56. Higham D.J.: An algorithmic introduction to numerical simulation of stochastic differential equations. SIAM Rev. **43** (2001)
57. Homer, S., Leibowitz, M.L.: Inside the Yield Book. John Wiley & Sons, Hoboken (2013)
58. Hull, J.C.: Options, Futures, and Other Derivatives. Pearson Prentice Hall, Uppersaddle River (1999)
59. IGEM:IMPERIAL/2006/project/Oscillator/Theoretical Analyses/Results/2D Modell
60. Investopedia on http://www.investopedia.com/financial-edge/0712/closed-end-vs.-open-end-funds.aspx. Accessed 11 April 2017
61. Isaacs R.: Differential Games. Dover (1999)
62. Isermann, R.: Digital Control Systems. Springer, Heidelberg (2013)
63. Kahneman, D., Krueger, A., Schkade, D., Schwarz, N., Stone, A.: Would you be happier if you were richer? A focusing illusion. Science **312** (2006)
64. Kailath, T.: Linear Systems. Prentice-Hall, Englewood Cliffs (1982)
65. Kalai, A., Kalai, E.: Cooperation in strategic games revisited. Q. J. Econ. **10** (2012)
66. Kallenberg, O.: Foundations of Modern Probability. Springer, New York (2002)
67. Kalman, R.E.: Mathematical description of linear dynamical system. J. SIAM Control **1** (1963)
68. Kennedy, J., Eberhardt, R.: Particle swarm optimization. In: Proceedings of the IEEE International Conference, Neural Networks, Piscataway (1995)
69. Kloeden, P.E., Platen, E.: Numerical Solution of Stochastic Differential Equations. Springer, Heidelberg (1999)
70. Kohn, R.V.: Lecture Notes PDE for Finance on http://www.math.nyu.edu/faculty/kohn/pde_finance.html. Accessed 20 Jan 2017
71. Kolda, T.G., Lewis, R.M., Torcson, V.: Optimization by direct search: new perspectives on some classical and modern methods. SIAM Rev. **45** (2005)
72. Kramer, P.R., Majda, A.J.: Stochastic mode reduction for immersed boundary. SIAM J. Appl. Math. **64** (2003)
73. Lauritzen, N.: Undergraduate Convexity: From Fourier and Motzkin to Kuhn and Tucker. World Scientific Publishing Co., Singapore (2013)
74. Lehmann, E.L., Casella, G.: Theory of Point Estimation. Springer, New York (1998)
75. Liberzon, D.: Calculus of Variations and Optimal Control Theory: A Concise Introduction. Princeton University Press, Princeton (2012)
76. Liu, S.W.: Adaptive multi-meshes in finite element approximation of optimal control. Contemp. Math. **383** (2005)
77. Llewellyn, D.T.: Principles of effective regulation and supervision of banks. J. Financ. Regul. Compliance **6** (1998)
78. Lotka, A.J.: Elements of Mathematical Biology. Dover Press, New York (1956)
79. Luenberger, D.G.: Investment Science. Oxford University Press, Oxford (1998)
80. Macaulay, F.R.: Some Theoretical Problems Suggested by the Movements of Interest Rates, Bond Yields and Stock Prices in the United States since 1856. National Bureau of Economic Research, New York (1938)
81. Malkiel, B.G.: Expectations, bond prices and the term structure of interest rates. Q. J. Econ. **76** (1962)
82. Malkiel, B.G.: The Term Structure of Interest Rates. McCaleb-Seiler, New York (1970)
83. Markowitz, H.: Portfolio selection. J. Financ. **7** (1952)
84. Martellini, L., Priaulet, P., Priaulet, S.: Fixed-Income Securities. John Wiley & Sons, Chichester (2003)
85. Marx, K.: A Contribution to the Critique of Political Economy. Progress Publishers, Moscow (1970)

86. Maruyama, G.: Continuous Markov processes and stochastic equations. Rendiconti del Circolo Matematico di Palermo **4** (1955)
87. McAndrew, C. (ed.): Fine Art and High Finance. Bloomberg Press, New York (2010)
88. McLean, B., Nocera, J.: All the Devils Are Here: The Hidden History of the Financial Crisis. Penguin Press, London (2010)
89. Merton, R.C.: Continuous-Time Finance. Blackwell, Malden (1992)
90. Meyer, C.D.: Matrix Analysis and Applied Linear Algebra. SIAM, Philadelphia (2000)
91. Meyers, R.A. (ed.): Encyclopedia of Complexity and Systems Science. Springer, Berlin (2009)
92. Miano, G., Maffuci, A.: Transmission Lines and Lumped Circuits. Academic Press, London (2001)
93. Mikosch, T.: Elementary Stochastic Calculus with Finance in View. World Scientific, Singapore (1998)
94. Milstein, G.: Numerical Integration of Stochastic Differential Equations. Kluwer, Dordrecht (1995)
95. Mlodinow, L.: The Drunkard's Walk: How Randomness Rules Our Lives. Pantheon Books, New York (2008)
96. Moler, C.B., van Loan, C.F.: Nineteen dubious ways to compute the exponential of a matrix, twenty-five years later. SIAM Rev. **45** (2003)
97. Möller, J.K., Madsen, H.: From State Dependent Diffusion to Constant Diffusion in Stochastic Differential Equations by the Lamperti Transform. Technical University of Denmark, IMM-Technical Report-2010-16, Lyngby (2010)
98. Murphey, J.E.: Bond Tables of Probable Future Yields. Crossgar Press, Crossgar (1997)
99. Neftci, S.N.: An Introduction to the Mathematics of Financial Derivatives. Academic Press, San Diego (2000)
100. Nocedal, J., Wright, S.J.: Numerical Optimization. Springer, Berlin (2006)
101. Øksendal, B.: Stochastic Differential Equations. Springer, Berlin (1998)
102. Petersen, K.B., Pedersen, M.S.: The matrix cookbook. www2.imm.dtu.dk/pubdb/views/edoc_download.php/3274/pdf/imm3274.pdf. Accessed 19 Jan 2017
103. Platen, E., Heath, D.: A Benchmark Approach to Quantitative Finance. Springer, Heidelberg (2010)
104. Pontryagin, L.S., Boltyanskii, V.G., Gamkrelidze, R.V., Mishchenko, E.F.: The mathematical theory of optimal processes. Interscience Publishers, New York (1962)
105. Pratt, J.W.: Risk aversion in the small and in the large. Econometrica **32** (1964)
106. Predko, M.: Programming Robot Controllers. McGraw-Hill, New York (2003)
107. Puhle, M.: Bond Portfolio Optimization. Springer, Berlin (2008)
108. Pukelsheim, F.: Optimal Design of Experiments. SIAM, Philadelphia (2006)
109. Reilly, F.K.: Investment Analysis and Portfolio Management. Dryden Press, Oak Brook (1985)
110. Robbins, L.: An Essay on the Nature and Significance of Economic Science. Ludwig von Mises Institute, Auburn (2007)
111. Rodgers, J.L., Nicewander, A.: Thirteen ways to look at the correlation coefficient. Am. Stat. **42** (1988)
112. Rush, R.H.: Antiques as an Investment. Bonanza Books, New York (1979)
113. Sage, A.P., White, C.C.: Optimum Systems Control. Prentice-Hall, Englewood Cliffs (1968)
114. Sauer, T.: Numerical Analysis. Pearson, Boston (2016)
115. Sarychev, A.V.: First- and second-order sufficient optimality conditions for bang-bang controls. SIAM J. Control Optim. **35** (1997)
116. Schumacher, H., Hanzon, B., Vellekoop, M.: Finance for control engineers: tutorial workshop. In: European Control Conference. Porto (2001)
117. Schwert, G.W.: Stock volatility and the crash of '87. Rev. Financ. Stud. **3** (1990)
118. Sethi, S.P., Thompson, G.L.: Optimal Control Theory: Applications to Management Science and Economics. Springer, New York (2003)
119. Sevinç C.: Estimation Algorithms for Portfolio Optimization, Bachelor Thesis, Boğaziçi University, Electrical and Electronics Engineering Department (2013)
120. Shiryaev, A.N.: Graduate Texts in Mathematics: Probability. Springer, Berlin (1996)

121. Sierksma, G., Zwols, Y.: Linear and Integer Optimization: Theory and Practice. Chapman & Hall, Boca Raton (2015)
122. Smolin, L.: Three Roads to Quantum Gravity. Basic Books, New York (2001)
123. Snyman, J.A.: Practical Mathematical Optimization: An Introduction to Basic Optimization Theory and Classical and New Gradient-Based Algorithms. Springer, Berlin (2005)
124. Speight, H.: Economics: The Science of Prices and Incomes. Methuen & Co., London (1960)
125. Steele, J.M.: Stochastic Calculus and Financial Applications. Springer, New York (2001)
126. Stringham E.P.: The extralegal development of securities trading in seventeenth century Amsterdam. Q. Rev. Econ. Financ.**43** (2003)
127. Sussmann, H.J.: Envelopes, conjugate points and optimal bang-bang extremals. In: Fliess, M., Hazewinkel, M. (eds.) Proceedings of the 1985 Paris Conference on Non-linear Systems. Reidel Publishers, Dordrecht (1987)
128. Swiss Structured Products Association: Swiss Derivative Map. Zurich (2016)
129. Taleb, N.N.: The Black Swan: The Impact of the Highly Improbable. Random House, New York (2007)
130. Thompson, D.E.: Design Analysis: Mathematical Modeling of Non-linear Systems. Cambridge University Press, Cambridge (1999)
131. Tversky, A., Kahneman, D.: The framing of decisions and the psychology of choice. Science **211** (1981)
132. Tu, P.N.V.: Introductory Optimization Dynamics. Springer, New York (1991)
133. Uyan, U.: Bond Portfolio Optimization, Bachelor Thesis, Boğaziçi University, Electrical and Electronics Engineering Department (2013)
134. Weatherall, J.O.: The Physics of Wall Street: A Brief History of Predicting the Unpredictable. Houghton Mifflin Harcourt Publishing, New York (2014)
135. Widnall, S.: Dynamics: MIT Open Course Ware (2009). http://ocw.mit.edu/courses/aeronautics-and-astronautics/16-07-dynamics-fall-2009/lecture-notes/MIT16_07F09_Lec20.pdf
136. Wiggins, S.: Introduction to Applied Non-linear Dynamical Systems and Chaos. Springer, New York (2003)
137. Wilhelm, J.: Fristigkeitsstruktur und Zinsänderungsrisiko, Zeitschrift für betriebswirtschaftliche Forschung **44** (1992)
138. Yates, R.D., Goodman, D.J.: Probability and Stochastic Processes: A Friendly Introduction for Electrical and Computer Engineers. Wiley & Sons, Hoboken (2005)
139. Yılmaz, E.D.: Empirical Analyses on Portfolio Optimization Strategies, Master Thesis, Boğaziçi University, Electrical and Electronics Engineering Department (2012)

Index

Index

301

Generalized boundary conditions, 89, 95, 97, 99
Golden section, 32, 37
Graham, Benjamin, 6, 219

H
Hamilton, William Rowan, 92
Hamiltonian, 92–95, 97, 98, 102, 106, 107, 132, 133
Hamilton-Jacobi equation, 103–105, 108–110, 261
Hayek, Friedrich, 217
Heaviside, Oliver, 12
Hedging, 244, 245, 257, 258, 261
Histogram correction, 57, 58
Hooke's law, 14
Hurwitz, Adolf, 276, 278

I
Identifiability, 23
Immunization strategy, 238
Inductive model, 10
Inflation, 14, 182, 199, 202, 204, 208, 210, 212
Insider trading, 200
Integrability, 77
Interest, 199, 201–204, 206, 209–211, 258
  accrued, 207
  rate, 5, 106, 141, 165, 178, 186, 201, 203–205, 207, 209, 210, 213, 233–235, 238, 254, 257, 258, 261
Inverted pendulum, 3, 17, 38
Investing, 185, 186, 188, 191, 192, 199, 206, 217, 219, 233, 239, 241, 242
IPO, 187
Istanbul Stock Exchange, 230
Itô
  calculus, 145, 150, 180
  integral, 139, 145, 146, 148, 149, 168, 176, 177
  lemma, 139, 152, 153, 158, 168, 169, 176–178, 252
Itô, Kiyoshi, 145

J
Jacobi, Carl Gustav Jacob, 103
Joint density function, 49

K
Kac, Mark, 166

Kahneman, Daniel, 182, 195
Kalman, Rudolf Emil, 180, 275, 277
Karush-Kuhn-Tucker conditions, 31, 225
Katona, George, 195
Kodak, 186
Kolmogorov, Andrey Nikolaevich, 42
Kolmogorov axioms, 4, 41, 42
Kuhn-Tucker conditions, 31, 225
Kurtosis, 52, 53, 78, 179

L
Lagrange, Joseph-Louis, 28
Lagrange multiplier, 28–30, 91, 135, 222
Lagrangian, 28, 29, 222
Lamperti transform, 171
Laplace, Pierre-Simon, 14, 24, 117, 118, 267–270, 274, 276
Least squares, 19, 23, 24
Lehman Brothers, 259
Leverage, 187, 189, 190, 241, 242, 247, 257, 258, 260
LIBOR, 204, 258
Limited control variable, 97, 100, 135, 136
Linear model, 13, 20, 35
Linear programming, 33
Linear system, 13, 24, 83, 108, 113, 117–120, 264–267, 269, 271, 273–278, 280
Llewellyn, David, 198
Log-normal distribution, 59, 80, 226, 251
Lorenz, Edward Norton, 11
Lotka-Volterra equations, 10
Lumped parameter model, 12
Lyapunov
  exponent, 11
  stability, 277–281
Lyapunov, Alexandr Mikhailovich, 56, 277, 279

M
Macaulay
  duration, 5, 201, 238
Macaulay, Frederick, 211
Macaulay duration, 211
Margin account, 5, 245–247
Markov, Andrey Andreyevich, 71
Markov process, 4, 41, 72, 74, 79, 82, 173
Markowitz, Harry, 216, 236
Martingale, 146, 148
Marx, Karl, 189
Mathematical model, 3, 7, 9–11, 15, 17, 35, 36, 42, 113, 180, 221

Printed in the United States
By Bookmasters